28th Annual Meeting of the North American Membrane Society (NAMS 2019)

Pittsburgh, Pennsylvania, USA
11-15 May 2019

ISBN: 978-1-7138-2787-0

Printed from e-media with permission by:

Curran Associates, Inc.
57 Morehouse Lane
Red Hook, NY 12571

Some format issues inherent in the e-media version may also appear in this print version.

Printed with permission by Curran Associates, Inc. (2021)

For permission requests, please contact North American Membrane Society (NAMS) at the address below.

North American Membrane Society (NAMS)
University of Arkansas
Ralph E. Martin Department of Chemical Engineering
3202 Bell Engineering Center
Fayetteville, AR 72701-1201
USA

Phone: +1 479 575-7926
Fax: +1 479 575-7926

names@uark.edu

Additional copies of this publication are available from:

Curran Associates, Inc.
57 Morehouse Lane
Red Hook, NY 12571 USA
Phone: 845-758-0400
Fax: 845-758-2633
Email: curran@proceedings.com
Web: www.proceedings.com

ORAL SESSIONS

Presenting authors are indicated by an asterisk*. See the most up-to-date version of the program with abstracts at **nams2019.org** or via the **Attendify (NAMS 2019)** app.

Plenary Session 1

7:45 am - 9:00 pm | Ballroom 1

Dr. Peter S. Fiske
Lawrence Berkeley National Laboratory

Innovation Pathways in the Water Ecosystem: The PAX Water Story

#1 Processes: Fundamentals of Predicting and Preventing Membrane Fouling ...pg. 1

9:30 am - 12:30 pm | Kings Garden 4

Co-Chairs: Manish Kumar (Penn State University), Davide Mattia (University of Bath), Saifur Rahaman (Concordia University)

9:30 am MONDAY

1a - Hydrodynamic-colloidal interactions of an oil droplet and a membrane surface

Mariano Galvagno (Technion - Israel Institute of Technology), Guy Z Ramon (Technion - Israel Institute of Technology)*

10:00 am MONDAY

1b - Do Graphene Oxide Nanostructured Coatings Mitigate Bacterial Adhesion to Membrane Interfaces?

Santiago Romero-Vargas Castrillon (University of Edinburgh)*, Sara BinAhmed (University of Minnesota), Karl Wuolo-Journey (University of Minnesota)

10:30 am MONDAY

1c - Fouling mechanisms in constant flux crossflow ultrafiltration

Alon Kirschner (The University of Texas at Austin)*, Yu-Heng Cheng (The University of Texas at Austin), Donald Paul (The University of Texas at Austin), Robert Field (University of Oxford), Benny Freeman (The University of Texas at Austin)

11:00 am MONDAY

1d - Fouling-Resistant, Multi-Functional Membranes with Ultra-Thin Hydrogel Selective Layers using Interfacially Initiated Free Radical Polymerization (IIFRP)

Ayse Asatekin (Tufts University)*, Ilin Sadeghi (Tufts University), Eric Liu (Tufts University), Alice Oliveira Aguiar (Tufts University), Hyunmin Yi (Tufts University)

11:30 am MONDAY

1e - Unravel the Gordian Knot of membrane desalination: Understanding and mitigating silica scaling in reverse osmosis

Tiezheng Tong (Colorado State University)*, Song Zhao (Tianjin University), Amanda Quay (Carnegie Mellon University), Yunlong Qi (Tianjin University)

12:00 pm MONDAY

1f - RO Mineral Scale Mitigation via Self-Adaptive Feed-Flow Reversal

Yeunha Kim (UCLA)*, Anditya Rahardianto (UCLA), Tae Lee (UCLA), Yoram Cohen (UCLA)

#2 Processes: Membrane Distillation and Pervaporation: Innovations in Process Design and Scalability ...pg 3

9:30 am - 12:30 pm | Ballroom 3

Co-Chairs: Lee Vane (Environmental Protection Agency), David Warsinger (Purdue University)

9:30 am MONDAY

2a - Continuous on-demand dehydration of solvents in flow chemistry manufacturing processes

Hannah Murnen (Compact Membrane Systems)*, Sudip Majumdar (Compact Membrane Systems)

10:00 am MONDAY

2b - How membrane technology has been successfully applied to ethanol processing

Stephan Blum (Whitefox Technologies)*, Trond Heggenhougen (Whitefox Technologies), Meridith K. Bridge (Whitefox Technologies)

10:30 am MONDAY

2c - Modeling the effect of different single-walled carbon nanotube orientations on permeation through mixed matrix membranes

Ali Zamani (University of Ottawa)*, Jules Thibault (University of Ottawa), Handan Tezel (University of Ottawa)

11:00 am MONDAY

2d - Unique 3D printed spacers for process enhancement in membrane distillation

Navya Thomas (Khalifa University)*, Nurshaun Sreedhar (Khalifa University), Oraib Al-Ketan (Khalifa University), Reza Rowshan (New York University Abu Dhabi), Rashid Abu Al-Rub (Khalifa University), Hassan Arafat (Khalifa University)

11:30 am MONDAY

2e - 2D boron nitride encapsulated nanofiber membranes for membrane distillation

Yunchul Woo (Korea Institute of Civil Engineering and Building Technology)*, Ho Kyong Shon (University of Technology Sydney), June-Seok Choi (Korea Institute of Civil Engineering and Building Technology)

12:00 pm MONDAY

2f - Impact of membrane properties on scaling-induced wetting in membrane distillation

Allyson L McGaughey (University of Southern California)*, Prathamesh Karandikar (University of Southern California), Malancha Gupta (University of Southern California), Amy Childress (University of Southern California)

#3 Applications: Seawater Desalination ...pg. 6

9:30 am - 12:30 pm | Kings Garden 3

Co-Chairs: Sunny Jiang (UC Irvine), Bill Phillip (University of Notre Dame)

9:30 am MONDAY

3a - Sacrificial Protective Layer for Fouling Control in Reverse Osmosis Desalination

Moon Son (Penn State University)*, Wulin Yang (Penn State University), Szilard S. Bucs (King Abdullah University of Science and Technology), Maria F. Nava-Ocampo (King Abdullah University of Science and Technology), Johannes S. Vrouwenvelder (King Abdullah University of Science and Technology), Bruce E. Logan (Penn State University)

10:00 am MONDAY

3b - Water Transport through Ultrathin Polyamide Nanofilms Used for Reverse Osmosis

Zhiwei Jiang (Imperial College London)*, Santanu Karan (Imperial College London), Andrew Livingston (Imperial College London)

10:30 am MONDAY

3c - Balancing Carbon, Nitrogen and Phosphorus Concentration in Seawater as a Strategy to Prevent Accelerated Membrane Biofouling

Siqian (Simon) Huang (University of Minnesota, Duluth), Nikolay Voutchkov (Water Globe Consultants, LLC), Sunny Jiang (UC Irvine)*

11:00 am MONDAY

3d - Flow Reversal for Increased Recovery of 2nd Pass Seawater RO

Tomer Osman (Rotec Water Ltd.), Dan Peled (AST-Water Ltd.), Noam Perlmutter (Rotec Water Ltd.), Arina Shulman (Mekorot Water Company), Riki Harpness (Mekorot Water Company), Sivan Bleich (Mekorot Water Company), Eli Korin (Ben-Gurion University of the Negev), Jack Gilron (Ben-Gurion University of the Negev)*

11:30 am MONDAY

3e - Zwitterionic poly(arylene ether sulfone) (PAES) copolymer for biofouling/chlorine resistant desalination membranes

Yi Yang (Arizona State University)*, Matthew Green (Arizona State University)

12:00 pm MONDAY

3f - Chlorine resistance of functionalized HNTs-based TFN membranes

Farhad Asempour (University of Ottawa), Somaye Akbari (Amirkabir University of Technology), Du Bai (University of Ottawa), Ramzi Aoun (University of Ottawa), Sofia Reyes Lombardo (University of Ottawa), Amirsajad Atashgar (University of Ottawa), Boguslaw Kruczek (University of Ottawa)*

#4 Materials: Selective Polymeric and Mixed-Matrix Materials - Liquid Separations ...pg. 9

9:30 am - 12:30 pm | Ballroom 4

Co-Chairs: Dan Miller (Lawrence Berkeley National Laboratory), Baoxia Mi (UC Berkeley)

9:30 am MONDAY

4a - Enantioselective Polymeric Membrane for Chiral Separation of Enantiomers

Marine Michel (Imperial College London)*

10:00 am MONDAY

4b - Tailored Synthesis of Inorganic-supported Polymer, Graphene, and Nanocomposite Membranes for Liquid-phase Separations

Michael Z Hu (Oak Ridge National Laboratory)*

10:30 am MONDAY

4c - Two-Dimensional (2D) Covalent Organic Framework Mixed-Matrix Membranes

Phuoc Duong (University of Wyoming), Valerie Kuehl (University of Wyoming), Bruce Mastorovich (University of Wyoming), John Hoberg (University of Wyoming), Bruce Parkinson (University of Wyoming), Katie D Li-Oakey (University of Wyoming)*

11:00 am MONDAY

4d - Enhancing water permeance in reduced graphene oxide membranes by chemical etching

Xiaoyi Chen (The State University of New York at Buffalo)*, Liang Huang (The State University of New York at Buffalo), Zhihao Feng (The State University of New York at Buffalo), Janavi Gohil (The State University of New York at Buffalo), Haiqing Lin (The State University of New York at Buffalo)

11:30 am MONDAY

4e - Nanocomposite RO Membranes with High Aspect Ratio Functional Nanoparticles for Controlled Interfacial Transport

Stephen M Martin (Virginia Tech)*, Ethan Smith (Virginia Tech), E. Johan Foster (Virginia Tech), Keith Hendren (Virginia Tech), James Haag (Virginia Tech)

12:00 pm MONDAY

4f - If one were to develop a new NF/RO membrane, specifically for uncharged organic solute rejection, how permeable could it be to water while meeting organic solute rejection goals?

Eric Hoek (UCLA)*

#5 Materials: Inorganic Materials...pg.11

9:30 am - 12:30 pm | Kings Garden 1

Co-Chairs: Jay Kniep (Membrane Technology & Research), Jerry Lin (Arizona State University)

9:30 am MONDAY

5a - Two-Dimensional Membranes for Gas Separation

Dan Zhao (National University of Singapore)*

10:00 am MONDAY

5b - Restricting Lattice Flexibility in Polycrystalline Metal-Organic Framework Membrane for Efficient Carbon Capture

Deepu Babu (L'Ecole polytechnique fédérale de Lausanne), Kumar Varoon Agrawal (L'Ecole polytechnique fédérale de Lausanne)*

10:30 am MONDAY

5c - New Asymmetric and Permselective Carbon Molecular Sieve Membranes

Jay (Junqiang) Liu (The Dow Chemical Company)*, Ted Calverley (The Dow Chemical Company), Dean Millar (The Dow Chemical Company), Mark Brayden (The Dow Chemical Company), Marcos Martinez (The Dow Chemical Company)

11:00 am MONDAY

5d - Hitz zeolite membrane separation system (HDS) for gas separation

Masaya Itakura (Hitachi Zosen Corporation)*, Satoshi Imasaka (Hitachi Zosen Corporation), Kentaro Shinoya (Hitachi Zosen Corporation), Masashi Okada (Hitachi Zosen Corporation)

11:30 am MONDAY

5e - Insights on the Carbon Molecular Sieve Structure from Stability and Rejuvenation Tendencies

Samuel Hays (Georgia Institute of Technology)*, Oishi Sanyal (Georgia Institute of Technology), Nicholas Leon (Georgia Institute of Technology), Chen Zhang (University of Maryland), Pezhman Arab (Georgia Institute of Technology), William Koros (Georgia Institute of Technology)

12:00 pm MONDAY

5f - Fabrication of zeolitic imidazolate framework ZIF-8 membranes by ligand-induced permselectivation in vapor phase

Xiaoli Ma (University of Wisconsin-Milwaukee)*, Prashant Kumar (University of Minnesota), Nitish Mittal (University of Minnesota), Alexandra Khlyustova (University of Minnesota), Prodromos Daoutidis (University of Minnesota), Andre Mkhoyan (University of Minnesota), Michael Tsapatsis (Johns Hopkins University)

#6 Applications: Assessing Performance, Robustness, and Scalability for Carbon Capture ...pg 13

9:30 am - 12:30 pm | Kings Garden 5

Co-Chairs: Dave Hopkinson (U.S. Department of Energy National Energy Technology Laboratory), Winston Ho (Ohio State University)

9:30 am MONDAY

6a - Polymers with Side Chain Porosity for Ultrapermeable and Plasticization Resistant Materials for Gas Separations

Sharon Lin (MIT)*, Francesco Benedetti (University of Bologna), Yuan He (MIT), Chao Liu (University of Chinese Academy of Sciences), Yanchuan Zhao (University of Chinese Academy of Sciences), Hong-Zhou Ye (MIT), Troy Van Voorhis (MIT), Maria Grazia De Angelis (University of Bologna), Timothy Swager (MIT), Zach Smith (MIT)

10:00 am MONDAY

6b - Bench scale testing of next generation hollow fiber membrane modules at subambient conditions at the National Carbon Capture Center

David Hasse (Air Liquide)*, Shilu Fu (Air Liquide), Sudhir Kulkarni (Air Liquide), Trapti Chaubey (Air Liquide), Alex Augustine (Air Liquide), Andrew Hamilton (Air Liquide), Monaca Mcnall (Air Liquide)

10:30 am MONDAY

6c - A Carbon Molecular Sieve Membrane-Based Reactive Separation Process for Pre-Combustion CO_2 Capture

Mingyuan Cao (Unversity of Southern California), Linghao Zhao (University of Southern California), Dongwan Xu (University of Southern California), Secgin Karagoz (UCLA), Patricia Pichardo (UCLA), Richard J. Ciora, Jr. (Media and Process Technology, Inc.), Paul Kt Liu (Media and Process Technology, Inc.), Vasilios I. Manousiouthakis (UCLA), Theodore T Tsotsis (University of Southern California)*

11:00 am MONDAY

6d - Synthesis of Polymers of Intrinsic Microporosity (PIM-1) with Poly(ethylene glycol) for Membrane Application

Volkan Filiz (Helmholtz-Zentrum Geesthacht)*, Gisela Bengtson (Helmholtz-Zentrum Geesthacht), Silvio Neumann (Helmholtz-Zentrum Geesthacht)

11:30 am MONDAY

6e - High-permeance polymer-functionalized graphene membranes that surpass the postcombustion carbon capture target

Guangwei He (L'Ecole polytechnique fédérale de Lausanne), Kumar Varoon Agrawal (L'Ecole polytechnique fédérale de Lausanne)*

12:00 pm MONDAY

6f - Fabrication of Thin Film Composite Membranes using Microporous Polymer Blends and Polybenzimidazole Nanoporous Supports for CO_2/N_2 Separation

Lingxiang Zhu (U.S. Department of Energy National Energy Technology Laboratory)*, Surendar Venna (U.S. Department of Energy National Energy Technology Laboratory), Ali Sekizkardes (Battelle/ U.S. Department of Energy National Energy Technology Laboratory), Victor A Kusuma (U.S. Department of Energy National Energy Technology Laboratory), David Hopkinson (U.S. Department of Energy National Energy Technology Laboratory)

#7 Applications: Assessing Performance, Robustness, and Scalability for Carbon Capture ...pg 16

2:00 pm - 5:00 pm | Kings Garden 5

Co-Chairs: Dave Hopkinson (U.S. Department of Energy National Energy Technology Laboratory), Winston Ho (Ohio State University)

2:00 pm MONDAY

7a - Field trial of spiral-wound facilitated transport membrane module for CO_2 capture from flue gas

Yang Han (Ohio State University)*, Witopo Salim (Ohio State University), Kai Chen (Ohio State University), Dongzhu Wu (Ohio State University), Winston Ho (Ohio State University)

2:30 pm MONDAY

7b - Gen-2 Proteus™ membrane and module development for CO_2 capture from gasification streams

Witopo Salim (Membrane Technology & Research Inc.)*, Karl Amo (Membrane Technology & Research Inc.), Trevor Carlisle (Oregon State University), Richard Baker (Membrane Technology & Research Inc.), Jenny He (Membrane Technology & Research Inc.), Jay Kniep (Membrane Technology & Research Inc.), Tim Merkel (Membrane Technology & Research Inc.), Vincent Nguyen (Membrane Technology & Research Inc.), Zhen Sun (Membrane Technology & Research Inc.), Jonathan Tan (Membrane Technology & Research Inc.), Bob Watson (Membrane Technology & Research Inc.), Erik Westling (Membrane Technology & Research Inc.)

3:00 pm MONDAY

7c - One-pot surface modification of nanocellulose fibrils and effect of manipulated fibril surfaces in Hybrid Facilitated Transport Membranes for CO_2 capture

Saravanan Janakiram (Norwegian University of Science and Technology)*, Xinyi Yu (Norwegian University of Science and Technology), Luca Ansaloni (SINTEF Industry), Zhongde Dai (Norwegian University of Science and Technology), Liyuan Deng (Norwegian University of Science and Technology)

3:30 pm MONDAY

7d - DDR-type zeolite membrane: The first opportunity in CO_2-EOR industry

Junya Okazaki (JGC Corporation)*, Hiroaki Hasegawa (JGC Corporation), Nobuyasu Chikamatsu (JGC Corporation), Kenji Yajima (NGK Insulators), Katsuya Shimizu (NGK Insulators), Makiko Niino (NGK Insulators)

4:00 pm MONDAY

7e - Coating of hollow fiber membrane for carbon molecular sieve membrane preparation

Yuhe Cao (Georgia Institute of Technology)*, Kuang Zhang (Georgia Institute of Technology), Oishi Sanyal (Georgia Institute of Technology), William Koros (Georgia Institute of Technology)

4:30 pm MONDAY

7f - Ultrapermeable benzotriptycene-based PIMs that redefine the upper bounds for CO_2 separations

Alessio Fuoco (National Research Council of Italy)*, Bibiana Comesaña-Gándara (University of Edinburgh), Jie Chen (University of Edinburgh), Grazia Bezzu (University of Edinburgh), Mariolino Carta (Swansea University), Ian Rose (University of Edinburgh), Maria-Chiara Ferrari (University of Edinburgh), Johannes Carolus (John) Jansen (National Research Council of Italy), Elisa Esposito (National Research Council of Italy), Marcello Monteleone (National Research Council of Italy) Neil McKeown (University of Edinburgh)

#8 Processes: Innovations in Module Modeling and Design ...pg. 19

2:00 pm - 5:00 pm | Ballroom 3

Co-Chairs: David Ladner (Clemson University), Grigorius Panagakos (U.S. Department of Energy National Energy Technology Laboratory)

2:00 pm MONDAY

8a - CFD simulation of crimped hollow fiber membranes for liquid separation processes

Mohammad Younas (University of Engineering and Technology, Peshawar)*, Amir Muhammad (University of Engineering and Technology, Peshawar), Glenn Lipscomb (University of Toledo)

2:30 pm MONDAY

8b - Computational Fluid Dynamics Modeling for the Investigation of Multi-Layer Spacer Effects

Zachary Binger (The University of Arizona), Andrea Achilli (The University of Arizona)*

3:00 pm MONDAY

8c - Direct numerical simulation of unsteady mixing in direct contact membrane distillation systems with membrane spacers

Jincheng Lou (Colorado School of Mines)*, Jacob Johnston (Colorado School of Mines), Nils Tilton (Colorado School of Mines)

3:30 pm MONDAY

8d - Concentration polarization modeling for high-pressure membranes with engineered surface features

Zuo Zhou (Clemson University)*, David Ladner (Clemson University), Sapna Sarupria (Clemson University), Steven Weinman (Clemson University), Scott Husson (Clemson University), Ilenia Battiato (Stanford University), Negin Kananizadeh (Clemson University)

4:00 pm MONDAY

8e - Analysis of Micromixers to Minimize Scaling Effects on Reverse Osmosis Membranes

Jeremy Walker (U.S. Army)*, Shawn McElmurry (Wayne State University), James Dusenbury (U.S. Army)

4:30 pm MONDAY

8f - Do Hydrodynamic Instabilities Cause Roughness in Thin Polyamide Films of Reverse Osmosis Membranes?

Akshay Deshmukh (Yale University)*, Puskar Mondal (Yale University), Subhajyoti Chaudhuri (Yale University), Menachem Elimelech (Yale University)

#9 Processes: Osmotically-Driven Membrane Processes for Water and Energy ...pg. 22

2:00 pm - 5:00 pm | Ballroom 4

Co-Chairs: Tony Straub (University of Colorado Boulder/MIT), Aaron Wilson (Idaho National Lab), Emily Tow (Olin College), Chris Gorski (Penn State University), Kitty Nijmeijer (Eijndhoven University of Technology)

2:00 pm MONDAY

9a - Structural and Transport Properties of Membranes in High-Salinity Desalination using Cascading Osmotically Mediated Reverse Osmosis

Xi Chen (Columbia University)*, Chanhee Boo (Columbia University), Ngai Yin Yip (Columbia University)

2:30 pm MONDAY

9b - Utilization of reverse solute diffusion to enhance membrane performance during osmotically driven processes - improved boron retention and silica scaling mitigation

Yining Wang (Nanyang Technological University)*, Weiyi Li (Southern University of Science and Technology), Chuyang Tang (University of New South Wales), Rong Wang (Nanyang Technological University)

3:00 pm MONDAY

9c - Osmotic Membrane Desalination Performance Governed by Molecular Reflection at the Liquid-Vapor Interface

Akshay Deshmukh (Yale University), Jongho Lee (University of British Columbia)*

3:30 pm MONDAY

9d - Characterization of External and Internal Fouling of Forward Osmosis (FO) Process Through Ultrasonic Time Domain Reflectometry (UTDR) Method

Li Lai (Nanyang Technological University)*, Tzyy Haur Chong (Singapore Membrane Technology Centre)

4:00 pm MONDAY

9e - Incorporating membrane deformation into the boundary layer equation to model water and reverse salt flux in osmotic processes

Jaime A Idarraga-Mora (Clemson University)*, Alton O'Neal (Clemson University), Morgan Pfeiler (Clemson University), David Ladner (Clemson University), Scott Husson (Clemson University)

4:30 pm MONDAY

9f - A Thin-Film Forward-Osmosis Membrane Made from a Graphene Oxide-Silver Metal-Organic Framework Nanocomposite: Antifouling and Antibiofouling Assessment

Mostafa Dadashi Firouzjaei (University of Alabama), Ahmad Arabi Shamsabadi (Drexel University), Mohammad Sharifian Gh (University of Virginia), Milad Rabbani Esfahani (University of Alabama)*, Ahmad Rahimpour (Babol Noshirvani University of Technology), Masoud Soroush (Drexel University)

#10 Applications: Water Reuse ...pg. 25

2:00 pm - 5:00 pm | Kings Garden 4

Co-Chairs: Pei Xu (New Mexico State University), Jack Gilron (Ben-Gurion University of the Negev)

2:00 pm MONDAY

10a - Tannic acid-iron network based green technique for enhanced membrane performance in water reuse

Hao Guo (The University of Hong Kong)*

2:30 pm MONDAY

10b - Membrane Distillation (MD) and Reverse Osmosis (RO) processes for water recovery from pre-treated high strength brewery wastewater

Nawrin Anwar (Concordia University)*, Mahbuboor Choudhury (Concordia University), Saifur Rahaman (Concordia University)

3:00 pm MONDAY

10c - Field Demonstration of Real-Time Colloidal Particle Monitoring to Improve MF and UF Membrane Performance

Jana Safarik (Orange County Water District)*, Ganesh Rajagopalan (Kennedy/Jenks Consultants), Helia Safaee (Kennedy/Jenks Consultants), Megan Plumlee (Orange County Water District), Zita Yu (West Basin Municipal Water District)

3:30 pm MONDAY

10d - Selective removal of phosphorus from wastewater using electromembrane process

Sanhita Chaudhury (Ben-Gurion University of the Negev)*, Oded Nir (Ben-Gurion University of the Negev)

4:00 pm MONDAY

10e - Water Recovery from Produced Water via Robust Membrane Distillation

Rong Wang (Nanyang Technological University)*

4:30 pm MONDAY

10f - Potable-quality water recovery from primary effluent through an integrated algal-osmosis membrane system

Lu Lin (New Mexico State University)*, Wenbin Jiang (New Mexico State University), Pei Xu (New Mexico State University)

#11 Materials: Ion-Exchange and Electrofunctional Materials ...pg. 28

2:00 pm - 5:00 pm | Kings Garden 1

Co-Chairs: Geoff Geise (University of Virginia), Orlando Coronell (University of North Carolina at Chapel Hill)

2:00 pm MONDAY

11a - Ion Transport Through Perforated Nanoporous Graphene

Mandakranta Ghosh (University of Twente)*, Koen F. A. Jorissen (University of Twente), Jeffery A. Wood (University of Twente), Rob G. H. Lammertink (University of Twente)

2:30 pm MONDAY

11b - Developing a new approach to describe ion sorption and transport in Nafion membranes

Rahul Sujanani (The University of Texas at Austin)*, Jovan Kamcev (UC Berkeley), Eui-Soung Jang (The University of Texas at Austin), Benny Freeman (The University of Texas at Austin), Donald Paul (The University of Texas at Austin)

3:00 pm MONDAY

11c - Co-transport of neutral solutes in Nafion 117 using In-situ ATR FTIR spectroscopy for multicomponent solution speciation

Breanna Dobyns (Auburn University)*, Bryan S Beckingham (Auburn University)

3:30 pm MONDAY

11d - Functionalized Nanoporous Ceramic Membranes Towards Low-Cost Electrodialysis

Gregory M Newbloom (Membrion, Inc.)*, Ryan Kingsbury (Membrion, Inc.), Rachel Malone (Membrion, Inc.)

4:00 pm MONDAY

11e - Polymerization of counterions in self-assembled, 1 nm pores of lyotropic liquid crystal anion exchange membrane to tune nanopore and ion transport properties

Michael J. McGrath (University of Colorado at Boulder)*, Samantha Hardy, Andrew Basalla (University of Colorado Boulder), Hans Funke (University of Colorado Boulder), Bryce Manubay (University of Colorado Boulder), Zhangxing Shi (University of Colorado Boulder), Douglas Gin (University of Colorado Boulder), Richard Noble (University of Colorado Boulder)

4:30 pm MONDAY

11f - Quantifying tortuosity and diffusion over multiple scales in ionic and non-ionic polymer membranes

Louis Madsen (Virginia Tech)*

#12 Materials: Membrane Material and Transport Simulation: Molecular & Process Modeling of Membranes ...pg. 30

2:00 pm - 5:00 pm | Kings Garden 3

Co-Chairs: Chris Wilmer (University of Pittsburgh), Abhishek Roy (The Dow Chemical Company)

2:00 pm MONDAY

12a - Expanding the chemical palette for reliable chemical separations using molecular simulations and machine learning

David Sholl (Georgia Institute of Technology)*, Farhad Gharagheizi (Georgia Institute of Technology), Dai Tang (Georgia Institute of Technology)

2:30 pm MONDAY

12b - Diffusion, Reaction, and Network Structure Modeling of Interfacial Polymerization of Polyamide Membranes

Jeffrey D Wilbur (DuPont Water Solutions)*, Toni Bechtel (DuPont Water Solutions), Matthew Jansma (DuPont Water Solutions), Steve Rosenberg (DuPont Water Solutions), Dan Arriola (DuPont Water Solutions)

3:00 pm MONDAY

12c - Optimization of Mixed Matrix Membrane Materials for Post-Combustion Carbon Capture

Janice A. Steckel (U.S. Department of Energy National Energy Technology Laboratory)*, Samir Budhathoki (U.S. Department of Energy National Energy Technology Laboratory), Miguel Zamarripa-Perez (U.S. Department of Energy National Energy Technology Laboratory), Paul Boone (University of Pittsburgh), Christopher Wilmer (University of Pittsburgh)

3:30 pm MONDAY

12d - Diffusivity in anion exchange membranes under low hydration: A molecular dynamics study

Dario Dekel (Technion - Israel Institute of Technology), Israel Zadok (Technion - Israel Institute of Technology), Srdan Pusara (Technion - Israel Institute of Technology), Simcha Srebnik (Technion - Israel Institute of Technology)*

4:00 pm MONDAY

12e - Modeling of polymer membrane formation via phase inversion by mesoscopic phase-field methods: Investigating the development of structure on multiple scales

Michael R Cervellere (University of Arkansas/EMD Millipore)*, Xianghong Qian (University of Arkansas), David Ford (University of Arkansas), Paul Millett (University of Arkansas)

4:30 pm MONDAY

12f - Multiscale Modeling of Gas Permeation Through Poly(Dimethyl Phenylene Oxide) (PPO) and Its Pre-Steady State Polymer Response

Marielle Soniat (Lawrence Berkeley National Laboratory)*, Meron Tesfaye (Lawrence Berkeley National Laboratory), Daniel Brooks (CalTech), Nicholas Humphrey (USC), Lien-Chun Weng (Lawrence Berkeley National Laboratory), Boris Merinov (CalTech), William Goddard (CalTech), Adam Weber (Lawrence Berkeley National Laboratory), Frances Houle (Lawrence Berkeley National Laboratory)

Plenary Session 2

7:45 am - 9:00 pm | Ballroom 1

Dr. Tim Merkel
Membrane Technology and Research, Inc.

Development of Membranes for CO_2 Capture

#13 Processes: Process Innovations in Electrofunctional and Electrocatalytic Membrane Processes ...pg. 33

9:30 am - 12:30 pm | Kings Garden 4

Co-Chairs: Charles-François de Lannoy (McMaster University), David Jassby (UCLA)

9:30 am TUESDAY

13a - Scaling mitigation in membrane distillation through electrokinetic mixing

Unnati Rao (UCLA)*, Guy Z Ramon (Technion - Israel Institute of Technology), Eric Hoek (UCLA), David Jassby (UCLA)

10:00 am TUESDAY

13b - Donnan dialysis desalination with thermolytic salts

Hanqing Fan (Columbia University)*, Ngai Yin Yip (Columbia University)

10:30 am TUESDAY

13c - Electrically Conductive Membranes as Sensors: Stability, Conductivity, and Sensitivity

Nan Zhang (McMaster University), Amin Halali (McMaster University), Charles-François de Lannoy (McMaster University)*

11:00 am TUESDAY

13d - Electrocatalytic Membrane Reactor-based Integrated Processes for Landfill Leachate Treatment

Jianxin Li (Tianjin Polytechnic University)*, Hong Wang (Tianjin Polytechnic University), Junhao Ding (Tianjin Polytechnic University)

11:30 am TUESDAY

13e - Nitrogen and Phosphorus Recovery from Agricultural Wastewater Effluents - Role of Electrically Conductive Membranes

Avner Ronen (Temple University)*, Kartikeya Kekre (Temple University)

12:00 pm TUESDAY

13f - Simulations of Highly Selective Separations Based on Electromigration

Merlin Bruening (University of Notre Dame)*, Andriy Yaroshchuk (Polytechnic University of Catalonia), Mykola Bondarenko (National Academy of Sciences of Ukraine), Chao Tang (University of Notre Dame), Muhammad Ahmad (University of Notre Dame)

#14 Applications: Contaminant Removal from Water Sources ...pg. 36

9:30 am - 12:30 pm | Ballroom 3

Co-Chairs: Andre da Costa (Michigan Technological University), Prakhar Prakash (Chevron - RPE)

9:30 am TUESDAY

14a - Oil drop behavior on model nanofiltration membrane surfaces under conditions of hydrodynamic shear

Emily Tummons (Michigan State University), Charifa Hejase (Michigan State University)*, Zhefei Yang (University of Massachusetts Amherst), Jia Wei Chew (Nanyang Technological University), Merlin Bruening (University of Notre Dame), Vlad Tarabara (Michigan State University)

10:00 am TUESDAY

14b - A case study review of the application of reverse osmosis membranes for water treatment and reagent recovery at mining and mineral processing operations

Chris Biederman (Hatch)*, Jeffrey Cobbledick (Hatch)

10:30 am TUESDAY

14c - Self-Cleaning Nanocomposite Membranes with Phosphorene-Based Pore Fillers for Water Treatment

Joyner Eke (University of Kentucky)*, Isabel Escobar (University of Kentucky)

11:00 am TUESDAY

14d - Optimization of nanofiltration process for treating industrial wastewater of time-varying composition

Salman Alizadeh Kordkandi (McMaster University); Ryan LaRue (McMaster University)*; Abhishek Premachandra (McMaster University); Jacob Sitko (McMaster University); David Latulippe (McMaster University)

11:30 am TUESDAY

14e - Adsorptive and Reactive Membranes for Remediation of PFOA, Chloro-organics, and RCRA Metals from Water

DB Bhattacharyya (University of Kentucky)*, Hongyi (Derek) Wan (University of Kentucky), Anthony Saad (University of Kentucky), Md. Saiful Islam (University of Kentucky), Ashish Aher (University of Kentucky), Rollie Mills (University of Kentucky), Lindell Ormsbee (University of Kentucky), Evan Hatakeyama (Chevron)

12:00 pm TUESDAY

14f - Selective separation of mono- and di-valent cations in electrodialysis during brackish water desalination: bench and pilot-scale studies

Xuesong Xu (New Mexico State University)*, Pei Xu (New Mexico State University), Guanyu Ma (New Mexico State University)

#15 Materials: Materials for Organic Solvent Separations ...pg. 39

9:30 am - 12:30 pm | Ballroom 4

Co-Chairs: Steve White (Membrane Technology & Research Inc.), Ryan Lively (Georgia Institute of Technology)

9:30 am TUESDAY

15a - Membrane solvent recovery in edible oil industry

Petrus Cuperus (SolSep BV)*, Ingrid Wienk (SolSep BV), Pieter Vandezande (VITO), Marco Pipolo (Bunge), Jan Kolijn (TUSTI), Will Fuchten (MTSA)

10:00 am TUESDAY

15b - Molecular Design of Polymeric Membranes for Organic Solvent Recovery

Tai-Shung Chung (National University of Singapore)*

10:30 am TUESDAY

15c - Epoxy-based solvent resistant nanofiltration membranes prepared through phase inversion

Maarten Bastin (KU Leuven)*, Jasper Raymenants (KU Leuven), Ivo Vankelecom (KU Leuven)

11:00 am TUESDAY

15d - Nanoporous Graphene Membranes for Organic Solvent Nanofiltration

David Cheng (MIT)*, Rohit Karnik (MIT)

11:30 am TUESDAY

15e - OSN Technology in Petrochemical Industry: Looking Back, Looking Forward

Dhaval Bhandari (ExxonMobil), JR Johnson* (ExxonMobil)

12:00 pm TUESDAY

15f - Enabling Organic Solvent Nanofiltration and Reverse Osmosis using Ceramic Supported TFC Membranes

Jeffrey R McCutcheon (University of Connecticut)*, Lingling Xia (University of Connecticut), Marcus Weyd (Fraunhofer Institute for Ceramic Technologies and Systems IKTS), Mi Zhang (University of Connecticut)

#16 Materials: Catalytic and Responsive Materials ...pg. 41

9:30 am - 12:30 pm | Kings Garden 3

Co-Chairs: Miao Yu (Rensselaer Polytechnic Institute), Ayse Asatekin (Tufts University)

9:30 am TUESDAY

16a - Responsive CNT Composite Membranes for Protection Against Chemical Warfare Agents

Melinda L Jue (Lawrence Livermore National Laboratory)*, Chiatai Chen (Cornell University), Yifan Li (MIT), Eric Meshot (Lawrence Livermore National Laboratory), Ngoc Bui (Lawrence Livermore National Laboratory), Rong Zhu (MIT), Myles Herbert (MIT), Sei Jin Park (Lawrence Livermore National Laboratory), Steven Buchsbaum (Lawrence Livermore National Laboratory), Kuang Jen Wu (Lawrence Livermore National Laboratory), Timothy Swager (MIT), Francesco Fornasiero (Lawrence Livermore National Laboratory)

10:00 am TUESDAY

16b - Highly Efficient Dimethyl Ether Production from CO_2 Hydrogenation in a Dehydration Membrane Reactor

Huazheng Li (Rensselaer Polytechnic Institute)*, Fanglei Zhou (Rensselaer Polytechnic Institute), Qiaobei Dong (Rensselaer Polytechnic Institute), Shoujie Ren (Missouri University of Science and Technology), Xinhua Liang (Missouri University of Science and Technology), Naomi Klinghoffer (Gas Technology Institute), Shiguang Li (Gas Technology Institute), Miao Yu (Rensselaer Polytechnic Institute)

10:30 am TUESDAY

16c - Poly-methacrylic acid Functionalized Membranes with Incorporated Reactive Pd/Fe Nanoparticles: Lab Scale to Groundwater Remediation Applications

Hongyi (Derek) Wan (University of Kentucky)*, Md. Saiful Islam (University of Kentucky), Nicolas Briot (Unuversity of Kentucky), Anthony Saad (University of Kentucky), Lindell Ormsbee (University of Kentucky), DB Bhattacharyya (University of Kentucky)

11:00 am TUESDAY

16d - Investigation of a smart electrically responsive ultrafiltration membrane

Chia Miang Khor (UCLA)*, David Jassby (UCLA)

11:30 am TUESDAY

16e - Catalytic Membranes for Groundwater Treatment

Alexander J Sutherland (McMaster University)*, Charles-François de Lannoy (McMaster University)

12:00 pm TUESDAY

16f - Plasmonic membrane catalytic activation of peroxide for quantized oxidation via residence time control

Hao Tang (University of Washington), Guozheng Shao (University of Washington), Bruce Hinds (University of Washington)*

#17 Materials: Selective Polymeric and Mixed-Matrix Materials - Gas Separations ...pg. 44

9:30 am - 12:30 pm | Kings Garden 5

Co-Chairs: Ben Sundell (Aramco Services Company), Zach Smith (MIT)

9:30 am TUESDAY

17a - Mixed-matrix membranes formed from imide-functionalized UiO-66-NH_2 for improved interfacial compatibility

Qihui Qian (MIT)*, Zach Smith (MIT)

10:00 am TUESDAY

17b - Janus polymers bearing tri(n-alkoxy)silyl side groups: glassy membrane materials with properties of rubbers

Yury P Yampolskii (A.V.Topchiev Institute of Petrochemical Synthesis, Russian Academy of Sciences)*, Ludmila Starannikova (A.V.Topchiev Institute of Petrochemical Synthesis, Russian Academy of Sciences), Dmitry Alentiev (A.V.Topchiev Institute of Petrochemical Synthesis, Russian Academy of Sciences), Maxim Bermeshev (A.V.Topchiev Institute of Petrochemical Synthesis, Russian Academy of

Sciences), Eugene Finkelshtein (A.V.Topchiev Institute of Petrochemical Synthesis, Russian Academy of Sciences)

10:30 am TUESDAY

17c - Solution processable metal organic frameworks for gas separations: from porous liquids to mixed matrix membranes

Anastasiya V Bavykina (King Abdullah University of Science and Technology)*, Alexander Knebel (Leibniz University Hannover Institute for Physical Chemistry and Electrochemistry), Shuvo Datta (King Abdullah University of Science and Technology), Magnus Rueping (King Abdullah University of Science and Technology), Mohamed Eddaoudi (King Abdullah University of Science and Technology), Juergen Caro (Leibniz University Hannover Institute for Physical Chemistry and Electrochemistry), Jorge Gascon (King Abdullah University of Science and Technology)

11:00 am TUESDAY

17d - Enhancing CO_2/N_2 Selectivity and Elimination of Langmuir Sorption within High Tg, Glassy Polynorbornene Membranes

Christopher Maroon (University of Tennessee), Jacob Townsend (University of Tennessee), Kevin Gmernicki (University of Tennessee), Daniel Harrigan (Aramco Services Company), Ben Sundell (Aramco Services Company), John Lawrence (Aramco), Shannon Mahurin (Oak Ridge National Laboratory), Konstantinos Vogiatzis (University of Tennessee), Brian Long (University of Tennessee)*

11:30 am TUESDAY

17e - Ultra-High Permeability Mixed-Matrix Membranes as a Next-Generation Carbon Capture Technology for Post-Combustion

Sameh K Elsaidi (U.S. Department of Energy National Energy Technology Laboratory)*, Surendar Venna (U.S. Department of Energy National Energy Technology Laboratory), Ali Sekizkardes (Battelle/ U.S. Department of Energy National Energy Technology Laboratory), Mona Mohamed (University of Pittsburgh), David Hopkinson (U.S. Department of Energy National Energy Technology Laboratory)

12:00 pm TUESDAY

17f - Composite gas separation membranes from metal-induced ordered polymeric frameworks

Zhihua Qiao (Tianjin University), Song Zhao (Tianjin University), Menglong Sheng (Tianjin University), Jixiao Wang (Tianjin University), Zhi Wang (Tianjin University), Chongli Zhong (Tianjin University), Michael D Guiver (Tianjin University)*

#18 Materials: Selective Polymeric and Mixed-Matrix Materials - Gas Separations ...pg.47

2:00 pm - 5:00 pm | Kings Garden 5

Co-Chairs: Ben Sundell (Aramco Services Company), Zach Smith (MIT)

2:00 pm TUESDAY

18a - High temperature gas separation properties of sub-micron polybenzimidazole membranes

Melanie M Merrick (The University of Texas at Austin)*, Benny Freeman (The University of Texas at Austin)

2:30 pm TUESDAY

18b - Thermally Rearranged Semi-interpenetrating Polymer Network (TR-SIPN) Membranes for Olefin/Paraffin Separations

Young Moo Lee (Hanyang University)*, Won Hee Lee (Hanyang University), Jong Geun Seong (Hanyang University), Ho Hyun Wang (Hanyang University), Sun Ju Moon (Hanyang University)

3:00 pm TUESDAY

18c - Conformation-controlled molecular sieving effect for membrane-based propylene/propane separation

Yang Liu (Georgia Institute of Technology)*, William Koros (Georgia Institute of Technology), Mohamed Eddaoudi (KAUST), Gongping Liu (Georgia Institute of Technology), Youssef Belmabkhout (KAUST), Zhijie Chen (KAUST)

3:30 pm TUESDAY

18d - Mixed matrix materials containing soluble discrete metal-organic polyhedra in rubbery polyethers for CO_2/gas separation

Haiqing Lin (The State University of New York at Buffalo)*, Junyi Liu (The State University of New York at Buffalo), Cressa Fulong (The State University of New York at Buffalo), Tim Cook (The State University of New York at Buffalo)

4:00 pm TUESDAY

18e - Accessing novel microporous polymers to enhance yield, stereochemical control, and membrane performance in the recovery of light hydrocarbons from natural gas

John Lawrence (Aramco Services Company)*

4:30 pm TUESDAY

18f - Insight into the transport in polymeric and mixed matrix membranes via analysis of unique mixed gas diffusion coefficients

Johannes Carolus (John) Jansen (National Research Council of Italy)*, Alessio Fuoco (National Research Council of Italy), Marcello Monteleone (National Research Council of Italy), Elisa Esposito (National Research Council of Italy)

#19 Processes: Process Scale-Up and Techno-Economic Assessment ...pg. 50

2:00 pm - 5:00 pm | Kings Garden 4

Co-Chairs: JR Johnson (ExxonMobil), Bharat Bhut (Merck Pharmaceuticals), Ivy Huang (Membrane Technology and Research), Albert Kim (University of Hawaii)

2:00 pm TUESDAY

19a - Minimizing the cost of membrane distillation

Timothy V Bartholomew (Carnegie Mellon University)*, Alexander V Dudchenko (Carnegie Mellon University), Nicholas Siefert (U.S. Department of Energy National Energy Technology Laboratory), Meagan S Mauter (Carnegie Mellon University)

2:30 pm TUESDAY

19b - Process Optimization Using Perturbation Expansion Solutions for Membrane Gas Separation Modules

Norfamila Che Mat (Universiti Malaysia Sarawak), Glenn Lipscomb (University of Toledo)*

3:00 pm TUESDAY

19c - Multi-physics Simulation of Hollow Fiber Vacuum Membrane Distillation Using OpenFoam

Albert Kim (University of Hawaii), Ho Ji (Korea Research Institute Of Ships And Ocean Engineering), Deok-Soo Moon (Korea Research Institute Of Ships And Ocean Engineering), Hyeon-Ju Kim (Korea Research Institute Of Ships And Ocean Engineering)

3:30 pm TUESDAY

19d - Scale-up of Facilitated Transport Membrane Module for CO_2 Capture from Flue Gas

Kai Chen (Ohio State University)*, Witopo Salim (Ohio State University), Yang Han (Ohio State University), Dongzhu Wu (Ohio State University), Winston Ho (Ohio State University)

4:00 pm TUESDAY

19e - Development of industrial scale polyvinylidene fluoride transfer membrane

Marta E Bojarska (GVS Filter Technology)*, Deb English (GVS Filter Technology), Gil Tavares (GVS Filter Technology)

4:30 pm TUESDAY

19f - Low energy seawater desalination using multistage electrodialysis

Gijs Doornbusch (Eindhoven University of Technology)*, Michele Tedesco (Wetsus), Jan Post (Wetsus), Zandrie Borneman (Eindhoven University of Technology), Kitty Nijmeijer (Eindhoven University of Technology)

#20 Processes: New Concepts in Hybrid Processes and Process Integration ...pg. 52

2:00 pm - 5:00 pm | Kings Garden 3

Co-Chairs: Jia Wei Chew (Nanyang Technological University), Ed Sanders (Air Liquide-Global Markets & Technologies)

2:00 pm TUESDAY

20a - Air Products Hybrid Membrane Adsorption

Cory E Sanderson (Air Products and Chemicals)*, Don Henry (Air Products and Chemicals)

2:30 pm TUESDAY

20b - Hybrid Distillation and Facilitated Transport Membrane Processes for C3 Splitter Debottlenecking

Kenneth Pennisi (Compact Membrane Systems), Christine Parrish (Compact Membrane Systems)*, Sudip Majumdar (Compact Membrane Systems)

3:00 pm TUESDAY

20c - Water Recovery via Forward Osmosis - Freeze Concentration

Georgios Kolliopoulos (University of Toronto), Jeffrey Martin (University of Toronto), Chenbo Xu (University of Toronto), Vladimiros Papangelakis (University of Toronto)*

3:30 pm TUESDAY

20d - Leverage membrane technology in process intensification and new application design - from early stage design to technical scale implementation

Peter Kreis (Evonik Technology & Infrastructure GmbH)*, Kah Peng Lee (Evonik Fibres GmbH), Lars Peters (Evonik Fibres GmbH), Erik Hoving (Evonik Cooperation), Steven Pedersen (Evonik Canada Inc.), Kumar Abhinava (Evonik Cooperation)

4:00 pm TUESDAY

20e - Carbon Capture utilizing Hybrid Membrane/Liquifaction

David Hasse (Air Liquide)*

4:30 pm TUESDAY

20f - Membrane distillation hybridized with a thermoelectric heat pump for energy-efficient water treatment and space cooling

Yong Zen Tan (Nanyang Technological University), Nick Guan Ping Chew (Nanyang Technological University), Wai Hoong Chow (Nanyang Technological University), Rong Wang (Nanyang Technological University), Jia Wei Chew (Nanyang Technological University)*

#21 Applications: Brine Treatment for Minimal and Zero Liquid Discharge ...pg 54

2:00 pm - 5:00 pm | Ballroom 3

Co-Chairs: Radisav Vidic (University of Pittsburgh), Kerri Hickenbottom (University of Arizona)

2:00 pm TUESDAY

21a - Reduction of Brackish Water Desalination Brine Volume using Membrane Evaporation Coupled with Activated Sludge Aeration Waste Heat

Drew W Johnson (The University of Texas at San Antonio)*

2:30 pm TUESDAY

21b - Smart High Recovery RO Boosts Recovery Rate to More Than 90% for Reuse Purposes

Ronit Erlitzki (AdEdge Water Technologies)*

3:00 pm TUESDAY

21c - Membrane Distillation for Enhanced Water Recovery by Inland Brackish Water Desalination Plants

Zhewei Zhang (University of Pittsburgh)*, Radisav Vidic (University of Pittsburgh)

3:30 pm TUESDAY

21d - Selective Separation of Industrial Relevant Metal Ions from High TDS Water using Wafer-Enhanced Electrodeionization (WE-EDI)

Humeyra B Ulusoy Erol (University of Arkansas)*, Christa Hestekin (University of Arkansas), Jamie Hestekin (University of Arkansas)

4:00 pm TUESDAY

21e - Pilot Evaluation of Closed Circuit Reverse Osmosis (CCRO) for Treatment of RO Concentrate for Potable Reuse

Han Gu (Orange County Water District)*, Megan Plumlee (Orange County Water District), Mike Boyd (Desalitech), Ran Nadav (Desalitech), Michael Hwang (Jacobs), Jim Lozier (Jacobs)

4:30 pm TUESDAY

21f - Brine Treatment for Minimal and ZLD

Malynda Cappelle (The University of Texas at El Paso)*, W. Shane Walker (The University of Texas at El Paso)

#22 Materials: Innovations in Membrane Synthesis and Casting ...pg. 57

2:00 pm - 5:00 pm | Ballroom 4

Co-Chairs: Rachel Dorin (Terapore Technologies), Ulrich Wiesner (Cornell University)

2:00 pm TUESDAY

22a - Novel Spirocyclic Polymers for Membrane-based Organic Solvent Separations

Ronita Mathias (Georgia Institute of Technology)*, Kirstie Thompson (Georgia Institute of Technology), Dhaval Bhandari (ExxonMobil), JR Johnson (ExxonMobil), Huaxing Zhou (ExxonMobil), M.G. Finn (Georgia Institute of Technology), Ryan Lively (Georgia Institute of Technology)

2:30 pm TUESDAY

22b - 3D Printed fouling resistant composite membranes

Davide Mattia (University of Bath)*

3:00 pm TUESDAY

22c - Next-Generation Membranes Using Thin-Film Lift Off

Mackenzie Anderson (UCLA)*, Brian McVerry (UCLA), Na He (UCLA), Eric Hoek (UCLA), Richard Kaner (UCLA)

3:30 pm TUESDAY

22d - Interfacial Junctions Control Electrolyte Transport Through Charge-Patterned Membranes

Feng Gao (University of Notre Dame), Bill Phillip (University of Notre Dame)*

4:00 pm TUESDAY

22e - Fabrication of Inside-Out Isoporous Hollow Fiber Membranes via Spinning and Coating Methods

Kirti Sankhala (Helmholtz-Zentrum Geesthacht, Geesthacht)*, Joachim Koll (Helmholtz-Zentrum Geesthacht, Geesthacht), Maryam Radjabian (Helmholtz-Zentrum Geesthacht, Geesthacht), Clarissa Abetz (Helmholtz-Zentrum Geesthacht, Geesthacht), Volker Abetz (Helmholtz-Zentrum Geesthacht, Geesthacht)

4:30 pm TUESDAY

22f - Empirical evidence suggests that polyamide film formation during interfacial polymerization in thin film composite (TFC) membranes is not self-limiting

Kasia Grzebyk (University of North Carolina at Chapel Hill)*, Orlando Coronell (University of North Carolina at Chapel Hill)

Plenary Session 3

7:45 am - 9:00 pm | Ballroom 1

Dr. Rachel Segalman
University of California, Santa Barbara

Polymeric Ionic Liquids: A New Platform for Materials and Energy

#23 Awards Session (Student Fellowship & Young Membrane Scientist Awards) ...pg. 59

9:30 am - 12:30 pm | Kings Garden 4

Chair: Caleb Funk (The Dow Chemical Company)

9:30 am WEDNESDAY

23a - Creation of well-defined "mid-sized" micropores in carbon molecular sieve membranes for organic solvent separations

Yao Ma (Georgia Institute of Technology)*, Dhaval Bhandari (ExxonMobil), JR Johnson (ExxonMobil), Ryan Lively (Georgia Institute of Technology)

10:00 am WEDNESDAY

23b - PoreDesigner for computational optimization of channel proteins and porous organic cages for highly selective membrane-based separations

Ratul Chowdhury (Penn State University)*

10:30 am WEDNESDAY

23c - Crack-free, large-area, single-layer graphene membranes with a record performance in gas mixture separation

Shiqi Huang (L'Ecole polytechnique fédérale de Lausanne)*, Kumar Varoon Agrawal (L'Ecole polytechnique fédérale de Lausanne), Jing Zhao (L'Ecole polytechnique fédérale de Lausanne)

11:00 am WEDNESDAY

23d - Unprecedented CO_2/H_2 Selectivity and CO_2 Permeance Demonstrated in Facilitated Transport Membranes with Tunable Amine-CO_2 Chemistry

Yang Han (Ohio State University)*, Winston Ho (Ohio State University)

11:30 am WEDNESDAY

23e - Engineering the Nanochannels in Reduced Graphene Oxide Membrane for Dye Desalination

Liang Huang (The State University of New York at Buffalo)*, Haiqing Lin (The State University of New York at Buffalo)

12:00 pm WEDNESDAY

23f - Impact of module design on heat transfer and Nusselt correlation selection in membrane distillation

Alexander V Dudchenko (Carnegie Mellon University)*, Meagan S Mauter (Carnegie Mellon University)

#24 Applications: Cell and Protein Purification, Harvesting, and Processing ...pg. 61

9:30 am - 12:30 pm | Kings Garden 3

Co-Chairs: Merlin Bruening (University of Notre Dame), Jessica Molek (GlaxoSmithKline), Prity Bengani-Lutz (Repligen Corporation), Sen Xu (Merck Pharmaceuticals)

9:30 am WEDNESDAY

24a - Experimental and Theoretical Investigations of the Loss of Protein Sieving Due to Fouling of ATF Microfilters

Alex Apostolidis (Amgen)*, Glen Bolton (Amgen)

10:00 am WEDNESDAY

24b - pH and Excipient Partitioning in Ultrafiltration/ Diafiltration Processes for Formulation of High Concentration Monoclonal Antibody (mAb) Products

Parinaz Emami (Penn State University)*, Youngbin Baek (Penn State University), Nripen Singh (Bristol-Myers Squibb), Andrew Zydney (Penn State University)

10:30 am WEDNESDAY

24c - Effect of membranes molecular weight cut-off on peptides migration and selectivity during electrodialysis with filtration membranes

Sabita Kadel (Laval University)*, Jacinthe Thibodeau (Laval University), Carole Laine (Amer-Sil), Laurent Bazinet (Laval University)

11:00 am WEDNESDAY

24d - Scale down depth filtration case study using Pall Supracap™ 50 to predict large scale STAX™ capacity and product quality

Patricia Rose (Merck Pharmaceuticals)*, Jennifer Pollard (Merck Pharmaceuticals)

11:30 am WEDNESDAY

24e - Unraveling the Plugging Mechanisms during a Combined Tangential-flow and Depth Filtration Process

Xianghong Qian (University of Arkansas)*, Da Zhang (University of Arkansas), Daniel Strauss (Asahi Kasei), Parag Patel (Asahi Kasei), Ranil Wickramasinghe (University of Arkansas)

12:00 pm WEDNESDAY

24f - Functionalized Microporous Membranes for Protein Capture and Analysis

Joshua Berwanger (University of Notre Dame)*, Merlin Bruening (University of Notre Dame), Liu Yang (University of Notre Dame), Hui Yin Tan (University of Notre Dame), Weijing Liu (University of Notre Dame)

#25 Processes: Innovations in Microfiltration and Ultrafiltration... pg. 64

9:30 am - 12:30 pm | Ballroom 3

Co-Chairs: Beth Goodrich (MilliporeSigma), Vlad Tarabara (Michigan State University)

9:30 am WEDNESDAY

25a - A systematic approach to remove bottlenecks in high concentration UF/DF for subcutaneous application of biological drug substances

Hasin M Feroz (Bristol-Myers Squib)*, Dongyoun Jang (Bristol-Myers Squibb), Jessica Hung (Bristol-Myers Squibb), Yan Chen (Bristol-Myers Squibb), Erinc Sahin (Bristol-Myers Squibb), Andrew Zydney (Penn State University), Melissa Holstein (Bristol-Myers Squibb), Sanchayita Ghose (Bristol-Myers Squibb)

10:00 am WEDNESDAY

25b - Continuous Diafiltration for cGMP Biomanufacturing

Akshat Gupta (MilliporeSigma)*, Beth Goodrich (MilliporeSigma), Herb Lutz (MilliporeSigma)

10:30 am WEDNESDAY

25c - Understanding the role that patterning plays on membrane biofouling by visualization using light imaging and simulation

Anna Malakian (Clemson University)*, Bowen Ling (Stanford University), Ilenia Battiato (Stanford University), Scott Husson (Clemson University)

11:00 am WEDNESDAY

25d - Treating Poultry Processing Wastewaters by Combined Electrocoagulation and Ultrafiltration

Ranil Wickramasinghe (University of Arkansas)*, Kamyar Sardari (University of Arkansas)

11:30 am WEDNESDAY

25e - Mass transfer assisted TIPS process to precisely tailor the hollow fiber membrane surface and sublayer structures

Hideto Matsuyama (Kobe University)*, Chuanjie Fang (Kobe University), Saeid Rajabzadeh (Kobe University)

12:00 pm WEDNESDAY

25f - Organic solvent resistant membranes obtain by using non-toxic solvents

Stefan Chisca (King Abdullah University of Science and Technology)*, Gheorghe Falca (King Abdullah University of Science and Technology), Tommaso Marchesi (University of Bologna), Valentina Elena Musteata (King Abdullah University of Science and Technology, Suzana Nunes (King Abdullah University of Science and Technology)

#26 Materials: Bio-inspired and Biomimetic Materials ...pg. 67

9:30 am - 12:30 pm | Ballroom 4

Co-Chairs: Chunlong Chen (Pacific Northwest National Laboratory), Yair Kaufman (Ben Gurion University of the Negev), DB Bhattacharyya (University of Kentucky)

9:30 am WEDNESDAY

26a - Artificial Water Channels - toward biomimetic membranes for desalination

Mihail Barboiu (Institut Europeen des Membranes)*

10:00 am WEDNESDAY

26b - Self-Assembly of Long-Lasting Lipid Bilayers and the Effect of Temperature of their Ion Rejection

Yair Kaufman (Ben Gurion University of the Negev)*, Shiju Abraham (Ben Gurion University of the Negev), Tabea Heckenthaler (Ben Gurion University of the Negev), Yacov Morgenstern (Ben Gurion University of the Negev)

10:30 am WEDNESDAY

26c - Peptoid-based membrane-mimetic 2D nanomaterials with incorporated natural and synthetic channels

Chunlong Chen (Pacific Northwest National Laboratory)*

11:00 am WEDNESDAY

26d - Scalable High-Performance Membranes with High Density Channel Protein-Polymer Nanosheets

Yu-Ming Tu (Penn State University)*, Woochul Song (Penn State University), Tingwei Ren (Penn State University), Yuexiao Shen (UC Berkeley), Ratul Chowdhury (Penn State University), Prasangi Rajapaksha (University of Kentucky), Drew Carson (Penn State University), Yuxuan Dai (Penn State University), Arwa Mukthar (Penn State University), Miaoci Zhang (Penn State University), Alina Thokkadam (Rutgers University), DB Bhattacharyya (University of Kentucky), Bill Phillip (University of Notre Dame), Robert Hickey (Penn State University), Yinai Wei (University of Kentucky), Manish Kumar (Penn State University)

11:30 am WEDNESDAY

26e - Fabrication of aquaporin-based biomimetic membrane for seawater desalination

Ye Li (Nanyang Technological University)*, Rong Wang (Nanyang Technological University)

12:00 pm WEDNESDAY

26f - Biomimetic channel membranes: can they realize their full potential upon scale-up?

Viatcheslav Freger (Technion - IIT)*

#27 Materials: Advances in Membrane and Materials Characterization ...pg. 69

9:30 am - 12:30 pm | Kings Garden 5

Co-Chairs: Weiyi Li (School of Environment Science and Engineering), Maria-Chiara Ferrari (The University of Edinburgh), Chuyang Tang (University of New South Wales)

9:30 am WEDNESDAY

27a - Surface Nano-structuring with Hydrophilic Polymer Brush Layers for Membrane Performance Tailoring and Optimization

Yian Chen (UCLA)*, Soomin Kim (UCLA), Anditya Rahardianto (UCLA), Yoram Cohen (UCLA)

10:00 am WEDNESDAY

27b - Roughness in Polyamide RO Membranes: Its Formation Mechanism, Control, and Implications

Chuyang Y Tang (University of Hong Kong / University of New South Wales)*

10:30 am WEDNESDAY

27c - Mechanism of chlorine attack on polyamide membranes studied by EIS

Mikhail Stolov (Technion - IIT), Viatcheslav Freger (Technion - IIT)*

11:00 am WEDNESDAY

27d - Physicochemical and electrochemical characterization of cation-exchange membranes modified with polyethyleneimine for elucidating enhanced monovalent permselectivity of electrodialysis

Wenbin Jiang (New Mexico State University)*, Lu Lin (New Mexico State University), Xuesong Xu (New Mexico State University), Pei Xu (New Mexico State University)

11:30 am WEDNESDAY

27e - Unraveling the Film-Formation Kinetics during Interfacial Polymerization via Low Coherence Interferometry

Xin Liu (Southern University of Science and Technology)*, Weiyi Li (Southern University of Science and Technology), Zhuo Li (Southern University of Science and Technology), Anthony Fane (Singapore Membrane Technology Centre), Baolin Deng (Southern University of Science and Technology)

12:00 pm WEDNESDAY

27f - Operando Membrane Spectroscopy for Elucidating Transport Mechanisms in Membranes

Casey O'Brien (University of Notre Dame)*

#28 Materials: Bio-inspired and Biomimetic Materials ...pg. 71

2:00 pm - 5:00 pm | Ballroom 4

Co-Chairs: Chunlong Chen (Pacific Northwest National Laboratory), Yair Kaufman (Ben Gurion University of the Negev), DB Bhattacharyya (University of Kentucky)

2:00 pm WEDNESDAY

28a - Biofouling resistant membranes for wound healing

Antoine Venault (Chung Yuan Christian University)*, Hao-Tung Lin (Chung Yuan Christian University), Yung Chang (Chung Yuan Christian University)

2:30 pm WEDNESDAY

28b - How pore hydrophilicity influences water permeability?

Mingjie Wei (Nanjing Tech University)*, Fang Xu (Nanjing Tech University), Yong Wang (Nanjing Tech University)

3:00 pm WEDNESDAY

28c - Biomimetic membrane systems utilizing electro-dynamic interfaces

Bruce Hinds (University of Washington)*

3:30 pm WEDNESDAY

28d - Responsiveness and function of DNA-gated membranes

Beñat Olave (Polymat, University of the Basque Country), Iliane Rafaniello (Polymat, University of the Basque Country), Thomas Schäfer (Polymat, University of the Basque Country)*

4:00 pm WEDNESDAY

28e - Nature-inspired Coating-free Membranes for Desalination

Himanshu Mishra (King Abdullah University of Science and Technology)*, Ratul Das (King Abdullah University of Science and Technology), Sankara Arunachalam (King Abdullah University of Science and Technology), Zain Ahmad (King Abdullah University of Science and Technology), Edelberto Manalastas (King Abdullah University of Science and Technology)

4:30 pm WEDNESDAY

28f - Carbon nanotube porins as model biomimetic membrane nanopores

Aleksandr Noy (Lawrence Livermore National Laboratory)*

#29 Materials: Advances in Membrane and Materials Characterization ...pg. 74

...pg. 74

2:00 pm - 5:00 pm | Kings Garden 5

Co-Chairs: Weiyi Li (School of Environment Science and Engineering), Maria-Chiara Ferrari (The University of Edinburgh), Chuyang Tang (University of New South Wales)

2:00 pm WEDNESDAY

29a - A novel method to distinguish between crystal and defect transport in zeolite membranes

David Carter (University of Ottawa), Shaaima Al Akwaa (University of Ottawa), Boguslaw Kruczek (University of Ottawa)*, Handan Tezel (University of Ottawa)

2:30 pm WEDNESDAY

29b - Study on antifouling behaviors of GO modified nanocomposite membranes through QCMD and surface energetics analysis

Amin Karkooti (University of Alberta), Neda Nazemifard (University of Alberta), Mohtada Sadrzadeh (University of Alberta)*

3:00 pm WEDNESDAY

29c - 3D characterization of polyamide reverse osmosis membranes

Tyler E Culp (Penn State University)*, Manish Kumar (Penn State University), Enrique Gomez (Penn State University)

3:30 pm WEDNESDAY

29d - Selective Ion Transport Properties of Membrane Materials

Anita Hill (Commonwealth Scientific and Industrial Research Organisation)*, Aaron Thornton (Commonwealth Scientific and Industrial Research Organisation), Cara Doherty (Commonwealth Scientific and Industrial Research Organisation)

4:00 pm WEDNESDAY

29e - Membrane-based oil-gas separation for dissolved gas-in-oil extraction: gas transport properties of ceramic supported Teflon membranes

Liang-Chih Ma (Arizona State University)*, Chuan Chen (Global Energy Interconnection Research Institute Co., Ltd.), Jerry Lin (Arizona State University)

4:30 pm WEDNESDAY

29f - Porous support of TFC membranes: Does it truly have negligible resistance?

Masoud Aghajani (University of Colorado Boulder)*, Alan R. Greenberg (University of Colorado Boulder), Yifu Ding (University of Colorado Boulder)

#30 Processes: Fundamentals of Predicting and Preventing Membrane Fouling ...pg. 76

2:00 pm - 5:00 pm | Kings Garden 4

Co-Chairs: Manish Kumar (Penn State University), Davide Mattia (University of Bath), Saifur Rahaman (Concordia University)

2:00 pm WEDNESDAY

30a - Numerical and experimental studies on the deposition of sticky particles near a membrane surface

Seon Yeop Jung (Seoul National University)*, Kyung Hyun Ahn (Seoul National University)

2:30 pm WEDNESDAY

30b - Characterization of the Striping Phenomenon during Membrane Fouling via Optical Coherence Tomography

Weiyi Li (Southern University of Science and Technology)*, Xin Liu (Southern University of Science and Technology), Anthony Fane (Singapore Membrane Technology Centre)

3:00 pm WEDNESDAY

30c - A closer look at biofouling: time-lapse optical microscopy of biofilm formation in wastewater reuse

Emily W Tow (Olin College)*, Behzad Rad (Lawrence Berkeley National Laboratory), Robert Kostecki (Lawrence Berkeley National Laboratory)

3:30 pm WEDNESDAY

30d - Improved graphene oxide membrane increases membrane distillation desalination of RO concentrate

Samuel Olatunji (Texas A&M University-Kingsville), Lucy M Camacho (Texas A&M University-Kingsville)*

4:00 pm WEDNESDAY

30e - Automated real-time membrane biofouling assessment using microbial enzyme activity

Babar K Khan (King Abdullah University of Science and Technology)*, Luca Fortunato (King Abdullah University of Science and Technology), TorOve Leiknes (King Abdullah University of Science and Technology)

4:30 pm WEDNESDAY

30f - Interactions between extracellular polymeric substances and zwitterionic hydrogels as a designing tool for antifouling surfaces

Angelin Ida Vedhamanickam (Ben-Gurion University of the Negev)*, Nir Goldberg (Ben-Gurion University of the Negev), Ronit Bitton (Ben-Gurion University of the Negev), Moshe Herzberg (Ben-Gurion University of the Negev)

#31 Applications: Purification of Non-Protein Biologics ...pg. 79

2:00 pm - 5:00 pm | Kings Garden 3

Co-Chairs: James McGrath (University of Rochester), Onur Kas (MilliporeSigma)

2:00 pm WEDNESDAY

31a - Optimizing Microfluidic Nanoparticle Capture from Biofluids on Ultrathin Silicon Nanomembranes: A Computational and Experimental Analysis of Tangential Flow Analyte Capture (TFAC)

Kilean Lucas (University of Rochester)*, Danial Ahmad (University of Rochester), Mehdi (Aslan) Dehghani (Rochester Institute of Technology), Thomas Gaborski (Rochester Institute of Technology), Richard Waugh (University of Rochester), James McGrath (University of Rochester)

2:30 pm WEDNESDAY

31b - Development and scale-up of laterally-fed membrane chromatography for the purification of therapeutic viruses

Karina Kawka (McMaster University)*, Pedram Madadkar (McMaster University), Shabnam Shoaebargh (McMaster University), Natasha Kazhdan (McMaster University), Maria Fe C. Medina (McMaster University), Brian Lichty (McMaster University), Raja Ghosh (McMaster University), David Latulippe (McMaster University)

3:00 pm WEDNESDAY

31c - Characterization of EV Secretion at Single Cell Resolution

Gerardo Mauleon Ramos (University of Chicago)*, Kilean Lucas (University of Rochester), Vladimir Riazanski (University of Chicago), James McGrath (University of Rochester), Deborah Nelson (University of Chicago)

3:30 pm WEDNESDAY

31d - Charge, size distribution and hydrophobicity of viruses: Effect of the virus purification method

Hang Shi (Michigan State University), Vlad Tarabara (Michigan State University)*

4:00 pm WEDNESDAY

31e - Risk Mitigation Strategies for the use of polymeric consumables for manufacturing of Anti-Sense Oligonucleotides

Anuradha Vaidya (Biogen), Bill Scott* (Biogen), Dave Kolwyck (Biogen)

4:30 pm WEDNESDAY

31f - Use of Nanosphere Self-Assembly to Pattern Nanoporous Membranes for the Study of Extracellular Vesicles

Marcela Mireles (University of Rochester)*, Cody Soule (Rochester Institute of Technology), Mehdi Dehghani (Rochester Institute of Technology), Thomas Gaborski (Rochester Institute of Technology)

#32 Applications: Fuel Cells and Batteries ...pg. 81

2:00 pm - 5:00 pm | Ballroom 3

Co-Chairs: Michael Guiver (Tianjin University), Mike Hickner (Penn State University), Peter Pintauro (Vanderbilt University)

2:00 pm WEDNESDAY

32a - A Holistic Approach to Explore Interfacially Confined Ionomers Designed for Energy Conversion Device

Shudipto K Dishari (University of Nebraska-Lincoln)*

2:30 pm WEDNESDAY

32b - Anion Conducting Membranes Based on Poly(norbornene): High Conductivity, Chemical Stability and Fuel Cell Performance >2.5 W/cm^2

Mrinmay Mandal (Georgia Institute of Technology)*, Garrett Huang (Georgia Institute of Technology), Wanting Chen (Georgia Institute of Technology), Paul A Kohl (Georgia Institute of Technology)

3:00 pm WEDNESDAY

32c - High Temperature Fuel Cells with Ion-Pair Membranes and Phosphonated Ionomers

Michael R Hibbs (Sandia National Laboratories)*, Cy Fujimoto (Sandia National Laboratories), Ehren Baca (Sandia National Laboratories), Albert Lee (Los Alamos National Laboratory), Yu Seung Kim (Los Alamos National Laboratory

3:30 pm WEDNESDAY

32d - High temperature, anhydrous proton conducting membranes and micropatterned bipolar membranes for electrochemical energy conversion and storage technologies

Christopher G Arges (Louisiana State University)*

4:00 pm WEDNESDAY

32e - Membrane through-plane alignment of ion-conducting channels

Xin Liu (Tianjin University), Yi Li (Tianjin University), Jiandang Xue (Tianjin University), Weikang Zhu (Tianjin University), Junfeng Zhang (Tianjin University), Yan Yin (Tianjin University), Michael D Guiver (Tianjin University)*

4:30 pm WEDNESDAY

32f - Novel sulfonated aromatic polymer membranes for breaking the proton selectivity-conductivity trade-off limitation in vanadium redox flow battery

Tongshuai Wang (University of Illinois, Chicago), JunYoung Han (Rensselaer Polytechnic Institute), Kihyun Kim (Rensselaer Polytechnic Institute), Andreas Muenchinger (Max-Planck-Institute for Solid State Research), Klaus-Dieter Kreuer (Max-Planck-Institute for Solid State Research), Chulsung Bae (Rensselaer Polytechnic Institute), Sangil Kim (University of Illinois, Chicago)*

Poster Presentations

Presenting authors are indicated by an asterisk*. See the most up-to-date version of the program with abstracts at **nams2019.org** or via the **Attendify (NAMS 2019)** app.

SESSION A - Innovation in Bio-Inspired and Bio-Applied Membranes ...pg. 84

A1 **High-Capacity Multimodal Anion-Exchange Membranes for Purification of Biologics,** Joshua Osuofa (Clemson University)*; Scott Husson (Clemson University)

A2 **High throughput screening of ultra-filtration membranes with mixed dextrans in 96 well plate,** Masha Khazan (Pall Corporation)*; James Hathcock (Pall Corporation); Jian Qiu (Pall Corporation)

A3 **Impacts of Bio-inspired Zwitterionic Membranes for Health Care Applications,** Yung Chang (R&D Center for Membrane Technology, CYCU)*

A4 **Enhanced Membranes for Biogas Upgrading,** Ning Shangguan (Compact Membrane Systems)*; Sudip Majumdar (Compact Membrane Systems); Kenneth Pennisi (Compact Membrane Systems)

A5 **Monitoring of Membrane Fouling by the Zeta Potential,** Thomas Luxbacher (Anton Paar GmbH)*; Hermina Buksek (University of Maribor); Irena Petrinic (University of Maribor); Elisa Innocenti (GSK Vaccines S.r.l.); Krzysztof Trzaskus (Aquaporin A/S); Maria Salud Camilleri-Rumbai (Aquaporin A/S)

A6 **Tangential flow microfluidics for the capture and release of nanoparticles and extracellular vesicles on conventional and ultrathin membranes,** Mehdi Dehghani (Rochester Institute of Technology)*; Kilean Lucas (University of Rochester); James McGrath (University of Rochester); Thomas Gaborski (Rochester Institute of Technology)

A7 **Low-cost ultrathin porous polymeric membranes with precise control over pore size and thickness gradient for investigation of leukocyte transmigration through barrier models,** Shayan Gholizadeh (Rochester Institute of Technology)*; Alec Salminen (University of Rochester); Zahra Allahyari (Rochester Institute of Technology); Robert Carter (Rochester Institute of Technology); Henry Chung (Rochester Institute of Technology); Marcela Mireles (University of Rochester); Thomas Gaborski (Rochester Institute of Technology)

A8 **Application of Novel Multimodal Anion-exchange Membrane Chromatography Columns in a Two-step mAb Purification Scheme - Aggregates and HCP Removal,** Daniel Henn (Purilogics, LLC)*; Anna Forsyth (Purilogics, LLC); Graham Temples (Purilogics, LLC); Jinxiang Zhou (Purilogics, LLC); Scott Husson (Clemson University)

A9 **Effect of Nanomembrane Orientation and Flow Modality on Benchtop Urea Clearance,** Sam Walker (The State University of New York at Buffalo)*; Kayli Hill (University of Rochester); Alec Salminen (University of Rochester); Dean G Johnson (University of Rochester)

A10 **Feed Spacer Modification for Listeria Control in Dairy Processing,** Mainara Costa Teixeira (The University of Alabama)*; Preston Richier (The University of Alabama); Jake Colburn (The University of Alabama); Shelby Brooks (The University of Alabama); Will Baker (The University of Alabama); Thomas Hendrich (The University of Alabama); Ryan Summers (The University of Alabama); Stephen Ritchie (The University of Alabama)

A11 **Critical Flux of Ultrathin Silicon Membranes in Tangential Flow Filtration of Protein Solutions,** Danial Ahmad (University of Rochester)*; Kilean Lucas (University of Rochester); Mehdi (Aslan) Dehghani (Rochester Institute of Technology); Thomas Gaborski (Rochester Institute of Technology); James McGrath (University of Rochester)

A12 Grafting Zwitterionic Polymer and Poly (amino acid) on Polyamide Membranes: "Defending and Attacking" Strategies for Biofouling Control, Wen Ma (Concordia University)*; Liuqing Yang (Concordia University); Saifur Rahaman (Concordia University)

A13 Monitoring of biological fluids with extraction membranes, Pinar Cay-Durgun (Arizona State University); Tianmiao Lai (Arizona State University); Nai-Yuan Liu (Arizona State University); Mark Sprowls (Arizona State University); Stewart Mann (Arizona State University); Leslie Thomas (Mayo Clinic in Arizona); Erica Forzani (Arizona State University); Mary Laura Lind (Arizona State University)*

~~**A14 Fine-Tuned Biohybrid Polymeric Particles Mimicking Biological Compartmentalization Prepared by Membrane Emulsification,** Emma Piacentini (National Research Council of Italy, Institute on Membrane Technology, CNR-ITM); Lidietta Giorno (National Research Council of Italy - Institute on Membrane Technology)*~~

A15 Bio-inspired immobilization of casein-coated silver nanoparticles on cellulose acetate ultrafiltration membranes: from laboratory to scale-up, Xiaobo Dong (University of Kentucky)*; Halle Shannon (University of Kentucky); Tequila Harris (Georgia Institute of Technology); Isabel Escobar (University of Kentucky)

A16 Surface-Specific Thermoresponsive Coating of Membranes, Marcela Mireles (University of Rochester)*; Cody Soule (Rochester Institute of Technology); Luis Delgadillo (University of Rochester); Thomas Gaborski (Rochester Institute of Technology)

A17 Pervaporation Separation of n-Butanol from Aqueous Solutions Using PDMS/Lignin Mixed Matrix Membranes, Ali Zamani (University of Ottawa); Handan Tezel (University of Ottawa)*; Jules Thibault (University of Ottawa)

A18 Rapid and Selective Water Permeation Across Biomimetic Membranes Through Artificial Water Channel Aggregates, Woochul Song (Penn State University)*; Ratul Chowdhury (Penn State University); Himanshu Joshi (University of Illinois at Urbana-Champaign); Joseph S. Najem (The University of Tennessee); Yue-xiao Shen (UC Berkeley); Chao Lang (Penn State University); Stephen Sarles (University of Tennesseee); Jun-li Hou (Fudan University); Aleksei Aksimentiev (University of Illinois at Urbana-Champaign); Manish Kumar (Penn State University)

A19 Can porous membranes be engineered to induce physiologically relevant cell-substrate interactions?, Zahra Allahyari (Rochester Institute of Technology)*; Stephanie Casillo (Rochester Institute of Technology); Shayan Gholizadeh (Rochester Institute of Technology); Henry Chung (Rochester Institute of Technology); Thomas Gaborski (Rochester Institute of Technology)

A20 High Productivity Harvest - Intensify Harvest and Displace Depth Filtration in Fed-Batch Cell Culture, Shashi Kudugunti (Repligen Corporation)*, Jyoti Amatya (Repligen Corporation), Jamie Peyser (Repligen Corporation)

A21 Membrane Filtration of Flexible Particles: Rejection of Single Stranded DNA, Hossein Nouri Alavijeh (Clarkson University)*; Ruth E. Baltus (Clarkson University)

SESSION B - Innovation in Membrane Materials, Synthesis, and Characterization: Gases ...pg. 90

B1 Covalently modified graphene oxide incorporated in PIM-1 mixed matrix thin film composite membranes for gas separation, Elvin Aliyev (Helmholtz-Zentrum Geesthacht)*; Volkan Filiz (Helmholtz-Zentrum Geesthacht); Sergey Shishatskiy (Helmholtz-Zentrum Geesthacht)

B2 Tunable Interlayer Channels in Graphene Oxide Membranes for Molecular Separations, Shaofei Wang (King Abdullah University of Science and Technology)*; Zhongyi Jiang (Tianjin University); Michael D Guiver (Tianjin University); Suzana Nunes (King Abdullah University of Science and Technology)

B3 Nanoporous two-dimensional nanosheets and their self-assembled gas-sieving membranes, Kumar Varoon Agrawal (École polytechnique fédérale de Lausanne)*; Mostapha Dakhchoune (École polytechnique fédérale de Lausanne); Luis Francisco Villalobos (École polytechnique fédérale de Lausanne)

B4 Thermally Rearranged Semi-interpenetrating Polymer Networks (TR-SIPNs) for Olefin/Paraffin Separation Applications, Won Hee Lee (Hanyang University)*; Young Moo Lee (Hanyang University); Jong Geun Seong (Hanyang University); Ho Hyun Wang (Hanyang University); Sun Ju Moon (Hanyang University)

B5 Design of new polymeric materials for membrane gas-separation based on norbornenes bearing hydrocarbonic groups, Alyona Wozniak (A.V. Topchiev Institute of Petrochemical Synthesis, Russian Academy of Sciences); Evgeniya Bermesheva (A.V. Topchiev Institute of Petrochemical Synthesis, Russian Academy of Sciences); Danila Bakhtin (A.V. Topchiev Institute of Petrochemical Synthesis, Russian Academy of Sciences); Ilya Borisov (A.V. Topchiev Institute of Petrochemical Synthesis, Russian Academy of Sciences); Alexey Volkov (A.V. Topchiev Institute of Petrochemical Synthesis, Russian Academy of Sciences); Eugene Finkelshtein (A.V. Topchiev Institute of Petrochemical Synthesis, Russian Academy of Sciences); Maxim Bermeshev (A.V. Topchiev Institute of Petrochemical Synthesis, Russian Academy of Sciences)*

B6 Cake layer characterization in Activated Sludge Membrane Bioreactor: Real-time analysis, Luca Fortunato (King Abdullah University of Science and Technology)*; TorOve Leiknes (King Abdullah University of Science and Technology)

B7 Facilitated Transport Membrane for CO_2 Capture from Flue Gas: Module Fabrication and Scale-Up, Kai Chen (Ohio State University)*; Witopo Salim (Ohio State University); Yang Han (Ohio State University); Dongzhu Wu (Ohio State University); Winston Ho (Ohio State University)

B8 Interfacial phenomenon-driven hierarchical structuring of hard templated asymmetric ultrathin carbon molecular sieves films, Megha Sharma (Lehigh University)*; Mark Snyder (Lehigh University)

B9 Layer-by-Layer Assembled Metal Organic Framework Nanosheets with Polymer, Christy Koerner (U.S. Department of Energy National Energy Technology Laboratory)*; Fangming Xiang (U.S. Department of Energy National Energy Technology Laboratory); Eric Popczun (U.S. Department of Energy National Energy Technology Laboratory); David Hopkinson (U.S. Department of Energy National Energy Technology Laboratory)

B10 Highly oxygen-rich rubbery polymers for membrane CO_2 /CH_4 separation, Krysta Clark (The State University of New York at Buffalo)*; Junyi Liu (Air Liquide. Inc); Gengyi Zhang (The State University of New York at Buffalo); Haiqing Lin (The State University of New York at Buffalo)

B11 Adsorptive and Destructive Mixed Matrix Membranes for Chemical Protection, Yufeng Song (New Jersey Institute of Technology)*; John Chau (New Jersey Institute of Technology); Kamalesh K Sirkar (New Jersey Institute of Technology); Gregory Peterson (U.S. Army Edgewood Chemical Biological Center); Uwe Beuscher (W.L. Gore & Associates Inc.)

B12 Maximizing Ether Oxygen Content in Polymers for Membrane CO_2 Removal from Natural Gas, Junyi Liu (American Air Liquide. Inc)*; Gengyi Zhang (The State University of New York at Buffalo); Krysta Clark (The State University of New York at Buffalo); Haiqing Lin (The State University of New York at Buffalo)

B13 Novel CO_2 gas separation membranes prepared by chemical modification of highly permeable polymers, Eugenia Mariana Sandru (SINTEF AS); Arpenik Kroyan (Norwegian University of Science and Technology); Nebosja Simic (Norwegian University of Science and Technology); Jing Deng (Norwegian University of Science and Technology); Liyuan Deng (Norwegian University of Science and Technology); Marius Sandru (SINTEF AS)*

B14 Intrinsically Microporous Pentiptycene-based Polymers for Enhanced Gas Separation Performance and Physical Aging Resistance, Tanner Corrado (University of Notre Dame)*; Ruilan Guo (University of Notre Dame)

B15 Structural Designs of Cross-linked Polymer Network and Ionic Liquids for Ion Gel Gas Separation Membranes, Victor A Kusuma (U.S. Department of Energy National Energy Technology Laboratory)*; James S Baker (U.S. Department of Energy National Energy Technology Laboratory); Megan Macala (U.S. Department of Energy National Energy Technology Laboratory); Samir Budhathoki (U.S. Department of Energy National Energy Technology Laboratory); David Hopkinson (U.S. Department of Energy National Energy Technology Laboratory)

B16 Scalable hydrogen-bonded polyimide/metal-organic framework hybrid membranes for ultra-fast separations of multiple gas pairs, Canghai Ma (Lawrence Berkeley National Laboratory)*; Jeffrey Urban (Lawrence Berkeley National Laboratory)

B17 Performance Testing of Polyphosphazene-Based Membranes for Post-combustion Carbon Capture with Humidified Gas, Zi Tong (U.S. Department of Energy National Energy Technology Laboratory)*; David Hopkinson (U.S. Department of Energy National Energy Technology Laboratory); Victor A Kusuma (U.S. Department of Energy National Energy Technology Laboratory)

B18 Systematic Optimization of MOF-based Mixed-Matrix Membranes: Surface Functionalization, Particle Size, and Loading, Patrick Muldoon (Oak Ridge Institute for Science and Education)*

~~**B19 Membrane heat exchanger for novel heat recovery in post-combustion carbon capture,** Shuaifei Zhao (Dalian Maritime University)*; Shuiping Yan (Huazhong Agricultural University); Qiufang Cui (Huazhong Agricultural University)~~

B20 CANAL Ladder Polymers for Membrane Gas Separation, Holden W. H. Lai (Stanford University)*; Francesco Benedetti (University of Bologna); Zach Smith (MIT); Yan Xia (Stanford University)

B21 Fabrication of CO_2 selective miscible polyimide blend membranes for gas separation applications, Chamaal Karunaweera (The University of Texas at Dallas)*; John Ferraris (The University of Texas at Dallas); Kenneth Balkus, Jr. (The University of Texas at Dallas); Inga Musselman (The University of Texas at Dallas); Samitha Panangala (The University of Texas at Dallas); Shahed Haghiri (The University of Texas at Dallas)

B22 On development of constant-volume permeation system for monitoring upstream pressure decay, Haoyu Wu (University of Ottawa)*; Jules Thibault (University Ottawa); Boguslaw Kruczek (University of Ottawa)

B23 Two-dimensional-material mixed-matrix membranes for gas separation, Gongping Liu (Nanjing Technical University)*

B24 Synthesis of crosslinked polyether-based membranes for gas separations, Malgorzata Chwatko (The University of Texas at Austin)*; Christina Rodriguez (The University of Texas at Austin); Caitlin Bentley (The University of Texas at Austin); Nathaniel Lynd (The University of Texas at Austin); Benny Freeman (The University of Texas at Austin)

B25 Novel polyimides with bulky tert-butyl and fluorine-containing side groups: gas permeation parameters and free volume, Susanta Banerjee (Indian Institutes of Information Technology); Nikolay Belov (A.V. Topchiev Institute of Petrochemical Synthesis, Russian Academy of Sciences); Roman Nikiforov (A.V. Topchiev Institute of Petrochemical Synthesis, Russian Academy of Sciences); Mikhail Mazo (Institute of Chemical Physics, Russian Academy of Sciences); Ivan Strelnikov (Institute of Chemical Physics, Russian Academy of Sciences); Yury P Yampolskii (A.V. Topchiev Institute of Petrochemical Synthesis, Russian Academy of Sciences)*

B26 New mixed matrix composite membrane PEBAX based membrane incorporated with metal organic framework for CO_2 separation, Tarik Eljaddi (Laboratoire Réactions et Génie des Procédés)*; Julien Bouillon (Normandie University); Laurent Lebrun (Normandie University)

B27 Dual-Layer MOF Composite Membranes with Tunable Gas Transport Properties for Post-combustion CO_2 Separation, Sameh K Elsaidi (U.S. Department of Energy National Energy Technology Laboratory)*; David Hopkinson (U.S. Department of Energy National Energy Technology Laboratory); Surendar Venna (U.S. Department of Energy National Energy Technology Laboratory); Mona Mohamed (University of Pittsburgh)

B28 Performance of Pd-Based Membranes and Effects of Various Gas Mixtures on H_2 Permeation, Kourosh Kian (Worcester Polytechnic Institute)*

SESSION C - Innovation in Membrane Materials, Synthesis, and Characterization: Liquids ...pg. 100

C1 Layer-by-layer modification of aliphatic polyamide anion-exchange membranes to increase chloride/sulfate selectivity, Muhammad Ahmad (University of Notre Dame)*; Chao Tang (University of Notre Dame); Liu Yang (University of Notre Dame); Andriy Yaroshchuk (Polytechnic University of Catalonia); Merlin Bruening (University of Notre Dame)

C2 Development of electrically conductive hollow fiber membranes, Melissa J Larocque (McMaster University)*, David R Latulippe (McMaster University), Charles-François de Lannoy (McMaster University)

C3 Structure Formation in Isoporous Hollow Fiber Membranes: An In Situ SAXS study, Kirti Sankhala (Helmholtz-Zentrum Geesthacht)*; D. C. Florian Wieland (Helmholtz-Zentrum Geesthacht); Joachim Koll (Helmholtz-Zentrum Geesthacht); Maryam Radjabian (Helmholtz-Zentrum Geesthacht); Clarissa Abetz (Helmholtz-Zentrum Geesthacht); Volker Abetz (Helmholtz-Zentrum Geesthacht)

C4 Microscopic membrane fouling characterization, Nandini Debnath (University of Alberta); Aloke Kumar (Indian Institute of Science); Thoams Thundat (The State University of New York at Buffalo); Mohtada Sadrzadeh (University of Alberta)*

C5 New Insights into the Impact of Nano-scale Surface Heterogeneity on the Wettability of Polymeric Membranes, Md Farhad Ismail (University of Alberta); Behnam Khorshidi (University of Alberta); Mohtada Sadrzadeh (University of Alberta)*

C6 A comparison of water and solute transport in ion exchange and desalination polymers, Ryan Kingsbury (Membrion, Inc.); Jingbo Wang (UCLA); Mikayla D Armstrong (University of North Carolina at Chapel Hill)*; Orlando Coronell (University of North Carolina at Chapel Hill)

C7 Water permeation through nanoporous particles in thin film nanocomposite membranes, Pinar Cay-Durgun (Arizona State University)*; Mary Laura Lind (Arizona State University)

C8 Printing Membranes: Enabling the Use of Both Conventional and Novel Polymers in High Performance Membranes, Jeffrey R McCutcheon (University of Connecticut); Ayse Asatekin (Tufts University); Maqsud Chowdhury (University of Connecticut); Xin Qian (University of Connecticut); Tulasi Ravindran (University of Connecticut)*; Samuel Loudner (Tufts University)

C9 Functionalization of PVDF Membranes with Thiol Groups for Heavy-Metal Capture, Ronald Vogler (University of Kentucky)*; Md. Saiful Islam (University of Kentucky); Evan Hatakeyama (Chevron Energy Technology Company); DB Bhattacharyya (University of Kentucky)

C10 Pure and mixed fluid sorption and transport in Celazole® polybenzimidazole: the effect of plasticization, Kelly P Bye (University of Oklahoma); Michele Galizia (University of Oklahoma)*

C11 Enhanced Water Interaction and Antifouling Compatible Zwitterion-PVDF Membrane Designs, Mahboobeh Maghami (University of Saskatchewan); Amira Abdelrasoul (University of Saskatchewan)*

C12 Novel strategy to develop a composite membrane for desalination by membrane distillation, Tarik Eljaddi (Laboratoire Réactions et Génie des Procédés - UMR 7274)*; Deisy Mejia Mendez (Laboratoire Réactions et Génie des Procédés - UMR 7274); Eric Favre (Laboratoire Réactions et Génie des Procédés - UMR 7274); Denis Roizard (Laboratoire Réactions et Génie des Procédés - UMR 7274)

C13 One-Step Tailoring Surface Roughness and Surface Energy to Prepared Superhydrophobic Polyvinylidene Fluoride (PVDF) Membranes for Enhanced Membrane Distillation Performances, Weihua Qing (The University of Hong Kong)*; Xiaonan Shi (New Jersey Institute of Technology); Chuyang Tang (University of New South Wales)

C14 High-Performance Thin Film Composite (TFC) Membranes: Design Consideration Underneath the Polyamide Thin Film, Wangxi Fang (Suzhou Institute of Nano-Tech and Nano-Bionics, Chinese Academy of Sciences)*; Jian Jin (Suzhou Institute of Nano-Tech and Nano-Bionics, Chinese Academy of Sciences); Rong Wang (Singapore Membrane Technology Centre, Nanyang Technological University)

C15 Wafer-Scale, Vertically-Aligned SWCNT Composite Membranes for Nanofiltration, Melinda L Jue (Lawrence Livermore National Laboratory)*; Chiatai Chen (Cornell University); Steven Buchsbaum (Lawrence Livermore National Laboratory); Eric Meshot (Lawrence Livermore National Laboratory); Sei Jin Park (Lawrence Livermore National Laboratory); Kuang Jen Wu (Lawrence Livermore National Laboratory); Francesco Fornasiero (Lawrence Livermore National Laboratory)

C16 Vapor Phase Infiltration of Metal Oxides into Microporous Polymers for Solvent Stable Nanofiltration Membranes, Fengyi Zhang (Georgia Institute of Technology)*; Emily McGuinness (Georgia Institute of Technology); Yao Ma (Georgia Institute of Technology); Mark Losego (Georgia Institute of Technology); Ryan Lively (Georgia Institute of Technology)

C17 Understanding the Interlayer-spacing and Mass Transport Nexus of Graphene Oxide Membrane for Organic Solvent Nanofiltration, Sunxiang Zheng (UC Berkeley)*; Baoxia Mi (UC Berkeley)

C18 Organic-inorganic hybrid separation membrane having high chemical tolerance, Koichi Takada (TORAY Industries, Inc.)*; Takahiro Tokuyama (TORAY Industries, Inc.); Hiroki Minehara (TORAY Industries, Inc.); Takafumi Ogawa (TORAY Industries, Inc.); Masahiro Kimura (TORAY Industries, Inc.)

C19 Magnetron Sputtering-based Synthesis of Pd film on UF membrane support for environmental catalysis, Michael J Detisch (University of Kentucky)*; DB Bhattacharyya (University of Kentucky); John Balk (University of Kentucky)

C20 Influences of polymeric additives in different solvent systems on membrane performance and fouling resistance, Catharina Kahrs (Leibniz University Hannover)*; Jan Schwellenbach (Sartorius Stedim Biotech GmbH)

C21 Porous membranes with synergic solvent and thermal resistance from polyoxindole derivatives, Bruno Pulido Ponce de Leon (King Abdullah University of Science and Technology)*; Suzana Nunes (King Abdullah University of Science and Technology)

C22 Functionalized random zwitterionic copolymers as chlorine- and fouling-resistant nanofiltration membranes, Samuel J Lounder (Tufts University)*; Ayse Asatekin (Tufts University)

C23 Deterioration of nanofiltration polyamide membrane by strong acid and its mechanism, Byung-Moon Jun (University of South Carolina); Hyung Kae Lee (Ulsan National Institute of Science and Technology); Young-Nam Kwon (Ulsan National Institute of Science and Technology)*

C24 Opening new doors: Epoxides as novel chemistry for interfacial polymerization, Rhea Verbeke (Katholieke Universiteit Leuven)*; Marijn Seynaeve (Katholieke Universiteit Leuven); Wouter Arts (Katholieke Universiteit Leuven); Elke Dom (Katholieke Universiteit Leuven); Ivo Vankelecom (Katholieke Universiteit Leuven)

C25 Solution-diffusion with ion association: a non-mean-field model that successfully describes NF of multi-ion mixtures, Viatcheslav Freger (Technion - Israel Institute of Technology)*

C26 Relative permittivity properties of hydrated polymer membranes for desalination applications, Kevin Chang (University of Virginia)*; Hongxi Luo (University of Virginia); Geoffrey M Geise (University of Virginia)

C27 Multi-solute transport behavior of aqueous mixtures through polyether-based membranes, Jung Min Kim (Auburn University)*; Bryan S Beckingham (Auburn University)

C28 Comparative analysis of novel covalent organic framework nanofiltration membranes synthesis, composition, thickness and substrate dependence, Gabrielle Levato (University of Illinois at Urbana-Champaign)*; David Burke (Northwestern University); Zhiwei Jiang (Imperial College London); Daniel Mosiman (University of Illinois at Urbana-Champaign); William Dichtel (Northwestern University); Andrew Livingston (Imperial College London); Benito Marinas (University of Illinois at Urbana-Champaign)

C29 Copper-MOF Functionalized Nanofiltration Membranes for Enhanced Dye Removal: Effects of Dip-Coating and In-Situ Modification Techniques, Mostafa Dadashi Firouzjaei (University of Alabama); Zane Joseph Parkerson (University of Alabama); Milad Rabbani Esfahani (University of Alabama)*

C30 Advancing Conductivity-Permselectivity Tradeoff of Ion-Exchange Membranes with Sulfonated CNT Nanocomposites, Hanqing Fan (Columbia University)*; Yuxuan Huang (Columbia University); Ngai Yin Yip (Columbia University)

C31 Ceramic-supported thin film composite membrane for organic solvent reverse osmosis, Mi Zhang (University of Connecticut)*; Lingling Xia (University of Connecticut); Jeffrey R McCutcheon (University of Connecticut); Marcus Weyd (Fraunhofer-Institut für Keramische Technologien und Systeme)

C32 Dynamic FO tests of commercial low pressure RO TFC membrane, Du Bai (University of Ottawa); Farhad Asempour (University of Ottawa); Boguslaw Kruczek (University of Ottawa)*

C33 New generation recycled polystyrene membranes for industrial water purification using membrane distillation, Machawe M Motsa (University of South Africa)*; Bhekie Mamba (University of South Africa)

C34 Advanced supported ionic liquid (SIL) membranes for Pi electron cloud mediated separation of aromatics, Mahmood Jebur (University of Arkansas); Arijit Sengupta (University of Arkansas); Ranil Wickramasinghe (University of Arkansas)*

C35 Graphene Oxide Membranes for Selective Molecular Separation of Lignin Model Compounds, Ashish Aher (University of Kentucky)*; DB Bhattacharyya (University of Kentucky)

C36 High Flux and Anti-fouling Electrospun Nylon 6 Ultrafiltration Membrane Coated with GO/N-carboxyethylchitosan (NCECS) Functionalized Polyvinyl Alcohol Hydrogel, Liuqing Yang (Concordia University)*; Tiantian Chen (Concordia University); Zhibin Ye (Concordia University); Benoit Barbeau (Polytechnique Montreal); Saifur Rahaman (Concordia University)

C37 Development and Characterization of nZVI/TiO_2 Supported Membranes for Photocatalytic Treatment of 1,4-Dioxane in Water, Larissa L. S. Silva (Federal University of Rio de Janeiro)*; Wael Abdelraheem (University of Cincinnati); Minghao Kon (University of Cincinnati); Mallikarjuna Nadagouda (Wright State University); Ana Maria Rocco (Federal University of Rio de Janeiro); Cristiano Borges (Federal University of Rio de Janeiro); Fabiana Fonseca (Federal University of Rio de Janeiro); Dionysios Dionysiou (University of Cincinnati)

C38 Oxone Mediated TEMPO-Oxidized Nanocellulose Form-I and Form-II: Materials, Membranes and Modeling, John P Moore (University of Arkansas)*

C39 Facile Fabrication of Amphiphobic Nanocomposite Membranes for Application in Membrane Distillation, Tiantian Chen (Concordia University); Liuqing Yang (Concordia University); Saifur Rahaman (Concordia University)*

C40 Graphene Oxide Nanoplatelets Embedded Polyamide Thin Films for Water Desalination, Mahsa Abbaszadeh (Mississippi State University)*; Santanu Kundu (Mississippi State University)

C41 Development of Isoporous 0.2 Micron Nanoslit Silicon Nitride Membranes for Sterile Filtration of Biological Therapeutics, Evan Wright (McMaster University), Joshua Miller (Simpore), Andrew Gosselin (Simpore), Jared Carter (Simpore), James McGrath (University of Rochester), David Latulippe (McMaster University), James Roussie (Simpore)*

SESSION D - Innovation in Membranes for Energy Applications ...pg. 113

D1 A membrane to separate light olefins and paraffins; an update on pilot trials, module design and C4 separations, Ken Loprete (Compact Membrane Systems)*; Bill Charlton (Compact Membrane Systems); Hannah Murnen (Compact Membrane Systems); Kenneth Pennisi (Compact Membrane Systems); Sudip Majumdar (Compact Membrane Systems)

D2 Cross-linked Polyethyleneimine (XLPEI) with superior H_2/CO_2 separation properties, Sankhajit Pal (The State University of New York at Buffalo)*; Weiguang Jia (The State University of New York at Buffalo); Haiqing Lin (The State University of New York at Buffalo)

D3 Self-Cleaning Nanocomposite Membranes with Phosphorene-Based Pore Fillers for Water Treatment, Joyner Eke (University of Kentucky)*; Isabel Escobar (University of Kentucky)

D4 Scale-up of A Catalytic Membrane Reactor for the Production of 5-Hydroxymethylfurfural from Lignocellulosic Biomass, Tanmoy Patra (University of Arkansas); Robert Beitle (University of Arkansas); Richard J. Ciora, Jr. (Media and Process Technology, Inc.); Ranil Wickramasinghe (University of Arkansas); Xianghong Qian (University of Arkansas)*

D5 Thin-Film Composite Hollow Fiber Membrane for Post-Combustion Carbon Capture, Shouliang Yi (U.S. Department of Energy National Energy Technology Laboratory)*; James S Baker (U.S. Department of Energy National Energy Technology Laboratory); Ali Sekizkardes (Battelle/U.S. Department of Energy National Energy Technology Laboratory); Surendar Venna (U.S. Department of Energy National Energy Technology Laboratory); Victor A Kusuma (U.S. Department of Energy National Energy Technology Laboratory); Fangming Xiang (U.S. Department of Energy National Energy Technology Laboratory); Lingxiang Zhu (U.S. Department of Energy National Energy Technology Laboratory); Hyuk Kwon (U.S. Department of Energy National Energy Technology Laboratory); David Hopkinson (U.S. Department of Energy National Energy Technology Laboratory); Kevin Resnik (U.S. Department of Energy National Energy Technology Laboratory)

D6 Salinity gradient heat engine: optimising reverse electrodialysis for thermal-to-electric, Anna Hulme (Cranfield University)*

D7 Use of desalination brine for low-energy concentration of orange juice via forward osmosis, Haley D White (Tennessee Technological University)*; Leif Templeton (Tennessee Technological University); Shelby Jones (Tennessee Technological University); Laura H Arias Chavez (Tennessee Technological University)

D8 High-performance SSZ-13 membranes for CO_2 and N_2 removals from natural gas, Rongfei Zhou (Nanjing Technical University)*; Xinping LI (Nanjing Technical University); Bin Wang (Nanjing Technical University)

D9 Highly efficient CO_2 capture by mixed matrix membranes containing three-dimensional covalent organic framework fillers, Youdong Cheng (National University of Singapore)*; Dan Zhao (National University of Singapore)

~~**D10 Industrially relevant membrane testing and novel mixed-matrix membranes for a brighter membrane future,** Raymond Thür (Katholicke Universiteit Leuven)*~~

D11 Torlon® Hollow Fiber Membranes for Organic Solvent Reverse Osmosis separation of Complex Aromatic Hydrocarbon Mixtures, Hye Youn Y Jang (Georgia Institute of Technology)*; Ryan Lively (Georgia Institute of Technology); JR Johnson (ExxonMobil); Dhaval Bhandari (ExxonMobil)

D12 Performance Investigation of Polymer Derived Ceramic Composite Membrane in Microbial Fuel Cell (MFC), Vignesh Ahilan (University of Bremen)*; Camila Cabral de Barros (University of Bremen); Gourav Dhar Bhowmick (Indian Institute of Technology); Makarand M Ghangrekar (Indian Institute of Technology); Michaela Wilhelm (University of Bremen); Kurosch Rezwan (University of Bremen)

D13 PIM-1 based polymeric blend membranes for gas separation applications, Ashley Miles (Oak Ridge Associated Universities/U.S. Department of Energy National Energy Technology Laboratory)*; Ali Sekizkardes (Battelle/U.S. Department of Energy National Energy Technology Laboratory); Janice Steckel (U.S. Department of Energy National Energy Technology Laboratory); David Hopkinson (U.S. Department of Energy National Energy Technology Laboratory)

D14 The dynamic fate of Organic micro-pollutants in MBR and their accumulation in fish tissue, Gal Dagan (Ben-Gurion University of the Negev)*; Inbal Zaibel (Ben-Gurion University of the Negev); Dina Zilberg (Ben-Gurion University of the Negev); Moshe Herzberg (Ben-Gurion University of the Negev)

D15 Improving H_2 Utilization of Solid Oxide Fuel Cell Using CO_2-Selective Membrane, Kai Chen (Ohio State University)*; Witopo Salim (Ohio State University); Yang Han (Ohio State University); Mike Gasda (Bloom Energy Corporation); Winston Ho (Ohio State University)

D16 Adapting a Blood CO_2 Removal Device for CO_2 Removal from Flue Gas, Katherine Hornbostel (University of Pittsburgh)*

D17 Experimental study of dehumidification potential of PDMS hollow fiber membrane for HVAC applications, Michelle K Croal (Oakland University); Jonathan Maisonneuve (Oakland University)*

D18 The Perfluoropolymer Upper Bound, Albert Wu (MIT)*; James Drayton (MIT); Zach Smith (MIT)

D19 Synthesis and characterization of catalytic PtNa-PdAu membranes for hydrogen production, Ana M Tarditi (Instituto de Investigaciones en Catálisis y Petroquímica)*; Laura Cornagllia (Instituto de Investigaciones en Catálisis y Petroquímica); Yohana Martínez Galeano Instituto de Investigaciones en Catálisis y Petroquímica)

D20 Hybrid facilitated transport membranes containing 2D GO-based nanoplatelets and scaffolds for CO_2 separation, Saravanan Janakiram (Norwegian University of Science and Technology)*; Juan Luis Martín Espejo (Norwegian University of Science and Technology); Zhongde Dai (Norwegian University of Science and Technology); Luca Ansaloni (SINTEF Industry); Liyuan Deng (Norwegian University of Science and Technology)

D21 Current Efforts in Harvesting Algae At-Scale For Biofuels to Meet the Dual Energy Challenge, Mark A Deimund (ExxonMobil Research and Engineering)*

D22 Atomically-Ordered Intermetallic Hydrogen Separation Membranes, Casey O'Brien (University of Notre Dame)*

D23 Molten Hydroxide Dual-Phase Membranes for Intermediate-temperature Fuel Cells, Vedasri Vedharathinam (Lawrence Livermore National Laboratory)*; Anna Ivanovskaya (Lawrence Livermore National Laboratory); Maira Ceron Hernandez (Lawrence Livermore National Laboratory); Patrick Campbell (Lawrence Livermore National Laboratory)

D24 High-Yield Recovery of Dissolved Methane Using Omniphobic Membranes, Xuesong Li (University of British Columbia); Abhishek Dutta (University of British Columbia); Qirong Dong (Fibracast); Sasha Rollings-Scattergood (Anaergia); Jongho Lee (University of British Columbia)*

D25 Aging minimization of CMSMs from the incorporation of copper pillars derived from metal organic polyhedra-18, Whitney Cosey (University of Texas at Dallas)*; Inga Musselman (University of Texas at Dallas); John Ferraris (University of Texas at Dallas); Kenneth Balkus (Univerisity of Texas at Dallas)

D26 (Cross-linked Poly(ionic liquid)-Ionic Liquid-Zeolite) Mixed-Matrix Membranes for CO_2/CH_4 Gas Separations Based on Curable Ionic Liquid Prepolymers, Collin A Dunn (University of Colorado Boulder)*; Zhangxing Shi (University of Colorado Boulder); Rongfei Zhou (Nanjing Technical University); Douglas Gin (University of Colorado Boulder); Richard Noble (University of Colorado Boulder)

D27 Integral hollow fiber membrane with chemical cross-linking for pressure retarded osmosis operated in the orientation of active layer facing feed solution, Ye Li (Nanyang Technological University)*; Rong Wang (Nanyang Technological University)

D28 GHG separation membrane application in KOGAS LNG regasification facilities, Kyung Taek Woo (Korea Gas Corporation)*; Hyun Seok You (Korea Gas Corporation); Jong Tae Chung (Korea Gas Corporation); Jae Dong Kim (Korea Gas Corporation)

D29 High performance electrochemical structures for power generation and oxygen separation, Ralph A Bauer (Global Research and Development Inc.)*

D30 Techno-economic comparison of RO and MD for small-scale brackish water desalination, Haamid S Usman (Concordia University)*; Khaled Touati (Concordia University); Saifur Rahaman (Concordia University)

D31 Tailor-made amphiphilic copolymer membranes for efficient gas and water vapor transport, Faheem Hassan Akhtar (King Abdullah University of Science and Technology)*; Klaus-Viktor Peinemann (King Abdullah University of Science and Technology)

D32 High-Temperature and High-Pressure Permeability-Selectivity Upper Bounds for Gas Separation Membranes Prepared From High Performance Polymers, Edson V Perez (The University of Texas at Dallas)*; Kenneth Balkus, Jr. (The University of Texas at Dallas); John Ferraris (University of Texas at Dallas); Inga Musselman (The University of Texas at Dallas)

D33 Synthesis and Characterization of Anion Exchange Membranes based on Semi-Crystalline Poly(Ethylene-block-Styrene-block-Ethylene) Triblock Copolymer, Carrie Trant (Rensselaer Polytechnic Institute)*; Chulsung Bae (Rensselaer Polytechnic Institute); Sangwoo Lee (Rensselaer Polytechnic Institute)

D34 Effect of Polyvinylamine Content on the CO_2 Separation Performance of Facilitated Transport Membranes, James S Baker (U.S. Department of Energy National Energy Technology Laboratory)*; Victor A Kusuma (U.S. Department of Energy National Energy Technology Laboratory); David Hopkinson (U.S. Department of Energy National Energy Technology Laboratory)

D35 Effect of organic ballast properties on the energy efficiency of a concentration gradient flow battery, Fei Liu (University of North Carolina at Chapel Hill)*; Ryan Kingsbury (Membrion, Inc.); Jeromy Rech (University of North Carolina at Chapel Hill); Wei You (University of North Carolina at Chapel Hill); Orlando Coronell (University of North Carolina at Chapel Hill)

SESSION E - Innovation in Membranes for Water Treatment ... pg. 124

E1 Feasibility of long-term treatment of malting wastewater in a submerged anaerobic membrane bioreactor (SAnMBR) at psychrophilic temperature conditions, Esmat Maleki (Lakehead University)*; Baoqiang Liao (Lakehead University)

E2 Biofouling in ultrafiltration process for drinking water treatment and its control by chlorinated-water and pure water backwashing, Shao Senlin (The University of Hong Kong)*

E3 Membrane Distillation of High Salinity Water by Induction Heated Thermally Conducting Membranes, Avner Ronen (Temple University)*; Arezou Anvari (Temple University)

E4 Bromide Removal from Drinking Water Source Impacted by Energy Wastewater Discharge Using Electrically Conductive Membranes, Avner Ronen (Temple University)*; Mei Sun (University of North Carolina at Charlotte); Kartikeya Kekre (Temple University)

E5 A surface modification of forward osmosis (FO) membranes to enhance dye retention and fouling resistance for the treatment of textile wastewater, Sara Azari (Sterlitech)*; Mohammd Karimi (Amirkabir University of Technology)

E6 Evaluation of surface, bulk and electrochemical properties of Ti3C2Tx membranes used for selective ion removal from brine., Yousuf Z Bootwala (Georgia Institute of Technology)*; Wahiduz Zaman (Vanderbilt University); Kelsey Hatzell (Vanderbilt University); Marta Hatzell (Georgia Institute of Technology)

E7 Microwave-enhanced Membrane Filtration for Water Treatment, Wanyi Fu (New Jersey Institute of Technology)*; Wen Zhang (New Jersey Institute of Technology)

E8 Application of Nanofiltration (NF) Membranes for Perfluoroalkyl and Polyfluoroalkyl substances (PFAS) Removal, Daniel G Kulas (Michigan Technological University)*; Andre da Costa (Michigan Technological University); Ananya Ananya (Michigan Technological University); Sam Jacobs (Michigan Technological University)

E9 Development of membrane synthesis and novel draw materials for forward osmosis process, Yusak Hartanto (Katholieke Universiteit Leuven)*

E10 3D Printed Self-Assembled Zwitterionic Copolymer Membranes, Xin Qian (University of Connecticut)*; Samuel J Lounder (Tufts University); Ayse Asatekin (Tufts University); Jeffrey R McCutcheon (University of Connecticut)

E11 Antifouling and Antiwetting Janus Membrane for Treating Hypersaline Oily Wastewater by Membrane Distillation, Chenxi Li (University of British Columbia); Xuesong Li (University of British Columbia); Xuewei Du (Colorado State University); Tiezheng Tong (Colorado State University); Tzahi Cath (Colorado School of Mines); Jongho Lee (University of British Columbia)*

E12 Treating Hydraulic Fracturing Produced Waters by Membrane Distillation, Ranil Wickramasinghe (University of Arkansas)*

E13 Synthesis of Polystyrene-b-poly(solketal methacrylate) (PS-b-PSMA) Polystyrene-b-poly(glyceryl methacrylate) (PS-b-PGMA) Block Copolymers for Isoporous Membranes, Sarah Saleem (Helmholtz-Zentrum Geesthacht); Sofia Rangou (Helmholtz-Zentrum Geesthacht); Clarissa Abetz (Helmholtz-Zentrum Geesthacht); Brigitte Lademann (Helmholtz-Zentrum Geesthacht); Volkan Filiz (Helmholtz-Zentrum Geesthacht)*; Volker Abetz (Helmholtz-Zentrum Geesthacht)

E14 Uranium isolation and concentration using reactive membranes for nuclear forensics applications, Abenazer W Darge (Clemson University); Timothy DeVol (Clemson University); Scott Husson (Clemson University)*

E15 Development of Nano-enabled Membrane Technology for Water Reuse in Agriculture, Xingmao Ma (Texas A&M University); Yinghao Wen (Texas A&M University)*; Jieming Yuan (Texas A&M University)

E16 Detection of Electrically Conductive Membrane Fouling by Impedance Spectroscopy, Nan Zhang (McMaster University)*; Charles-François de Lannoy (McMaster University)

E17 Stability of Polyamide Nanofiltration Membranes with Peracetic Acid/Hydrogen Peroxide Disinfection, Mohsen Ghafari (State University of New York at Buffalo)*; Boya Xiong (MIT); Ning Dai (State University of New York at Buffalo)

E18 Fouling-Resistant Two-Dimensional (2D) Covalent Organic Framework Membranes for Industrial Water Reuse, Phuoc Duong (University of Wyoming); Valerie Kuehl (University of Wyoming); Mohammad Afroz (University of Wyoming); John Hoberg (University of Wyoming); Bruce Parkinson (University of Wyoming); Katie D Li-Oakey (University of Wyoming)*

E19 Reduced-holy graphene oxide(r-HGO) membranes with enhanced water permeance for water purification, Xiaoyi Chen (The State University of New York at Buffalo)*; Liang Huang (The State University of New York at Buffalo); Zhihao Feng (The State University of New York at Buffalo); Janavi Gohil (The State University of New York at Buffalo); Haiqing Lin (The State University of New York at Buffalo)

E20 A Comparison of PolarClean, Gamma-Valerolactone and their Mixture as Bio-derived Solvents for Polysulfone Membrane Fabrication, Xiaobo Dong (University of Kentucky)*; Halle Shannon (University of Kentucky); Isabel Escobar (University of Kentucky)

E21 Filterability of crude oil emulsions stabilized by COREXIT 9500 dispersant, Seyma Kucuk (Michigan State University); Charifa Hejase (Michigan State University)*; Iryna Kolesnyk (National University of Kyiv-Mohyla Academy); Jia Wei Chew (Nanyang Technological University); Vlad Tarabara (Michigan State University)

~~**E22 Polyamide/CNT Thin Film Nanocomposite (TFN) Seawater RO Membranes with Improved Boron Rejection,** Aysa Guvensoy (Istanbul Technical University); Süer Kurklu-Kocaoglu (Istanbul Technical University); Cansu Yildirim (Istanbul Technical University); Sadiye Velioglu (Istanbul Technical University); S. Birgül Tantekin-Ersolmaz (Istanbul Technical University)*~~

E23 New generation of patterned membranes for water treatment, Asad Asad (University of Alberta); Dan Sameoto (University of Alberta); Mohtada Sadrzadeh (University of Alberta)*

E24 Enhancing boron rejection on electrically conducting reverse osmosis membranes through local electrochemical pH modification, Bongyeon Jung (UCLA)*; David Jassby (UCLA)

E25 Cellulose Triacetate Membrane Parameter Estimation Under Osmotic and Osmotic-Assisted Process Conditions, Jacob L Weidman (U.S. Department of Energy National Energy Technology Laboratory/Oak Ridge Institute for Science and Education)*; Sara Osipi (Federal University of Rio de Janeiro); Alexander V Dudchenko (Carnegie Mellon University); Meagan S Mauter (Carnegie Mellon University); Nicholas Siefert (U.S. Department of Energy National Energy Technology Laboratory)

E26 Porous hydrophobic-hydrophilic composite flat and hollow fiber membranes for direct contact membrane distillation, Aishwarya Puranik (New Jersey Institute of Technology); Lydia Rodrigues (New Jersey Institute of Technology); John Chau (New Jersey Institute of Technology); Lin Li (New Jersey Institute of Technology); Kamalesh K Sirkar (New Jersey Institute of Technology)*; Ashok Sharma (Applied Membrane Technology Inc); Stephen Conover (Applied Membrane Technology Inc); Adam Juelfs (Applied Membrane Technology Inc); Connor Colling (Applied Membrane Technology Inc); Saket Sharma (Applied Membrane Technology Inc)

E27 Polyamide-imide: a new platform for fabrication of oil/water separation membranes, Nusrat Helali (University of Alberta); Masoud Rastgar (University of Alberta); Mohtada Sadrzadeh (University of Alberta)*

E28 Monolayer Graphene transfer onto polyvinyl alcohol (PVA) microfiltration membrane for water desalination., Mansour Saberi (Clemson University)*

E29 Membrane Distillation-Crystallization (MDC) for Maximum Water Recovery in Inland Desalination, Evangelos Balis (University of Nevada, Reno); Sage Hiibel (University of Nevada, Reno)*

E30 Insights regarding thermomechanical bonding between porous membranes and thermoplastic polymers, Masoud Aghajani (MAST Center, University of Colorado Boulder)*; Adrienne Blevins (University of Colorado Boulder); Jason P. Killgore (National Institute of Standards and Technology); Ryan Sylvia (MilliporeSigma); Christina Carbrello (MilliporeSigma); Alan R. Greenberg (MAST Center, University of Colorado Boulder); Rong Long (MAST Center, University of Colorado Boulder); Yifu Ding (MAST Center, University of Colorado Boulder)

E31 Ion exchange polymer coatings enhance solute rejection of polyamide thin-film composite membranes, Mikayla D Armstrong (University of North Carolina at Chapel Hill)*; Ryan Kingsbury (Membrion, Inc.); Kasia Grzebyk (University of North Carolina at Chapel Hill); Orlando Coronell (University of North Carolina at Chapel Hill)

E32 Mechanical behaviour of the porous support of TFC membranes and its influence on the overall resistance, Masoud Aghajani (University of Colorado Boulder)*; Alan R. Greenberg (University of Colorado Boulder); Yifu Ding (University of Colorado Boulder)

E33 Development of Membrane Distillation through Tuning Membrane Surface Properties and Module Design, Siamak Nejati (University of Nebraska-Lincoln)*; Mahdi Mohammadi (University of Nebraska-Lincoln); Mona Bavarian (University of Nebraska-Lincoln)

E34 Integration of Forward Osmosis in the Treatment of Sewage by Chlorella vulgaris: Comparison between External and Immersed Systems, Xue Jin (Oregon State University)*; Mathieu Larronde-Larretche (University of Glasgow)

E35 Nanoengineered condensation surfaces for air gap membrane distillation, Yashwant S. Yogi (Purdue University); Sina Nejati (Purdue University); Akshay K. Rao (Purdue University); Rishav Roy (Purdue University); Abhimanyu Das, Longnan Li (University of Illinois, Urbana-Champaign); Soumyadip Sett (University of Illinois, Urbana-Champaign); John Lienhard (MIT); Nenad Miljkovic (University of Illinois, Urbana-Champaign); Justin A. Weibel (Purdue University); Jaichander Swaminathan (Massachusetts Institute of Technology); David Warsinger (Purdue University)*

E36 Fouling of Anion Exchange Membranes in Ferric Chloride Solutions: mechanism and the role of ion exchange capacity, water content, and pore structure, Michael J. McGrath (University of Colorado Boulder)*; Nicholas Patterson (UC San Diego); Bryce Manubay (University of Colorado Boulder); Hans Funke (University of Colorado); Samantha Hardy (University of Colorado Boulder); Andrew Basalla (University of Colorado Boulder); Xiujun Yue (UC San Diego); Ping Liu (UC San Diego); Douglas Gin (University of Colorado Boulder); Richard Noble (University of Colorado Boulder)

SESSION F - Innovation in Modeling of Membrane Fundamentals and Processes ...pg 135

F1 Molecular Dynamics Processing Tool for Transmembrane Simulations, Jia Lin Cheoh (Purdue University)*; Abhimanyu Das (Purdue University); David Warsinger (Purdue Unversity)

F2 **Low energy-consumption pre-treatment system for integrated forward osmosis - reverse osmosis desalination process,** Yunchul Woo (Korea Institute of Civil Engineering and Building Technology)*; June-Seok Choi (Korea Institute of Civil Engineering and Building Technology); Kwang-Duk Park (Korea Institute of Civil Engineering and Building Technology)

F3 **Arrested mobility effects on the spinodal decomposition of ternary polymer solutions for immersion precipitation,** Jan Garcia (UC Santa Barbara)*; Douglas Tree (Brigham Young University); Kris Delaney (UC Santa Barbara); Glenn Fredrickson (UC Santa Barbara)

F4 **Hydro- and perfluoro-carbon sorption, diffusion and permeation modeling in poly(dimethylsiloxane) using a PC-SAFT Equation of State,** Liang Liu (The University of Melbourne)*

F5 **The Potential of Low-Temperature Forward Osmosis for Water Recovery,** Jeffrey Martin (University of Toronto); Georgios Kolliopoulos (University of Toronto); Vladimiros Papangelakis (University of Toronto)*

F6 **Fouling mechanisms in constant flux crossflow ultrafiltration,** Alon Kirschner (The University of Texas at Austin)*; Yu-Heng Cheng (The University of Texas at Austin); Donald Paul (The University of Texas at Austin); Robert Field (University of Oxford); Benny Freeman (The University of Texas at Austin)

F7 **Hitz zeolite membrane separation system (HDS) for liquid separation,** Akira Hamasaki (Hitachi Zosen Corporation)*

F8 **Effect of short-term contact with C1-C4 alcohols on the water permeance and salt rejection of MPD-TMC thin-film composite reverse osmosis membranes,** Jaime A Idarraga-Mora (Clemson University)*; Michael Lemelin (Clemson University); Steven Weinman (Clemson University); Scott Husson (Clemson University)

F9 **Whey recovery using forward osmosis with hollow fiber configuration - Evaluating the factors limiting the flux performance,** YiNing Wang (Nanyang Technological University)*; Rong Wang (Nanyang Technological University)

F10 **Bimetallic Nanoparticles Incorporated Poly(methacrylic acid) Functionalized Membranes for Water Remediation: Synthesis, Advanced Characterization and Reactive Properties,** Hongyi (Derek) Wan (University of Kentucky)*; Md. Saiful Islam (University of Kentucky); Nicolas Briot (University of Kentucky); Anthony Saad (University of Kentucky); Lindell Ormsbee (University of Kentucky); DB Bhattacharyya (University of Kentucky)

F11 **Membrane compaction in forward osmosis process,** Daniel Yee Fan Ng (Nanyang Technological University); Yunfeng Chen (Nanyang Technological University); Rong Wang (Nanyang Technological University)*; Zhili Dong (Nanyang Technological University)

F12 **Novel Isothermal Acid Sweep Membrane Distillation for Selective and Energy Efficient Removal and Recovery of Ammonia,** Stephanie N McCartney (Columbia University)*; Natalie Williams (Columbia University); Chanhee Boo (Columbia University); Ngai Yin Yip (Columbia University)

F13 **In-situ micro-rheology of a foulant layer,** Jose Agustin Epstein (Technion - Israel Institute of Technology); Guy Z Ramon (Technion - Israel Institute of Technology)*

F14 **Mixed-gas permeation in thermally rearranged poly(benzoxazole-co-imide) polymer membranes,** Adele Brunetti (Institute on Membrane Technology of Italian National Research Council); Elena Tocci (Institute on Membrane Technology of Italian National Research Council); Jong Geun Seong (Hanyang University); Young Moo Lee (Hanyang University)*; Enrico Drioli (Institute on Membrane Technology of Italian National Research Council); Giuseppe Barbieri (Institute on Membrane Technology of Italian National Research Council)

F15 **Plug flow reactor IEMB - performance and modelling,** Shalom Fox (Ben-Gurion University of the Negev); Kristina Stadnik (Ben-Gurion Univ.); Amit Thakur (Ben-Gurion University of the Negev); Zeev Ronen (Ben-Gurion Univ.); Yoram Oren (Ben-Gurion University of the Negev); Jack Gilron (Ben-Gurion University of the Negev)*

F16 **Isolation of Extracellular Vesicles from a Biofluid Using Ultrathin Microslit and Nanopore Silicon Membranes,** Jared Carter (SiMPore inc.)*; Cassandra Walinski (SiMPore Inc.); Akash Patel (SiMPore Inc.); Joshua Miller (SiMPore Inc.); Jon-Paul DesOrmeaux (SiMPore Inc.); James A Roussie (SiMPore Inc.)

F17 **High-Pressure Reverse Osmosis for Energy-Efficient Desalination of High Salinity Brines,** Douglas M Davenport (Yale University)*; Akshay Deshmukh (Yale University); Jay Werber (Yale University); Yuhao Du (Yale University); Menachem Elimelech (Yale University)

F18 **Computationally Efficient Methods of Simulating Membrane Spacers,** Jacob Johnston (Colorado School of Mines)*; Jincheng Lou (Colorado School of Mines); Nils Tilton (Colorado School of Mines)

F19 **Dissipative Particle Dynamics Simulations of PES membrane via NIPS,** Eric Ledieu (University of Arkansas)*; Yuanhui Tang (University of Arkansas); Rosario Cervellere (University of Arkansas); Paul Millett (University of Arkansas); David Ford (University of Arkansas); Xianghong Qian (University of Arkansas)

F20 **Leveraging Surplus Energy from Nuclear Power Generation for Seawater Desalination: A Process Modelling Investigation,** J. Hunter Himes (Tennessee Tech University); Mary Adkisson (Tennessee Tech University); Laura H Arias Chavez (Tennessee Tech University)*

F21 **Maxwell-Stefan characterization of surface diffusion flow through defect containing silicalite membranes.,** David Carter (University of Ottawa); Shaaima Al Akwaa (University of Ottawa); Boguslaw Kruczek (University of Ottawa); Handan Tezel (University Ottawa)*

F22 **Self-Regulating FO-RO Hybrid Systems,** Noah Ferguson (University of Connecticut)*; Nicole Beauregard (University of Connecticut); Jeffrey R McCutcheon (University of Connecticut)

F23 **Estimating salt diffusion coefficients in polyamide active layers of reverse osmosis membranes using microscale continuum modeling,** Riley Vickers (University of North Carolina at Chapel Hill)*; Jingbo Wang (UCLA); Timothy Weigand (University of North Carolina at Chapel Hill); Casey Miller (University of North Carolina at Chapel Hill); Orlando Coronell (University of North Carolina at Chapel Hill)

F24 **Energy-efficient Biogas Upgrading Membrane Process,** Chung-Seop Lee (Airrane Co. Ltd); Jinhyuk Lim (Airrane Co. Ltd); Sang Hoon Han (Airrane Co. Ltd.)*; Sejong Kim (Airrane Co. Ltd); Seong Yong Ha (Airrane Co. Ltd); Won Seok Chang (Korea District Heating Corporation); Moon Sei Oh (Korea District Heating Corporation)

F25 **Multi-stage Membrane Process for CO_2 Separation from LNG flue gas,** Chung-Seop Lee (Airrane Co. Ltd)*; Sejong Kim (Airrane Co. Ltd); Chung-Seop Lee (Airrane Co. Ltd); Won Seok Chang (Korea District Heating Corporation); Moon Sei Oh (Korea District Heating Corporation)

F26 **Cost Optimization of Pressure Retarded Osmosis,** Alexandra N Newby (Carnegie Mellon University)*; Timothy V Bartholomew (Carnegie Mellon University); Meagan S Mauter (Carnegie Mellon University)

F27 **Tearing propagation of an airship envelope fabric based on FEM and X-FEM methods,** Chen Yonglin (Shanghai Jiaotong University); Wang Fengxin (Shanghai Jiaotong University)*

F28 **Modelling of the effect of current density and contact time in electro-coagulation on membrane fouling,** In-Soung Chang (Hoseo University)*

F29 **Characterization and Molecular Dynamics Simulation of Poly(arylethersulfone) Ultrafiltration Hemodialysis Membrane,** Arash Mollahosseini (University of Saskatchewan); Amira Abdelrasoul (University of Saskatchewan)*

F30 **Humidified gas permeabilities in polybenzimidazole gas separation membranes,** Joshua D Moon (The University of Texas at Austin)*; Benny Freeman (The University of Texas at Austin); Donald Paul (The University of Texas at Austin)

F31 **Hollow Fiber Countercurrent Dialysis for Continuous Buffer Exchange of High Value Biotherapeutics,** Christopher Yehl (Penn State University)*

F32 **Multiscale modeling of microporous polymer blends for CO_2 separation,** Alec Kaija (U.S. Department of Energy National Energy Technology Laboratory)*; Ali Sekizkardes (Battelle/U.S. Department of Energy National Energy Technology Laboratory); Samir Budhathoki (U.S. Department of Energy National Energy Technology Laboratory); Janice Steckel (U.S. Department of Energy National Energy Technology Laboratory)

North American Membrane Society

North American Membrane Society 28th Annual Meeting

Membrane Separations for Emerging Water, Energy, and Health Applications

Program Book

Conference Chairs:
David Latulippe, McMaster University
Meagan Mauter, Carnegie Mellon University
Andrew Zydney, Penn State University

www.nams2019.org
@NAMS2019

May 11-15, 2019
Pittsburgh, PA

Program at a Glance

Date	Time	Event
Saturday May 11	8:00 am - 5:00 pm	Workshop 1: Measurement Methods for Membranes (Benedum Room)
	8:00 am - 5:00 pm	Workshop 2: Membranes for Water Treatment Applications (Duquesne Room)
Sunday May 12	8:00 am - 5:00 pm	Workshop 3: Membranes for Gas Separations (Birmingham/Smithfield Room)
	8:00 am - 5:00 pm	Workshop 4: Polymeric and Inorganic Membrane Materials and Membrane Formation (Benedum Room)
	8:00 am - 5:00 pm	Workshop 5: Membranes for Bioprocessing (Duquesne Room)
	12:00 pm - 6:00 pm	Registration (Lobby)
	1:30 pm - 4:30 pm	NAMS Student Workshop (Sterlings Room)
	6:00 pm - 9:00 pm	Welcome Reception (Carnegie Science Center, 1 Allegheny Avenue)
Monday May 13	7:00 am - 7:00 pm	Registration (Ballroom Foyer)
	7:00 am - 9:00 pm	Exhibitors (Ballroom Foyer)
	7:45 am - 9:00 am	Plenary Lecture - Dr. Peter Fiske (Ballroom 1)
	9:00 am - 9:30 am	Coffee Break
	9:30 am - 12:30 pm	Parallel Technical Sessions I
	12:30 pm - 2:00 pm	Lunch With Legends (Commonwealth Room) – Pre-registration required
	2:00 pm - 5:00 pm	Parallel Technical Sessions II
	7:00 pm - 10:00 pm	Poster Session (Ballrooms 1 & 2)
Tuesday May 14	7:30 am - 7:00 pm	Registration (Ballroom Foyer)
	7:00 am - 9:00 pm	Exhibitors (Ballroom Foyer)
	7:45 am - 9:00 am	Plenary Lecture - Dr. Tim Merkel (Ballroom 1)
	9:00 am - 9:30 am	Coffee Break
	9:30 am - 12:30 pm	Parallel Technical Sessions III
	2:00 pm - 5:00 pm	Parallel Technical Sessions IV
	5:30 pm - 6:30 pm	NAMS Business Meeting (Kings Garden 5)
	6:30 pm - 7:30 pm	Happy Hour (Ballroom Foyer)
	7:30 pm - 11:00 pm	Banquet & Awards Ceremony (Ballroom 1)
Wednesday May 15	7:30 am - 12:00 pm	Registration (Ballroom Foyer)
	7:00 am - 5:00 pm	Exhibitors (Ballroom Foyer)
	7:45 am - 9:00 am	Plenary Lecture - Dr. Rachel Segalman (Ballroom 1)
	9:00 am - 9:30 am	Coffee Break
	9:30 am - 12:30 pm	Parallel Technical Sessions V
	2:00 pm - 5:00 pm	Parallel Technical Sessions VI

Get Connected!

Wyndham WiFi – Username: **NAMS1** Password: **CONF19** **#NAMS2019**

7:45	**Plenary Session 1 - Dr. Peter S. Fiske, Innovation Pathways in the Water Ecosystem: The PAX Water Story (Ballroom 1)**					
Parallel Sessions	**BALLROOM 3** **#2 Processes: Membrane Distillation and Pervaporation: Innovations in Process Design and Scalability**	**BALLROOM 4** **#4 Materials: Selective Polymeric and Mixed-Matrix Materials - Liquid Separations**	**KINGS GARDEN 1** **#5 Materials: Inorganic Materials**	**KINGS GARDEN 3** **#3 Applications: Seawater Desalination**	**KINGS GARDEN 4** **#1 Processes: Fundamentals of Predicting and Preventing Membrane Fouling**	**KINGS GARDEN 5** **#6 Applications: Assessing Performance, Robustness, and Scalability for Carbon Capture**
9:30	**2a - Continuous on-demand dehydration of solvents in flow chemistry manufacturing processes** Hannah Murnen (Compact Membrane Systems)	**4a - Enantioselective Polymeric Membrane for Chiral Separation of Enantiomers** Marine Michel (Imperial College London)	**5a - Two-Dimensional Membranes for Gas Separation** Dan Zhao (National University of Singapore)	**3a - Sacrificial Protective Layer for Fouling Control in Reverse Osmosis Desalination** Moon Son (Penn State University)	**1a - Hydrodynamic-colloidal interactions of an oil droplet and a membrane surface** Guy Z Ramon (Technion - Israel Institute of Technology)	**6a - Polymers with Side Chain Porosity for Ultrapermeable and Plasticization Resistant Materials for Gas Separations** Sharon Lin (MIT)
10:00	**2b - How membrane technology has been successfully applied to ethanol processing** Stephan Blum (Whitefox Technologies)	**4b - Tailored Synthesis of Inorganic-supported Polymer, Graphene, and Nanocomposite Membranes for Liquid-phase Separations** Michael Z Hu (Oak Ridge National Laboratory)	**5b - Restricting Lattice Flexibility in Polycrystalline Metal-Organic Framework Membrane for Efficient Carbon Capture** Kumar Varoon Agrawal (L'Ecole polytechnique fédérale de Lausanne)	**3b - Water Transport through Ultrathin Polyamide Nanofilms Used for Reverse Osmosis** Zhiwei Jiang (Imperial College London)	**1b - Do Graphene Oxide Nanostructured Coatings Mitigate Bacterial Adhesion to Membrane Interfaces?** Santiago Romero-Vargas Castrillon (University of Edinburgh)	**6b - Bench scale testing of next generation hollow fiber membrane modules at subambient conditions at the National Carbon Capture Center** David Hasse (Air Liquide)
10:30	**2c - Modeling the effect of different single-walled carbon nanotube orientations on permeation through mixed matrix membranes** Ali Zamani (University of Ottawa)	**4c - Two-Dimensional (2D) Covalent Organic Framework Mixed-Matrix Membranes** Katie D Li-Oakey (University of Wyoming)	**5c - New Asymmetric and Permselective Carbon Molecular Sieve Membranes** Jay (Junqiang) Liu (The Dow Chemical Company)	**3c - Balancing Carbon, Nitrogen and Phosphorus Concentration in Seawater as a Strategy to Prevent Accelerated Membrane Biofouling** Sunny Jiang (UC Irvine)	**1c - Fouling mechanisms in constant flux crossflow ultrafiltration** Alon Kirschner (The University of Texas at Austin)	**6c - A Carbon Molecular Sieve Membrane-Based Reactive Separation Process for Pre-Combustion CO_2 Capture** Theodore T Tsotsis (University of Southern California)
11:00	**2d - Unique 3D printed spacers for process enhancement in membrane distillation** Navya Thomas (Khalifa University)	**4d - Enhancing water permeance in reduced graphene oxide membranes by chemical etching** Xiaoyi Chen (The State University of New York at Buffalo)	**5d - Hitz zeolite membrane separation system (HDS) for gas separation** Masaya Itakura (Hitachi Zosen Corporation)	**3d - Flow Reversal for Increased Recovery of 2nd Pass Seawater RO** Jack Gilron (Ben-Gurion University)	**1d - Fouling-Resistant, Multi-Functional Membranes with Ultra-Thin Hydrogel Selective Layers using Interfacially Initiated Free Radical Polymerization (IIFRP)** Ayse Asatekin (Tufts University)	**6d - Synthesis of Polymers of Intrinsic Microporosity (PIM-1) with Poly(ethylene glycol) for Membrane Application** Volkan Filiz (Helmholtz-Zentrum Geesthacht)
11:30	**2e - 2D boron nitride encapsulated nanofiber membranes for membrane distillation** Yunchul Woo (Korea Institute of Civil Engineering and Building Technology)	**4e - Nanocomposite RO Membranes with High Aspect Ratio Functional Nanoparticles for Controlled Interfacial Transport** Stephen M Martin (Virginia Tech)	**5e - Insights on the Carbon Molecular Sieve Structure from Stability and Rejuvenation Tendencies** Samuel Hays (Georgia Institute of Technology)	**3e - Zwitterionic poly(arylene ether sulfone) (PAES) copolymer for biofouling/chlorine resistant desalination membranes** Yi Yang (Arizona State University)	**1e - Unravel the Gordian Knot of membrane desalination: Understanding and mitigating silica scaling in reverse osmosis** Tiezheng Tong (Colorado State University)	**6e - High-permeance polymer-functionalized graphene membranes that surpass the post-combustion carbon capture target** Kumar Varoon Agrawal (L'Ecole polytechnique fédérale de Lausanne)
12:00	**2f - Impact of membrane properties on scaling-induced wetting in membrane distillation** Allyson L McGaughey (University of Southern California)	**4f - If one were to develop a new NF/RO membrane, specifically for uncharged organic solute rejection, how permeable could it be to water while meeting organic solute rejection goals?** Eric Hoek (UCLA)	**5f - Fabrication of zeolitic imidazolate framework ZIF-8 membranes by ligand-induced permselectivation in vapor phase** Xiaoli Ma (University of Wisconsin-Milwaukee)	**3f - Chlorine resistance of functionalized HNTs-based TFN membranes** Boguslaw Kruczek (University of Ottawa)	**1f - RO Mineral Scale Mitigation via Self-Adaptive Feed-Flow Reversal** Yeunha Kim (UCLA)	**6f - Fabrication of Thin Film Composite Membranes using Microporous Polymer Blends and Polybenzimidazole Nanoporous Supports for CO_2/N_2 Separation** Lingxiang Zhu (U.S. Department of Energy National Energy and Technology Laboratory)

Parallel Sessions	BALLROOM 3 **#8 Processes: Innovations in Module Modeling and Design**	BALLROOM 4 **#9 Processes: Osmotically Driven Membrane Processes for Water and Energy**	KINGS GARDEN 1 **#11 Materials: Ion-Exchange and Electrofunctional Materials**	KINGS GARDEN 3 **#12 Materials: Membrane Material and Transport Simulation: Molecular & Process Modeling of Membranes**	KINGS GARDEN 4 **#10 Applications: Water Reuse**	KINGS GARDEN 5 **#7 Applications: Assessing Performance, Robustness, and Scalability for Carbon Capture**
2:00	**8a - CFD simulation of crimped hollow fiber membranes for liquid separation processes** Mohammad Younas (University of Engineering and Technology, Peshawar)	**9a - Structural and Transport Properties of Membranes in High-Salinity Desalination using Cascading Osmotically Mediated Reverse Osmosis** Xi Chen (Columbia University)	**11a - Ion Transport Through Perforated Nanoporous Graphene** Mandakranta Ghosh (University of Twente)	**12a - Expanding the chemical palette for reliable chemical separations using molecular simulations and machine learning** David Sholl (Georgia Institute of Technology)	**10a - Tannic acid-iron network based green technique for enhanced membrane performance in water reuse** Hao Guo (The University of Hong Kong)	**7a - Field trial of spiral-wound facilitated transport membrane module for CO_2 capture from flue gas** Yang Han (Ohio State University)
2:30	**8b - Computational Fluid Dynamics Modeling for the Investigation of Multi-Layer Spacer Effects** Andrea Achilli (The University of Arizona)	**9b - Utilization of reverse solute diffusion to enhance membrane performance during Osmotically-Driven processes - improved boron retention and silica scaling mitigation** YiNing Wang (Nanyang Technological University)	**11b - Developing a new approach to describe ion sorption and transport in Nafion membranes** Rahul Sujanani (The University of Texas at Austin)	**12b - Diffusion, Reaction, and Network Structure Modeling of Interfacial Polymerization of Polyamide Membranes** Jeffrey D Wilbur (DuPont Water Solutions)	**10b - Membrane Distillation (MD) and Reverse Osmosis (RO) processes for water recovery from pre-treated high strength brewery wastewater** Nawrin Anwar (Concordia University)	**7b - Gen-2 Proteus™ membrane and module development for CO CO_2 capture from gasification streams** Witopo Salim (Membrane Technology & Research Inc.)
3:00	**8c - Direct numerical simulation of unsteady mixing in direct contact membrane distillation systems with membrane spacers** Jincheng Lou (Colorado School of Mines)	**9c - Osmotic Membrane Desalination Performance Governed by Molecular Reflection at the Liquid-Vapor Interface** Jongho Lee (University of British Columbia)	**11c - Co-transport of neutral solutes in Nafion 117 using In-situ ATR FTIR spectroscopy for multicomponent solution speciation** Breanna Dobyns (Auburn University)	**12c - Optimization of Mixed Matrix Membrane Materials for Post-Combustion Carbon Capture** Janice A. Steckel (U.S. Department of Energy National Energy and Technology Laboratory)	**10c - Field Demonstration of Real-Time Colloidal Particle Monitoring to Improve MF and UF Membrane Performance** Jana Safarik (Orange County Water District)	**7c - One-pot surface modification of nano-cellulose fibrils and effect of manipulated fibril surfaces in Hybrid Facilitated Transport Membranes for CO_2 capture** Saravanan Janakiram (Norwegian University of Science and Technology)
3:30	**8d - Concentration polarization modeling for high-pressure membranes with engineered surface features** Zuo Zhou (Clemson University)	**9d - Characterization of External and Internal Fouling of Forward Osmosis (FO) Process Through Ultrasonic Time Domain Reflectometry (UTDR) Method** LI Lai (Nanyang Technological University)	**11d - Functionalized Nanoporous Ceramic Membranes Towards Low-Cost Electrodialysis** Gregory M Newbloom (Membrion, Inc.)	**12d - Diffusivity in anion exchange membranes under low hydration: A molecular dynamics study** Simcha Srebnik (Technion - Israel Institute of Technology)	**10d - Selective removal of phosphorus from wastewater using electromembrane process** Sanhita Chaudhury (Ben-Gurion University of the Negev)	**7d - DDR-type zeolite membrane: The first opportunity in CO_2-EOR industry** Junya Okazaki (JGC Corporation)
4:00	**8e - Analysis of Micromixers to Minimize Scaling Effects on Reverse Osmosis Membranes** Jeremy Walker (U.S. Army)	**9e - Incorporating membrane deformation into the boundary layer equation to model water and reverse salt flux in osmotic processes** Jaime A Idarraga-Mora (Clemson University)	**11e - Polymerization of counterions in self-assembled, 1 nm pores of lyotropic liquid crystal anion exchange membrane to tune nanopore and ion transport properties** Michael J. McGrath (University of Colorado at Boulder)	**12e - Modeling of polymer membrane formation via phase inversion by mesoscopic phase-field methods: Investigating the development of structure on multiple scales** Michael R Cervellere (University of Arkansas/EMD Millipore)	**10e - Water Recovery from Produced Water via Robust Membrane Distillation** Rong Wang (Nanyang Technological University)	**7e - Coating of hollow fiber membrane for carbon molecular sieve membrane preparation** Yuhe Cao (Georgia Institute of Technology)
4:30	**8f - Do Hydrodynamic Instabilities Cause Roughness in Thin Polyamide Films of Reverse Osmosis Membranes?** Akshay Deshmukh (Yale University)	**9f - A Thin-Film Forward-Osmosis Membrane Made from a Graphene Oxide-Silver Metal-Organic Framework Nanocomposite: Antifouling and Antibiofouling Assessment** Milad Rabbani Esfahani (University of Alabama)	**11f - Quantifying tortuosity and diffusion over multiple scales in ionic and non-ionic polymer membranes** Louis Madsen (Virginia Tech)	**12f - Multiscale Modeling of Gas Permeation Through Poly(Dimethyl Phenylene Oxide) (PPO) and Its Pre-Steady State Polymer Response** Marielle Soniat (Lawrence Berkeley National Laboratory)	**10f - Potable-quality water recovery from primary effluent through an integrated algal-osmosis membrane system** Lu Lin (New Mexico State University)	**7f - Ultrapermeable benzotriptycene-based PIMs that redefine the upper bounds for CO_2 separations** Alessio Fuoco (CNR)
	POSTER SESSION – 7:00 pm - 10:00 pm (Ballrooms 1 & 2)					

7:45	**Plenary Session 2 - Dr. Tim Merkel, Development of Membranes for CO_2 Capture (Ballroom 1)**				
Parallel Sessions	**BALLROOM 3** **#14 Applications: Contaminant Removal from Water Sources**	**BALLROOM 4** **#15 Materials: Materials for Organic Solvent Separations**	**KINGS GARDEN 3** **#16 Materials: Catalytic and Responsive Materials**	**KINGS GARDEN 4** **#13 Processes: Process Innovations in Electrofunctional and Electrocatalytic Membrane Processes**	**KINGS GARDEN 5** **#17 Materials: Selective Polymeric and Mixed-Matrix Materials - Gas Separations**
9:30	**14a - Oil drop behavior on model nanofiltration membrane surfaces under conditions of hydrodynamic shear** Charifa Hejase (Michigan State University)	**15a - Membrane solvent recovery in edible oil industry** Petrus Cuperus (SolSep BV)	**16a - Responsive CNT Composite Membranes for Protection Against Chemical Warfare Agents** Melinda L Jue (Lawrence Livermore National Laboratory)	**13a - Scaling mitigation in membrane distillation through electrokinetic mixing** Unnati Rao (UCLA)	**17a - Mixed-matrix membranes formed from imide-functionalized UiO-66-NH_2 for improved interfacial compatibility** Qihui Qian (MIT)
10:00	**14b - A case study review of the application of reverse osmosis membranes for water treatment and reagent recovery at mining and mineral processing operations** Chris Biederman (Hatch)	**15b - Molecular Design of Polymeric Membranes for Organic Solvent Recovery** Tai-Shung Chung (National University of Singapore)	**16b - Highly Efficient Dimethyl Ether Production from CO_2 Hydrogenation in a Dehydration Membrane Reactor** Huazheng Li (Rensselaer Polytechnic Institute)	**13b - Donnan dialysis desalination with thermolytic salts** Hanqing Fan (Columbia University)	**17b - Janus polymers bearing tri(n-alkoxy)silyl side groups: glassy membrane materials with properties of rubbers** Yury P Yampolskii (A.V.Topchiev Institute of Petrochemical Synthesis, Russian Academy of Sciences)
10:30	**14c - Self-Cleaning Nanocomposite Membranes with Phosphorene-Based Pore Fillers for Water Treatment** Joyner Eke (University of Kentucky)	**15c - Epoxy-based solvent resistant nanofiltration membranes prepared through phase inversion** Maarten Bastin (KU Leuven)	**16c - Poly-methacrylic acid Functionalized Membranes with Incorporated Reactive Pd/Fe Nanoparticles: Lab Scale to Groundwater Remediation Applications** Hongyi (Derek) Wan (University of Kentucky)	**13c - Electrically Conductive Membranes as Sensors: Stability, Conductivity, and Sensitivity** Charles-François de Lannoy (McMaster University)	**17c - Solution processable metal organic frameworks for gas separations: from porous liquids to mixed matrix membranes** Anastasiya V Bavykina (KAUST)
11:00	**14d - Optimization of nanofiltration process for treating industrial wastewater of time-varying composition**, Ryan LaRue (McMaster University)	**15d - Nanoporous Graphene Membranes for Organic Solvent Nanofiltration** David Cheng (MIT)	**16d - Investigation of a smart electrically responsive ultrafiltration membrane** Chia Miang Khor (UCLA)	**13d - Electrocatalytic Membrane Reactor-based Integrated Processes for Landfill Leachate Treatment** Jianxin Li (Tianjin Polytechnic University)	**17d - Enhancing CO_2/N_2 Selectivity and Elimination of Langmuir Sorption within High Tg, Glassy Polynorbornene Membranes** Brian Long (University of Tennessee)
11:30	**14e - Adsorptive and Reactive Membranes for Remediation of PFOA, Chloro-organics, and RCRA Metals from water** DB Bhattacharyya (University of Kentucky)	**15e - OSN Technology in Petrochemical Industry: Looking Back, Looking Forward** JR Johnson (ExxonMobil)	**16e - Catalytic Membranes for Groundwater Treatment** Alexander J Sutherland (McMaster University)	**13e - Nitrogen and Phosphorus Recovery from Agricultural Wastewater Effluents - Role of Electrically Conductive Membranes** Avner Ronen (Temple University)	**17e - Ultra-High Permeability Mixed-Matrix Membranes as a Next-Generation Carbon Capture Technology for Post-Combustion** Sameh K Elsaidi (U.S. Department of Energy National Energy Technology Laboratory)
12:00	**14f - Selective separation of mono- and di-valent cations in electrodialysis during brackish water desalination: bench and pilot-scale studies** Xuesong Xu (New Mexico State University)	**15f - Enabling Organic Solvent Nanofiltration and Reverse Osmosis using Ceramic Supported TFC Membranes** Jeffrey R McCutcheon (University of Connecticut)	**16f - Plasmonic membrane catalytic activation of peroxide for quantized oxidation via residence time control** Bruce Hinds (University of Washington)	**13f - Simulations of Highly Selective Separations Based on Electromigration** Merlin Bruening (University of Notre Dame)	**17f - Composite gas separation membranes from metal-induced ordered polymeric frameworks** Michael D Guiver (Tianjin University)

Parallel Sessions	BALLROOM 3 **#21 Applications: Brine Treatment for Minimal and Zero Liquid Discharge**	BALLROOM 4 **#22 Materials: Innovations in Membrane Synthesis and Casting**	KINGS GARDEN 3 **#20 Processes: New Concepts in Hybrid Processes and Process Integration**	KINGS GARDEN 4 **#19 Processes: Process Scale-up and Techno-economic Assessment**	KINGS GARDEN 5 **#18 Materials: Selective Polymeric and Mixed-Matrix Materials - Gas Separations**
2:00	**21a - Reduction of Brackish Water Desalination Brine Volume using Membrane Evaporation Coupled with Activated Sludge Aeration Waste Heat** Drew W Johnson (The University of Texas at San Antonio)	**22a - Novel Spirocyclic Polymers for Membrane-based Organic Solvent Separations** Ronita Mathias (Georgia Institute of Technology)	**20a - Air Products Hybrid Membrane Adsorption** Cory E Sanderson (Air Products and Chemicals)	**19a - Minimizing the cost of membrane distillation** Timothy V Bartholomew (Carnegie Mellon University)	**18a - High temperature gas separation properties of sub-micron polybenzimidazole membranes** Melanie M Merrick (The University of Texas at Austin)
2:30	**21b - Smart High Recovery RO Boosts Recovery Rate to More Than 90% for Reuse Purposes** Ronit Erlitzki (AdEdge Water Technologies)	**22b - 3D Printed fouling resistant composite membranes** Davide Mattia (University of Bath)	**20b - Hybrid Distillation and Facilitated Transport Membrane Processes for C3 Splitter Debottlenecking** Christine Parrish (Compact Membrane Systems)	**19b - Process Optimization Using Perturbation Expansion Solutions for Membrane Gas Separation Modules** Glenn Lipscomb (University of Toledo)	**18b - Thermally Rearranged Semi-interpenetrating Polymer Network (TR-SIPN) Membranes for Olefin/Paraffin Separations** Young Moo Lee (Hanyang University)
3:00	**21c - Membrane Distillation for Enhanced Water Recovery by Inland Brackish Water Desalination Plants** Zhewei Zhang (University of Pittsburgh)	**22c - Next-Generation Membranes Using Thin-Film Lift Off** Mackenzie Anderson (UCLA)	**20c - Water Recovery via Forward Osmosis - Freeze Concentration** Vladimiros Papangelakis (University of Toronto)	**19c - Multi-physics Simulation of Hollow Fiber Vacuum Membrane Distillation Using OpenFoam** Albert Kim (University of Hawaii)	**18c - Conformation-controlled molecular sieving effect for membrane-based propylene/propane separation** Yang Liu (Georgia Institute of Technology)
3:30	**21d - Selective Separation of Industrial Relevant Metal Ions from High TDS Water using Wafer-Enhanced Electrodeionization (WE-EDI)** Humeyra B Ulusoy Erol (University of Arkansas)	**22d - Interfacial Junctions Control Electrolyte Transport Through Charge-Patterned Membranes** Bill Phillip (University of Notre Dame)	**20d - Leverage membrane technology in process intensification and new application design - from early stage design to technical scale implementation** Peter Kreis (Evonik Technology & Infrastructure GmbH)	**19d - Scale-up of Facilitated Transport Membrane Module for CO_2 Capture from Flue Gas** Kai Chen (Ohio State University)	**18d - Mixed matrix materials containing soluble discrete metal-organic polyhedra in rubbery polyethers for CO_2/gas separation** Haiqing Lin (The State University of New York at Buffalo)
4:00	**21e - Pilot Evaluation of Closed Circuit Reverse Osmosis (CCRO) for Treatment of RO Concentrate for Potable Reuse** Han Gu (Orange County Water District)	**22e - Fabrication of Inside-Out Isoporous Hollow Fiber Membranes via Spinning and Coating Methods** Kirti Sankhala (Helmholtz-Zentrum Geesthacht, Geesthacht)	**20e - Carbon Capture utilizing Hybrid Membrane/Liquifaction** David Hasse (Air Liquide)	**19e - Development of industrial scale polyvinylidene fluoride transfer membrane** Marta E Bojarska (GVS Filter Technology)	**18e - Accessing novel microporous polymers to enhance yield, stereochemical control, and membrane performance in the recovery of light hydrocarbons from natural gas** John Lawrence (Aramco)
4:30	**21f - Brine Treatment for Minimal and ZLD** Malynda Cappelle (The University of Texas at El Paso)	**22f - Empirical evidence suggests that polyamide film formation during interfacial polymerization in thin film composite (TFC) membranes is not self-limiting** Kasia Grzebyk (University of North Carolina at Chapel Hill)	**20f - Membrane distillation hybridized with a thermoelectric heat pump for energy-efficient water treatment and space cooling** Jia Wei Chew (Nanyang Technological University)	**19f - Low energy seawater desalination using multistage electrodialysis** Gijs Doornbusch (Technical University Eindhoven/ Wetsus)	**18f - Insight into the transport in polymeric and mixed matrix membranes via analysis of unique mixed gas diffusion coefficients** Johannes Carolus (John) Jansen (National Research Council of Italy)

7:45	**Plenary Session 3 - Dr. Rachel Segalman, Polymerice Ionic Liquids: A New Platform for Materials and Energy**				
Parallel Sessions	**BALLROOM 3** **#25 Processes: Innovations in Microfiltration and Ultrafiltration**	**BALLROOM 4** **#26 Materials: Bio-inspired and Biomimetic Materials**	**KINGS GARDEN 3** **#24 Applications: Cell and Protein Purification, Harvesting, and Processing**	**KINGS GARDEN 4** **#23 Awards Session (Student Fellowship & Young Membrane Scientist Awards)**	**KINGS GARDEN 5** **#27 Materials: Advances in Membrane and Materials Characterization**
9:30	**25a - A systematic approach to remove bottlenecks in high concentration UF/DF for subcutaneous application of biological drug substances** Hasin M Feroz (Bristol-Myers Squib)	**26a - Artificial Water Channels - toward biomimetic membranes for desalination** Mihail Barboiu (Institut Europeen des Membranes)	**24a - Experimental and Theoretical Investigations of the Loss of Protein Sieving Due to Fouling of ATF Microfilters** Alex Apostolidis (Amgen)	**23a - Creation of well-defined "mid-sized" micropores in carbon molecular sieve membranes for organic solvent separations** Yao Ma (Georgia Institute of Technology)	**27a - Surface Nano-structuring with Hydrophilic Polymer Brush Layers for Membrane Performance Tailoring and Optimization** Yian Chen (UCLA)
10:00	**25b - Continuous Diafiltration for cGMP Biomanufacturing** Akshat Gupta (MilliporeSigma)	**26b - Self-Assembly of Long-Lasting Lipid Bilayers and the Effect of Temperature of their Ion Rejection** Yair Kaufman (Ben Gurion University of the Negev)	**24b - pH and Excipient Partitioning in Ultrafiltration / Diafiltration Processes for Formulation of High Concentration Monoclonal Antibody (mAb) Products** Parinaz Emami (Penn State University)	**23b - PoreDesigner for computational optimization of channel proteins and porous organic cages for highly selective membrane-based separations** Ratul Chowdhury (Penn State University)	**27b - Roughness in Polyamide RO Membranes: Its Formation Mechanism, Control, and Implications** Chuyang Y Tang (University of Hong Kong / University of New South Wales)
10:30	**25c - Understanding the role that patterning plays on membrane biofouling by visualization using light imaging and simulation** Anna Malakian (Clemson University)	**26c - Peptoid-based membrane-mimetic 2D nanomaterials with incorporated natural and synthetic channels** Chunlong Chen (Pacific Northwest National Laboratory)	**24c - Effect of membranes molecular weight cut-off on peptides migration and selectivity during electrodialysis with filtration membranes** Sabita Kadel (Laval University)	**23c - Crack-free, large-area, single-layer graphene membranes with a record performance in gas mixture separation** Shiqi Huang (L'Ecole polytechnique fédérale de Lausanne)	**27c - Mechanism of chlorine attack on polyamide membranes studied by EIS** Viatcheslav Freger (Technion - IIT)
11:00	**25d - Treating Poultry Processing Wastewaters by Combined Electrocoagulation and Ultrafiltration** Ranil Wickramasinghe (University of Arkansas)	**26d - Scalable High-Performance Membranes with High Density Channel Protein-Polymer Nanosheets** Yu-Ming Tu (Penn State University)	**24d - Scale down depth filtration case study using Pall Supracap™ 50 to predict large scale STAX™ capacity and product quality** Patricia Rose (Merck Pharmaceuticals)	**23d - Unprecedented CO_2/H_2 Selectivity and CO_2 Permeance Demonstrated in Facilitated Transport Membranes with Tunable Amine-CO_2 Chemistry** Yang Han (Ohio State University)	**27d - Physicochemical and electrochemical characterization of cation-exchange membranes modified with polyethyleneimine for elucidating enhanced monovalent permselectivity of electrodialysis** Wenbin Jiang (New Mexico State University)
11:30	**25e - Mass transfer assisted TIPS process to precisely tailor the hollow fiber membrane surface and sublayer structures** Hideto Matsuyama (Kobe University)	**26e - Fabrication of aquaporin-based biomimetic membrane for seawater desalination** Ye Li (Nanyang Technological University)	**24e – Unraveling the Plugging Mechanisms during a Combined Tangential-flow and Depth Filtration Process** Xianghong Qian (University of Arkansas)	**23e - Engineering the Nanochannels in Reduced Graphene Oxide Membrane for Dye Desalination** Liang Huang (The State University of New York at Buffalo)	**27e - Unraveling the Film-Formation Kinetics during Interfacial Polymerization via Low Coherence Interferometry** Xin Liu (Southern University of Science and Technology)
12:00	**25f - Organic solvent resistant membranes obtain by using non-toxic solvents** Stefan Chisca (King Abdullah University of Science and Technology)	**26f - Biomimetic channel membranes: can they realize their full potential upon scale-up?** Viatcheslav Freger (Technion - IIT)	**24f - Functionalized Microporous Membranes for Protein Capture and Analysis** Joshua Berwanger (University of Notre Dame)	**23f - Impact of module design on heat transfer and Nusselt correlation selection in membrane distillation** Alexander V Dudchenko (Carnegie Mellon University)	**27f - Operando Membrane Spectroscopy for Elucidating Transport Mechanisms in Membranes** Casey O'Brien (University of Notre Dame)

Parallel Sessions	BALLROOM 3 **#32 Applications: Fuel Cells and Batteries**	BALLROOM 4 **#28 Materials: Bio-inspired and Biomimetic Materials**	KINGS GARDEN 3 **#31 Applications: Purification of Non-Protein Biologics**	KINGS GARDEN 4 **#30 Processes: Fundamentals of Predicting and Preventing Membrane Fouling**	KINGS GARDEN 5 **#29 Materials: Advances in Membrane and Materials Characterization**
2:00	**32a - A Holistic Approach to Explore Interfacially Confined Ionomers Designed for Energy Conversion Device** Shudipto K Dishari (University of Nebraska-Lincoln)	**28a - Biofouling resistant membranes for wound healing** Antoine Venault (Chung Yuan Christian University)	**31a - Optimizing Microfluidic Nanoparticle Capture from Biofluids on Ultrathin Silicon Nanomembranes: A Computational and Experimental Analysis of Tangential Flow Analyte Capture (TFAC)** Kilean Lucas (University of Rochester)	**30a - Numerical and experimental studies on the deposition of sticky particles near a membrane surface** Seon Yeop Jung (Seoul National University)	**29a - A novel method to distinguish between crystal and defect transport in zeolite membranes** Boguslaw Kruczek (University of Ottawa)
2:30	**32b - Anion Conducting Membranes Based on Poly(norbornene): High Conductivity, Chemical Stability and Fuel Cell Performance >2.5 W/cm^2** Mrinmay Mandal (Georgia Institute of Technology)	**28b - How pore hydrophilicity influences water permeability?** Mingjie Wei (Nanjing Tech University)	**31b - Development and scale-up of laterally-fed membrane chromatography for the purification of therapeutic viruses** Karina Kawka (McMaster University)	**30b - Characterization of the Striping Phenomenon during Membrane Fouling via Optical Coherence Tomography** Weiyi Li (Southern University of Science and Technology)	**29b - Study on antifouling behaviors of GO modified nanocomposite membranes through QCMD and surface energetics analysis** Mohtada Sadrzadeh (University of Alberta)
3:00	**32c - High Temperature Fuel Cells with Ion-Pair Membranes and Phosphonated Ionomers** Michael R Hibbs (Sandia National Laboratories)	**28c - Biomimetic membrane systems utilizing electro-dynamic interfaces** Bruce Hinds (University of Washington)	**31c - Characterization of EV Secretion at Single Cell Resolution** Gerardo Mauleon Ramos (University of Chicago)	**30c - A closer look at biofouling: time-lapse optical microscopy of biofilm formation in wastewater reuse** Emily W Tow (Olin College)	**29c - 3D characterization of polyamide reverse osmosis membranes** Tyler E Culp (Penn State University)
3:30	**32d - High temperature, anhydrous proton conducting membranes and micropatterned bipolar membranes for electrochemical energy conversion and storage technologies** Christopher G Arges (Louisiana State University)	**28d - Responsiveness and function of DNA-gated membranes** Thomas Schäfer (Polymat, University of the Basque Country)	**31d - Charge, size distribution and hydrophobicity of viruses: Effect of the virus purification method** Vlad Tarabara (Michigan State University)	**30d - Improved graphene oxide membrane increases membrane distillation desalination of RO concentrate** Lucy M Camacho (Texas A&M University-Kingsville)	**29d - Selective Ion Transport Properties of Membrane Materials** Anita Hill (CSIRO)
4:00	**32e - Membrane through-plane alignment of ion-conducting channels** Michael D Guiver (Tianjin University)	**28e - Nature-inspired Coating-free Membranes for Desalination** Himanshu Mishra (King Abdullah University of Science and Technology)	**31e - Risk Mitigation Strategies for the use of polymeric consumables for manufacturing of Anti-Sense Oligonucleotides** Bill Scott (Biogen)	**30e - Automated real-time membrane biofouling assessment using microbial enzyme activity** Babar K Khan (King Abdullah University of Science and Technology)	**29e - Membrane-based oil-gas separation for dissolved gas-in-oil extraction: gas transport properties of ceramic supported Teflon membranes** Liang-Chih Ma (Arizona State University)
4:30	**32f - Novel sulfonated aromatic polymer membranes for breaking the proton selectivity-conductivity trade-off limitation in vanadium redox flow battery** Sangil Kim (University of Illinois, Chicago)	**28f - Carbon nanotube porins as model biomimetic membrane nanopores** Aleksandr Noy (Lawrence Livermore National Laboratory)	**31f - Use of Nanosphere Self-Assembly to Pattern Nanoporous Membranes for the Study of Extracellular Vesicles** Marcela Mireles (University of Rochester)	**30f - Interactions between extracellular polymeric substances and zwitterionic hydrogels as a designing tool for antifouling surfaces** Angelin Ida Vedhamanickam (Ben-Gurion University of the Negev)	**29f - Porous support of TFC membranes: Does it truly have negligible resistance?** Masoud Aghajani (University of Colorado Boulder)

ORAL PRESENTATIONS

MONDAY 9:30 am - Kings Garden 4

1a - Hydrodynamic-colloidal interactions of an oil droplet and a membrane surface

Mariano Galvagno (Technion - Israel Institute of Technology), Guy Z Ramon (Technion - Israel Institute of Technology)*

Membranes offer a superior separation capacity for stable emulsions, which are otherwise difficult to process. However, membrane fouling is always prevalent and requires extensive backwashing and cleaning as part of the operating regime. Oil droplets deform when in close proximity to a membrane, primarily due to permeation drag, which can lead to irreversible deposition on the membrane. Understanding the influence of hydrodynamic stresses and colloidal interactions between droplets and membranes will inform better engineered membrane surfaces and process conditions. Using a long-wave hydrodynamic description that incorporates wettability via a Derjaguin (disjoining) pressure and electrostatic repulsion via electric double layer (EDL) contributions, we derive a set of equations for the thickness and pressure in the liquid film separating the droplet and the permeable membrane. Numerical simulations are used to analyse equilibrium states where a zero net force acts on the droplet, to investigate the influence of droplet shape, distance from the membrane, membrane permeability and characteristic parameters of the colloidal interactions on the overall repulsion / attraction towards the membrane surface. A stability phase diagram is constructed in terms of the modified capillary number - accounting for the ratio of the viscous and surface tension forces - and a parameter representing the ratio of the hydrodynamic pressure scale over the electrostatic interaction intensity. These calculations are used to classify stable (upward deflection, increasing distance from the membrane) vs. unstable drop shapes (downward deflection bringing the droplet closer to the membrane), and investigate conditions leading to irreversible deposition onto the membrane. In particular, the model can describe how membrane properties and emulsion characteristics dictate a 'critical' flux.

MONDAY 10:00 am - Kings Garden 4

1b - Do Graphene Oxide Nanostructured Coatings Mitigate Bacterial Adhesion to Membrane Interfaces?

Santiago Romero-Vargas Castrillon (University of Edinburgh)*, Sara BinAhmed (University of Minnesota), Karl Wuolo-Journey (University of Minnesota)

Graphene oxide (GO) is a one-atom thick, two-dimensional sheet of hexagonally arranged carbon atoms, decorated with carboxyl, hydroxyl, and epoxide functional groups. GO nanosheets possess exceptional physical and chemical properties, as well as antimicrobial activity against both gram-negative and -positive bacteria, rendering them a promising building block for anti-biofouling coatings. Extensive research has been performed to explore the mechanisms underlying GO's biocidal activity. Yet, to date, a fundamental understanding of the behavior of bacterial cells at GO-functionalized aqueous interfaces is largely lacking. Elucidation of the interfacial activity of GO is essential for the development of graphene-based environmental applications, such as membrane materials for water purification. In this study, we use single-cell force spectroscopy to investigate the mechanism of microbial adhesion of single P. fluorescens cells to polyethersulfone ultrafiltration (UF) membranes. We find that, in addition to possessing antimicrobial activity, GO nanostructured coatings result in a statistically significant decrease in the bioadhesion forces compared to GO-free control UF membranes. Given that GO does not significantly modify the hydrophilicity or zeta potential of the UF membrane surfaces, we speculate that GO mitigates bioadhesion due to steric repulsion, i.e., through a mechanism akin to that of antifouling polymer brushes. This investigation therefore provides insight into GO's properties at biological interfaces, and provides novel nanoscale insights into its antifouling functionality.

MONDAY 10:30 am - Kings Garden 4

1c - Fouling mechanisms in constant flux crossflow ultrafiltration

Alon Kirschner (The University of Texas at Austin)*, Yu-Heng Cheng (The University of Texas at Austin), Donald Paul (The University of Texas at Austin), Robert Field (University of Oxford), Benny Freeman (The University of Texas at Austin)

Four fouling models due to Hermia (complete pore blocking, intermediate pore blocking, cake filtration and standard pore blocking), have long been used to describe membrane filtration and fouling in constant transmembrane pressure operation of membranes. A few studies apply these models to constant flux dead-end filtration systems. However, these models have not been reported for constant flux crossflow filtration, despite the frequent use of this mode of membrane operation in practical applications. We report derivation of these models for constant flux crossflow filtration. Of the four models, complete pore blocking and standard pore blocking were deemed inapplicable due to contradicting assumptions and relevance, respectively. Constant flux crossflow fouling experiments of dilute latex bead suspensions and soybean oil emulsions were conducted on commercial poly (ether sulfone) flat sheet ultrafiltration membranes to explore the models' abilities to describe such data. A model combining intermediate pore blocking and cake filtration appeared to give the best agreement with the experimental data. Below the threshold flux, both the intermediate pore blocking model and the combined model fit the data well. As permeate flux approached and passed the threshold flux, the combined model was required for accurate fits. Based on this observation, a physical interpretation of the threshold flux is proposed: the threshold flux is the flux below which cake buildup is negligible and above which cake filtration becomes the dominant fouling mechanism.

MONDAY 11:00 am - Kings Garden 4

1d - Fouling-Resistant, Multi-Functional Membranes with Ultra-Thin Hydrogel Selective Layers using Interfacially Initiated Free Radical Polymerization (IIFRP)

Ayse Asatekin (Tufts University)*, Ilin Sadeghi (Tufts University), Eric Liu (Tufts University), Alice Oliveira Aguiar (Tufts University), Hyunmin Yi (Tufts University)

Hydrogels are promising materials as membrane selective layers due to their fouling-resistance, tunability, and functionalizability. Yet, the fabrication of thin, uniform hydrogel membrane selective layers using a simple, scalable process is a challenge. We developed a novel method, interfacially initiated free radical polymerization (IIFRP), for the fabrication of ultra-thin hydrogel selective layers on porous supports in a simple and reproducible process. This method utilizes the segregation of the monomer and the photoinitiator into two separate, immiscible phases to form a very thin, uniform and defect-free hydrogel layer at the interface upon photopolymerization. The resulting hydrogel-coated membranes have selective layers as thin as <100 nm, and can separate the proteins based on size. The method is readily tunable for a broader range of separations by altering experimental parameters or the addition of inert porogens or comonomers. IIFRP can also be used to prepare membranes that simultaneously remove large organic molecules and catalytically treat smaller contaminants by incorporating catalytic nanoparticles (NPs) into the membrane selective layer. We prepared hybrid water filtration membranes incorporating catalytic NPs immobilized on tobacco mosaic virus (TMV) nanotemplates within their ultrathin hydrogel selective layers by adding the TMV-templated NPs to the monomer solution during IIFRP. These membranes catalytically reduce Cr(VI), a toxic and carcinogenic contaminant, to non-toxic Cr(III), achieving up to 98% conversion in a single pass. Catalytic activity is fully retained with continuous filtration for at least 3 days. These results illustrate a versatile approach for integration of various nanomaterials within membrane selective layers, enabling the creation of advanced membranes with new and versatile functionalities for next-generation applications.

MONDAY 11:30 am - Kings Garden 4

1e - Unravel the Gordian Knot of membrane desalination: Understanding and mitigating silica scaling in reverse osmosis

Tiezheng Tong (Colorado State University)*, Song Zhao (Tianjin University), Amanda Quay (Carnegie Mellon University), Yunlong Qi (Tianjin University)

Mineral scaling, a largely unexplored problem compared to organic and biological fouling, significantly constrains water recovery of membrane desalination. Silica scaling is among the most challenging types of scaling due to the complex nature of scaling mechanisms and the lack of effective anti-scalants. In this presentation, we will present our research on understanding the mechanisms as well as developing novel scaling-resistant membranes associated with silica scaling in reverse osmosis (RO). First, surface properties of thin-film composite (TFC) polyamide membranes were correlated with silica scaling in RO desalination. By developing a battery of ten membranes grafted with a variety of polymers, we were able to demonstrate a strong correlation of membrane surface charge with the extent of silica scaling, which was independent of both membrane hydrophilicity and surface energy. A more negative surface charge reduced silica scaling, suggesting the important role of negatively charged silica species in scale formation. Further, we investigated combined silica scaling and protein fouling. Our results showed that water flux decline was synergistically enhanced when silica and protein coexisted in the feedwater. The silica-protein interactions were elucidated by using a series of microscopic and spectroscopic analytical tools, which revealed a facilitating effect of proteins in the formation of aggregates with silica. Lastly, a novel TFC polyamide membrane modified with layer-by-layer (LbL) assembly was developed to improve resistance to combined silica scaling and protein fouling. Our systematic work provides fundamental insights into the mechanisms governing silica scaling in RO, and highlights the potential of membrane surface modification as a strategy to reduce silica scaling and improve RO water recovery.

MONDAY 12:00 pm - Kings Garden 4

1f - RO Mineral Scale Mitigation via Self-Adaptive Feed-Flow Reversal

Yeunha Kim (UCLA)*, Anditya Rahardianto (UCLA), Tae Lee (UCLA), Yoram Cohen (UCLA)

Reverse osmosis (RO) membrane desalination is increasingly used for production of potable water from seawater and brackish water and in municipal and industrial water reuse applications. However, RO processes are adversely impacted by membrane mineral scaling especially at high recovery. Membrane surface scaling decreases membrane water permeability, increase operating cost and reduce membrane longevity. Membrane scaling has been partially mitigated via antiscalant dosing of the RO feed, but this is at an added cost. An alternate approach to scale mitigation can be achieved via RO Feed flow reversal (FFR) which is a chemical-free method. By periodically reversing the direction of the raw RO feed, mineral scale that developed at the membrane stage exit can be removed via dissolution and the membrane permeability is recovered. An evaluation of the FFR process in a spiral-wound RO pilot system was undertaken whereby flow reversal was triggered by a novel membrane scale monitoring (MSM) system. The RO pilot was fitted with an arrangement of automated valves and piping for self-adaptive automated switching of the feed flow direction. The MSM system contains a membrane channel that mimics the level of concentration polarization of the spiral-wound RO element being monitored. Studies with gypsum as a model scalant demonstrated that continuous RO desalination was feasible without scaling and without antiscalant dosing up to twice the saturation limit of gypsum.

MONDAY 9:30 am - Ballroom 3

2a - Continuous on-demand dehydration of solvents in flow chemistry manufacturing processes

Hannah Murnen (Compact Membrane Systems)*, Sudip Majumdar (Compact Membrane Systems)

Anhydrous solvents can be critical in the development and production of new pharmaceutical molecules, affecting yield, product quality and process efficiency to name just a few key elements. Production and management of those anhydrous solvents is not always trivial given the temperature, quality and process requirements of APIs, particularly in flow chemistry systems. Membrane dehydration and purification can provide an avenue for efficient and cost-effective, on-demand production of these anhydrous solvents. Compact Membrane Systems has installed several fluoropolymer membrane pervaporation systems into flow chemistry manufacturing processes to allow inline dehydration of solvents. One example of this work was published in OPR&D along with authors from Pfizer and Asymchem and provides a new route to manage solvents in flow chemistry processes. This presentation will discuss solvent purification and management in flow chemistry processes using pervaporation membranes and will also cover specific operational issues such as how to manage low flow rates or small hold up volumes.

MONDAY 10:00 am - Ballroom 3

2b - How membrane technology has been successfully applied to ethanol processing

Stephan Blum (Whitefox Technologies)*, Trond Heggenhougen (Whitefox Technologies), Meridith K Bridge (Whitefox Technologies)

Whitefox's Integrated Cartridge Efficiency (ICE) is a membrane-based dehydration technology that removes bottlenecks in biofuels processes. The ICE solution dehydrates water-rich recycle streams that significantly reduces the load on existing equipment and frees up capacity in distillation and dehydration sections, enabling producers to increase ethanol output and revenues, as well as improve margins by reducing energy costs, improve carbon intensity (CI) scores, reduce cooling water costs year-round, and reduce operation and maintenance costs. Existing plants uses Molecular Sieve Units (MSU) to dehydrate ethanol. Recycle streams generated during distillation and MSU dehydration create bottlenecks and limit plant expansion. These regenerate streams are therefore recycled to distillation that increases load and energy consumption in the distillation columns. The ICE solution is a stand-alone membrane system that removes existing bottlenecks by dehydrating water-rich recycle streams (up to 50 wt%) to target moisture in one pass. This significantly reduces the load on distillation columns and MSU allowing for plant expansion, energy and cooling water savings. Whitefox is currently installed at four U.S. plants with several more installations planned for 2019. Benefits seen from an ICE installation include: load reduction on rectifier/side stripper and molecular sieves, removal of fluctuation in columns, reduction in steam and improved Carbon Intensity levels, reduction in cooling water recirculation rate and electrical load. Whitefox will share information about its vapor permeation membranes, module configurations and process diagrams before and after an ICE installation. Whitefox will also take the audience through the customer journey from baselining to complete installation, demonstrating the savings by providing operating data from its existing U.S. installations to show tangible benefits that can contribute to profitable operations.

MONDAY 10:30 am - Ballroom 3

2c - Modeling the effect of different single-walled carbon nanotube orientations on permeation through mixed matrix membranes

Ali Zamani (University of Ottawa)*, Jules Thibault (University of Ottawa), Handan Tezel (University of Ottawa)

Pervaporation, a membrane-based process, is considered to be a promising technique for the separation of organic compounds from aqueous solutions. Different types of polymeric membranes have been used for this process due to their wide range of separation properties, but these types of membranes are suffering from the trade-off between permeability and selectivity. Mixed matrix membranes (MMMs), comprising an inorganic filler phase embedded into a polymer matrix, have emerged in an attempt to partly overcome some limitations of conventional polymer and inorganic membranes. Among them, membranes incorporating single walled carbon nanotubes (SWCNTs) having the potential for transcending Robeson's upper bound. Recent studies tried to align SWCNTs in the host polymer matrix and results showed an improvement in membrane permeability. However, while much effort has been devoted to SWCNTs mixed matrix membrane in practice, their modeling is mostly based on early theories for transport in composites. These theories consider ideal morphology, which is very challenging to reach because of the imperfect filler-polymer adhesion. Also, they neglect the orientation of the nanotubes. Therefore, models for the permeability in MMMs often do not perform well when compared to experimental permeation data. These non-idealities can be classified into the following three major categories: (i) interface voids, (ii) rigidified polymer layer around the fillers, and (iii) particle pore blockage. Therefore, non-ideal morphology for three-phase systems containing filler-polymer interface defects, while considering SWCNTs orientation has been proposed. The Fickian diffusion equation was used to calculate the concentration profile of the component through the block of the mixed matrix membrane. Simulations showed that the membrane performance is highly dependent on the filler orientation and MMMs effective permeability would be increased up to 35% by aligning the SWCNTs vertical to the surface.

MONDAY 11:00 am - Ballroom 3

2d - Unique 3D printed spacers for process enhancement in membrane distillation

Navya Thomas (Khalifa University)*, Nurshaun Sreedhar (Khalifa University), Oraib Al-Ketan (Khalifa University), Reza Rowshan (New York University Abu Dhabi), Rashid Abu Al-Rub (Khalifa University), Hassan Arafat (Khalifa University)

Spacers, an essential component of an effective module design, influence the hydrodynamics in the flow channels of membrane distillation (MD) units, thus inducing multifaceted impact on the process performance. In this study, we investigated the performance of innovative 3D printed spacer designs based on mathematically modelled triply periodic minimal surfaces (TPMS), which was benchmarked against a conventional commercial mesh spacer (CS). The TPMS structures are particularly interesting because of their interconnected maze-like tortuous pathways that creates increased turbulence, and zero mean curvature that abates dead zone formation. Performance evaluation was based on flux and flux stability, heat transfer, feed channel pressure drop and fouling. Several TPMS architectures were evaluated, including Schwarz CLP, tCLP, P; Schoen Gyroid and Fisher Koch S surface. The best performing TPMS spacer design, the tCLP, resulted in flux and overall film heat transfer coefficient enhancement of 60% and 63%, respectively, compared to the CS in direct contact MD. The improved performance was further elucidated through temperature polarization modelling. However, the tCLP spacer also exhibited higher channel pressure drop compared to CS. For further improvement, a hybrid spacer design which combines two TPMS architectures was conceived, and resulted in high flux, on par with tCLP, but at a lower pressure drop penalty. Under fouling inducing conditions, processing RO brine with continuous feed concentration build-up (75,000-100,000 ppm), the use of TPMS spacers resulted in significant flux stabilization over time, as compared to CS. Additionally, inorganic scaling tests using calcium sulfate as the feed solution were conducted and a new technique for scale mapping using alizarin red S staining was used. The latter technique showed that membrane areas in contact with CS had a two-fold higher scalant deposition than those in contact with the TPMS spacers.

MONDAY 11:30 am - Ballroom 3

2e - 2D boron nitride encapsulated nanofiber membranes for membrane distillation

Yunchul Woo (Korea Institute of Civil Engineering and Building Technology)*, Ho Kyong Shon (University of Technology Sydney), June-Seok Choi (Korea Institute of Civil Engineering and Building Technology)

The instability of PVDF membranes in membrane distillation (MD) for seawater desalination still remains a problem awaiting solution, despite tremendous efforts. Here we propose simple and feasible approach for improving desalination performance through incorporation of two-dimensional boron nitride nanosheets (BNNSs) in PVDF-co-HFP electrospun nanofiber membrane (BNs-PH) as well as demonstrate its origin for fundamental understanding. The BNs-PH membrane exhibits stable water vapor flux (18 LMH) and superior salt rejection (99.99%), even after operation for 280 h (Commercial PVDF: steep decay within 28 h, and neat PH: wetting within 4h). From structural/chemical analysis, the BNNSs play a crucial role in forming favorable phase of the PH polymer crystal structure, inducing superhydrophobic surface with the larger nanoporosity and higher heterogeneity as well as enhanced mechanical properties for long-term operation. Furthermore, theoretical modelling for air gap MD (AGMD) system is clearly consistent with the experimental data. Approach introduced in this study has great potential for broad application with other desalination system, which will boost various water treatment applications.

MONDAY 12:00 pm - Ballroom 3

2f - Impact of membrane properties on scaling-induced wetting in membrane distillation

Allyson L McGaughey (University of Southern California)*, Prathamesh Karandikar (University of Southern California), Malancha Gupta (University of Southern California), Amy Childress (University of Southern California)

Membrane distillation (MD) relies on a hydrophobic, unwetted membrane to reject solutes. However, at high salinities, salt precipitation (scaling) on the membrane can induce pore wetting. Implicit assumptions are often made about pore wetting: that feed-side hydrophobicity and pore size (and feed stream properties) control wetting. Neither the role of asymmetric nor internal membrane properties in maintaining membrane hydrophobicity have been studied. In this study, we utilized initiated chemical vapor deposition to deposit a hydrophobic polymer (pPFDA) on a hydrophilic substrate (CA), resulting in symmetric superhydrophobicity (161° average contact angle on both sides of membrane) without impacting porosity. However, when pPFDA was deposited on a hydrophobic substrate (PTFE), hydrophobicity was unchanged and an asymmetric, low-porosity (5%) surface layer formed - likely due to low surface energy of PTFE preventing spreading. In short-term performance with scaling, both materials behaved similarly with high wetting resistance. However, in longer-term testing, the superhydrophobic pPFDA-CA membrane wetted much faster than the low-porosity ppFDA-PTFE membrane. Superhydrophobic membranes are frequently used in MD for their wetting resistance; they have also been suggested as anti-scaling materials. Because the energy barrier to crystal nucleation (ΔGM) increases with contact angle, superhydrophobicity is expected to reduce scaling. However, despite the higher ΔGM of the superhydrophobic membrane, nucleation eventually occurred and wetting followed. The low surface porosity membrane provided a more robust barrier. The reduced pore size may have prevented crystal growth into pores by size exclusion. Also, internal hydrophobicity (PTFE pore walls) may have inhibited liquid progression. In ongoing work, we are investigating surfactant-induced wetting; results will be presented.

MONDAY 9:30 am - Kings Garden 3

3a - Sacrificial Protective Layer for Fouling Control in Reverse Osmosis Desalination

Moon Son (Penn State University)*, Wulin Yang (Penn State University), Szilard S. Bucs (King Abdullah University of Science and Technology), Maria F. Nava-Ocampo (King Abdullah University of Science and Technology), Johannes S. Vrouwenvelder (King Abdullah University of Science and Technology), Bruce E. Logan (Penn State University)

A new approach was examined to minimize inevitable membrane fouling in reverse osmosis (RO) desalination based on adding a sacrificial coating of a protective layer (PL) using two polyelectrolytes, poly(diallyldimethylammonium chloride) and poly(sodium-4-styrenesulfonate), to the membrane surface. After the PL is fouled, this coating was removed by flushing with a highly saline solution such as an RO brine, and a new coating was regenerated in situ to provide a fresh PL coating on the membrane surface. The utility of this approach was demonstrated by conducting four consecutive dead-end filtration experiments using a model foulant (alginate, 200 ppm) in a synthetic brackish water (2000 ppm of NaCl). Brine removal and regeneration of the PL coating restored the water flux to an average of 97 ± 3% of its initial flux, compared to only 83 ± 3% for the pristine membrane. The average water flux for the PL-coated membranes was 15.5 ± 0.6 $L \cdot m^{-2} \cdot h^{-1}$ until the flux was decreased by 10% versus its initial flux, compared to 13.4 ± 0.5 $L \cdot m^{-2} \cdot h^{-1}$ for the non-treated control. Applicability and stability of the PL coating under cross-flow conditions was also further examined. These results suggest that the use of a sacrificial PL coating could provide a more sustainable approach for addressing RO membrane fouling. In addition, this approach could be also applied on other surfaces such as feed spacers, reservoirs, and pipes and released as needed to help remove biofilms or accumulated organic matter.

MONDAY 10:00 am - Kings Garden 3

3b - Water Transport through Ultrathin Polyamide Nanofilms Used for Reverse Osmosis

Zhiwei Jiang (Imperial College London)*, Santanu Karan (Imperial College London), Andrew Livingston (Imperial College London)

Thin film composite membranes comprising polyamide films as separating layers have been widely used in reverse osmosis (RO) applications for decades. These membranes have become the most attractive option for desalination due to the energy efficiency of reverse osmosis compared to other separation techniques such as multiple effect evaporation. Conventionally, polyamide films are fabricated via interfacial polymerisation (IP) between m-phenylenediamine (MPD) in an aqueous phase and trimesoyl chloride (TMC) in an organic phase on an ultrafiltration support. This results in the formation of crumpled polyamide films with apparent thickness in the range of ~50 to 300 nm. Such crumpled features created on ultrafiltration supports present a challenge to investigation of the active thickness of the polyamide layer and to understanding of the transport mechanism through the membranes. To tackle this challenge, herein we report ultrathin polyamide films (down to ~6 nm) fabricated at a free liquid-liquid interface, independently from any support materials. Polymerisation at a free interface created ultrathin free-standing polyamide nanofilms which could be transferred independently onto various supports. Interestingly, the pristine support with fast water transport exhibited >4 times higher permeance of the resulting composite membrane while identical nanofilms were attached on. In addition, by controlling the interfacial reaction conditions, we manipulated the intrinsic thickness of polyamide nanofilms down to ~6 nm. Composite membranes comprising such smooth films showed comparable water permeances (2.7 $L \cdot m^{-2} \cdot h^{-1} \cdot bar^{-1}$) to commercial and reported RO membranes with crumpled features. Considering that the ultra-smooth surfaces of the polyamide films provide less than one quarter of the permeable area of that provided by crumpled features, the water permeance per area of nanofilm is more than four times higher than for conventional RO membranes.

MONDAY 10:30 am - Kings Garden 3

3c - Balancing Carbon, Nitrogen and Phosphorus Concentration in Seawater as a Strategy to Prevent Accelerated Membrane Biofouling

Siqian (Simon) Huang (University of Minnesota, Duluth), Nikolay Voutchkov (Water Globe Consultants, LLC), Sunny Jiang (UC Irvine)*

Membrane biofouling remains a significant challenge in sweater reverse osmosis desalination for drinking water production. This study investigated nutrient imbalance as the cause of biofouling in lab scale experiments and carried out a year-long field-testing at a seawater desalination pilot plant. Lab experiments showed that growth medium with excess organic carbon (C) but low nitrogen (N) and phosphorus (P) accelerated the production of bacterial biofilm. Balancing C to N and P ratio by adding N and P to growth medium increased the cell proliferation but reduced biofilm production. The cell excretion of excess C in the form of extracellular polysaccharides (EPS) was considered as a strategy for nutrient storage for future use. Cell enzyme activity assays indicated some of the bacteria had enhanced enzyme activity to degrade polysaccharides in the absence of organic carbon in growth medium, possibly using EPS in the biofilm. But others do not show enhanced enzyme activity. A year-long field study indicated that accelerated biofouling of membrane was associated with the elevation of total organic carbon (TOC) in the intake water. Dosing of intake stream with addition of N and P to balance the increase of TOC resulted in reduction of membrane biofouling. Microbial community analysis of the biofouling layer using 16s rRNA gene sequencing indicated biofouling communities varied with seasonal changes. Dosing of N and P did not induced dramatic changes in fouling community recovered from membrane surface. The outcomes of this work imply that accelerated membrane biofouling associated with elevated TOC in intake water is likely due to imbalance of C:N:P. Addition of N and P to rebalance the nutrients can alleviate the accelerate biofouling.

MONDAY 11:00 am - Kings Garden 3

3d - Flow Reversal for Increased Recovery of 2nd Pass Seawater RO

Tomer Osman (Rotec Water Ltd.), Dan Peled (AST-Water Ltd.), Noam Perlmutter (Rotec Water Ltd.), Arina Shulman (Mekorot Water Company), Riki Harpness (Mekorot Water Company), Sivan Bleich (Mekorot Water Company), Eli Korin (Ben-Gurion University), Jack Gilron (Ben-Gurion University)*

A partial 2nd pass is needed on SWRO permeate when strict boron limits are imposed for agricultural needs. The pH of the 2nd pass feed is increased above pH=10 to maximize boron rejection, since Mg2+ and Ca2+ levels are quite low in the SWRO 1st pass permeate. The drive to reach maximum recovery from 1ST PASS permeate means that at recoveries above 90-92% become Mg2+ concentrations are high enough to create large supersaturations of magnesium hydroxide (brucite) which can foul the membrane. The flow reversal technique invented by BGU scientists periodically switches the position of the last membrane element in the train with the first membrane element in the train so that the tail element which saw the most supersaturated conditions will now see the lease supersaturated conditions of the feed element thus preventing onset of scaling. A planned expansion by Mekorot of the SWRO Sabha C plant near Eilat requires expanding the second pass as well to keep boron levels low. The feasibility of incorporating flow reversal into this expansion was evaluated in a pilot experiment using feeds and rejects from the existing second pass LPRO plant. To simulate a 4 stage second pass operation, Rotec implemented a pilot including four pressure vessels (PV) of four 4-inch elements divided into a 1st stage block with three pressure vessels in parallel and a 4th stage block with one pressure vessel. The feed blocks were fed by the feed to the existing second pass unit. The 4th stage block was fed by the concentrate leaving the existing three-stage second pass unit. The block in the 4th stage was periodically switched with a 1st stage PV while flipping the feed and concentrate connections. Non-increase of the axial pressure drops in the 4th stage were used as an online indicator of scale free operation. During pilot operation recoveries were increased to 97% and then to 98% for the final month with no sign of brucite precipitation.

MONDAY 11:30 am - Kings Garden 3

3e - Zwitterionic poly(arylene ether sulfone) (PAES) copolymer for biofouling/chlorine resistant desalination membranes

Yi Yang (Arizona State University)*, Matthew Green (Arizona State University)

Amphiphilic poly(arylene ether sulfone)s with covalently bonded sulfobetaine groups (PAES-co-SBAES) show great potential for sea water desalination membranes. This family of polymers showed impressive performances on transport properties, bio-fouling/chlorine resistance and strong mechanical stabilities. Here we demonstrated a novel synthesis method to prepare the controlled zwitterionic charge content PAES polymers. The polymer chemical structures were analyzed by 1H NMR spectroscopy and the molecular weight of polymers was identified by size exclusion chromatography (SEC). Our study indicated that with the introduction of zwitterionic charge in a rigid PAES matrix, the water permeance and antifouling ability were both remarkably improved to 2.5 $L \cdot m^{-2} \cdot h^{-1} \cdot bar^{-1}$ and 94% of flux recovery ratio, respectively, while salt rejection remained at a high level (98%) even under the high exposure to chlorine (12,000 ppm·h). Specifically, the zwitterion functionalized free-standing PAES membranes were fabricated via the non-solvent induced phase separation (NIPS) with incorporation of solvent evaporation process. The membrane composition, surface morphology (roughness), cross-section morphology, and surface hydrophilicity were determined by fourier transform infrared (FT-IR) spectrum, atomic force microscopy (AFM), scanning electron microscopy (SEM), and water contact angle measurements, respectively. The results indicated that both the porosity of the support layer and surface hydrophilicity increased drastically due to the incorporation of hydrophilic SBAES segments. This work provided a valuable and scalable strategy to fabricate free-standing and bio-fouling/chlorine resistant desalination membranes, and further performance optimization and fundamental understandings on water/salt transport behaviors in this family of polymers are under investigation.

MONDAY 12:00 pm - Kings Garden 3

3f - Chlorine resistance of functionalized HNTs-based TFN membranes

Farhad Asempour (University of Ottawa), Somaye Akbari (Amirkabir University of Technology), Du Bai (University of Ottawa), Ramzi Aoun (University of Ottawa), Sofia Reyes Lombardo (University of Ottawa), Amirsajad Atashgar (University of Ottawa), Boguslaw Kruczek (University of Ottawa)*

Thin Film Nanocomposite (TFN) Reverse Osmosis (RO) membranes are reported to have promising improvements compared to conventional Thin Film Composite (TFC) membranes in terms of separation characteristics for desalination. In this work, we investigate enhancing TFN membrane'S stability and chlorine resistance by focusing on the functionalization of nanoparticles. Fully aromatic RO membranes were synthesized by in situ interfacial polymerization of TMC and MPD; Halloysite NanoTubes (HNTs) were used as the additive nanoparticles. The HNTs were functionalized by three different generations of amine terminated poly(amidoamine) PAMAM dendrimers. Resulting membranes, were treated by passive batch chlorination using 50 ppm NaOCl solution for 24 hr of exposure time. Physiochemical properties of modified HNTs and membranes were characterized by SEM, FTIR and water contact angle measurements. The water permeability and salt rejection of the membranes were evaluated by cross-flow RO experiments with brackish water. FTIR analysis reviled formation of amide bonds between dendrimers from functionalized HNTs with TMC monomer. The exposure to batch chlorination of both the reference TFC membrane and the modified HNT-based TFN membrane led to a decrease in salt rejection and an increase in the water flux. However, the respective changes in the performance of TFN membrane were smaller than those in the reference TFC membrane. Therefore, increasing the number of amine groups by increasing the generation of PAMAM dendrimers on HNTs improves chlorine resistance of the resulting TFN membrane.

MONDAY 9:30 am - Ballroom 4

4a - Enantioselective Polymeric Membrane for Chiral Separation of Enantiomers

Marine Michel (Imperial College London)*

Chiral separation of enantiomers (optical isomers) is attracting interest as the demand for enantiopure pharmaceuticals is growing dramatically. Many pharmaceuticals exist as a mixture of two enantiomers which can present different biological response such as toxicological behaviour, therapeutic activity, or immune response. As they have the same physico-chemical properties, their separation is a challenging task but it is crucial as the incorrect enantiomer of the drug can offer no curative effect or even be detrimental. Optical resolution of racemic mixtures has been broadly studied by various methods including chromatography, preferential crystallization or membrane processes. The latest is seen as serious candidate to established chiral separation technologies, especially since it requires lower energy costs, is continuous, eco-friendly and easy to scale up. In optical resolution, the membrane acts as a selective barrier and transports one of the paired enantiomers preferentially because of a stereo-specific interaction between enantiomers and chiral recognition sites present in the membrane. Most studies have been performed using dialysis method, where the driving force is the concentration difference between feed and permeate. With this method the concentration of the final product is more dilute than that of the feed solution, and permeation is extremely slow. This work focuses on membrane-based enantioseparation technique, performed in pressure driven separation mode, which has the potential for large-scale production of enantiopure compounds and could pave the way for many more commercial applications satisfying the considerable demand for large-scale chiral separation techniques.

MONDAY 10:00 am - Ballroom 4

4b - Tailored Synthesis of Inorganic-supported Polymer, Graphene, and Nanocomposite Membranes for Liquid-phase Separations

Michael Z Hu (Oak Ridge National Laboratory)*

Molecular separations of liquid mixture solutions represent a separation challenge that demands the nanoscale control of materials structures and tailoring of surface/interfacial functionality chemistry. Synthesis capability of nanostructured functional materials is foundational to advance separation membranes with better flux and selectivity. For biomass conversion to biofuel/biochemical processing, recovery of biooil and water-soluble oxygenated hydrocarbon molecules is critical to ensuring the carbon efficiency and bioeconomy. In this work, surface-engineered (superhydrophobic) nanoporous inorganic membranes were developed to provide several orders of magnitude higher flux in phase separation biooil-water emulsions. Ceramic tube-supported polymer membranes were developed to dewater and concentrate the aqueous solutions that contain oxygenated hydrocarbon molecules (such as carboxylic acid). Thin film nanocomposite membranes that contain embedded nanoparticles (such as graphene oxide or covalent organic frameworks) are also discussed for their promise in dewatering, desalination, and organic solvent separation.

MONDAY 10:30 am - Ballroom 4

4c - Two-Dimensional (2D) Covalent Organic Framework Mixed-Matrix Membranes

Phuoc Duong (University of Wyoming), Valerie Kuehl (University of Wyoming), Bruce Mastorovich (University of Wyoming), John Hoberg (University of Wyoming), Bruce Parkinson (University of Wyoming), Katie D Li-Oakey (University of Wyoming)*

Mixed-matrix ultrafiltration (UF) membranes containing a new two-dimensional (2D) carboxyl-functionalized covalent organic framework (COF) as a nanofiller and polyacrylonitrile (PAN) as a polymer matrix have been investigated. The effect of COF concentration has been evaluated on the membrane morphology, intrinsic and mechanical properties, separation performance, and fouling propensity. Increasing COF concentration from 0 to 0.8 wt% could significantly enhance the rejection rate from 3.5 to 81.9 wt% for bovine serum albumin (BSA) and 65.0 to 99.4 wt% for gamma-globulin. Moreover, membrane hydrophilicity greatly increased, with the decrease of water contact angle from 64.4° to 45.3°. Overall, compared to the pristine PAN membrane, the mixed-matrix PAN membrane with 0.8 wt% COF demonstrated high water flux of 940 $Lm^{-2} h^{-1} bar^{-1}$, advantageous protein retention (81.9% for BSA and 99.4% gamma-globulin), enhanced thermal stability, mechanical properties, and fouling resistance.

MONDAY 11:00 am - Ballroom 4

4d - Enhancing water permeance in reduced graphene oxide membranes by chemical etching

Xiaoyi Chen (The State University of New York at Buffalo)*, Liang Huang (The State University of New York at Buffalo), Zhihao Feng (The State University of New York at Buffalo), Janavi Gohil (The State University of New York at Buffalo), Haiqing Lin (The State University of New York at Buffalo)

Reduced-graphene oxide (r-GO) membranes have emerged as a potential candidate for next-generation nanofiltration (NF) applications. However, previously reported r-GO membranes have lower water permeance than polyamide-based NF membranes because of the small 2D channels required to achieve good size-sieving ability. We report here an effective approach in enhancing the water permeance by creating nano-pores on the r-GO nanosheets. Specifically, GO sheets are chemically etched using H_2O_2 to form holey GO sheets, which are then deposited on porous supports before being reduced by HI. The prepared membranes were thoroughly characterized using SEM, Raman Spectroscopy, FTIR, and XRD. The effect of the etching and reduction conditions on the membrane performance is systematically investigated. Increasing the etching time increases the water permeance with slight decrease in the salt rejection rate, while increasing the reduction degree decreases the water permeance and increases the salt rejection. For instance, increasing the etching time from 0 h to 4 h improves the water permeance from 0.4 $L\ m^{-2} \cdot h^{-1} \cdot bar^{-1}$ to 6.6 $L\ m^{-2} \cdot h^{-1} \cdot bar^{-1}$ by over 15 times, while the Na_2SO_4 rejection remains at 98.5%. The r-HGO membranes exhibit NF performance comparable with the commercial polyamide-based membranes. For instance, when r-HGO with 4 h etching and a reduction using 34 wt% HI was fabricated into a 20 nm-thick membrane, it exhibits water permeance of 12 L $m^{-2} \cdot h^{-1} \cdot bar^{-1}$ and Na_2SO_4 rejection of 91.4% at 4 bar and 22°C. We have also prepared multi-layer r-HGO membranes with various degrees of etching and evaluated for NF applications. We will also present the long-term stability test results of these membranes using a crossflow system.

MONDAY 11:30 am - Ballroom 4

4e - Nanocomposite RO Membranes with High Aspect Ratio Functional Nanoparticles for Controlled Interfacial Transport

Stephen M Martin (Virginia Tech)*, Ethan Smith (Virginia Tech), E. Johan Foster (Virginia Tech), Keith Hendren (Virginia Tech), James Haag (Virginia Tech)

Thin film nanocomposite membranes (TFNs) are a recent class of materials that use nanoparticles to provide improvements over traditional thin film composite (TFC) reverse osmosis membranes. Prior work in the our lab has demonstrated that significant improvements in flux can be achieved through the inclusion of functionalized carbon nanotubes (CNT) in the polyamide selective layer. We hypothesized that the rapid water transport occurs not in the internal pore of the CNT, but at the interface between the CNTs and PA matrix. Thus, careful control of surface functionality and polymerization conditions can be used to tune the membrane performance. In this study, TFNs were produced using high aspect ratio cellulose nanocrystals (CNCs) and TEMPO-oxidized cellulose nanocrystals (TOCNs) as nanoparticle additives. CNCs are broadly interesting due to their low cost, sustainability, and potential for surface modification. Two methods of membrane fabrication were used in order to study the effects of nanoparticle dispersion on membrane flux and salt rejection: a vacuum filtration method and a monomer dispersion method. In both cases, various quantities of CNCs and TOCNs were incorporated into a polyamide TFC membrane via in-situ interfacial polymerization. The flux and rejection performance of the resulting membranes was evaluated, and the membranes were characterized via ATR-FTIR, TEM, and AFM. The best improvement was obtained via the monomer dispersion method at 0.5 wt% TOCN loading resulting in a 260% increase in water flux and an increase in salt rejection to $98.98 \pm 0.41\%$ compared to $97.53 \pm 0.31\%$ for the plain polyamide membrane. The increased flux is attributed to the formation of nanochannels at the interface between the high aspect ratio nanocrystals and the polyamide matrix. These nanochannels serve as rapid transport pathways through the membrane, and can be used to tune selectivity via control of particle/polymer interactions.

MONDAY 12:00 pm - Ballroom 4

4f - If one were to develop a new NF/RO membrane, specifically for uncharged organic solute rejection, how permeable could it be to water while meeting organic solute rejection goals?

Eric Hoek (UCLA)*

Today, we have commercially available NF/RO membranes with separation performance ranging from high-divalent/low-monovalent rejecting NF membranes up to seawater RO membranes exhibiting >99.85% TDS rejection, and everywhere between. However, few NF or RO membranes were specifically developed with the removal of uncharged organic solutes, which may be present at trace levels in drinking water sources and higher levels in wastewater effluents. For over a decade, we have explored the transport of environmentally-relevant uncharged organics (NDMA, 1,4-dioxane, methanol, etc.) through theoretically-modeled, hand-cast, commercially-available NF/RO membranes. We now have a solid practical understanding of the fundamental relationships between membrane structural (or "physical") descriptors such as apparent pore radius and structure factor (which are determined from thin film thickness and free volume along with support membrane pore structure), in addition to the thermodynamic (or "chemical") descriptors such as the solute-membrane partition coefficient (which derives from the free energies of interaction of the solute with water, the solute with the membrane and the membrane with water). Herein, we present new results - both experimental and theoretical - on organic transport through commercial NF/RO membranes in the hopes of answering the question, "If one were to develop a new NF/RO membrane, specifically for uncharged organic solute rejection, how permeable could it be to water while meeting organic solute rejection goals?".

MONDAY 9:30 am - Kings Garden 1

5a - Two-Dimensional Membranes for Gas Separation

Dan Zhao (National University of Singapore)*

It is highly desirable to reduce the membrane thickness in order to maximize the throughput and break the trade-off limitation for membrane-based gas separation. Two-dimensional (2D) membranes composed of ultrathin nanosheets have been attracting much attention recently because of their ultrasmall thickness that may offer interesting gas separation performance. In this talk, I will introduce our study on the fabrication of 2D membranes for gas separation. Our candidate materials include graphene oxide (GO) and metal-organic frameworks (MOFs). We have identified some important steps in fabricating and testing 2D membranes. In addition, our 2D MOF membranes exhibit responsive gas separation behaviors under the stimuli of temperature or pressure, paving the way for the rational design of smart membranes for H_2 purification and CO_2 capture.

MONDAY 10:00 am - Kings Garden 1

5b - Restricting Lattice Flexibility in Polycrystalline Metal-Organic Framework Membrane for Efficient Carbon Capture

Deepu Babu (L'Ecole polytechnique fédérale de Lausanne), Kumar Varoon Agrawal (L'Ecole polytechnique fédérale de Lausanne)*

Polycrystalline membranes based on metal-organic frameworks (MOFs) are advantageous to their nanoporous counterparts owing to a high degree of structural tunability, relatively fast crystallization kinetics, and higher reproducibility [1]. However, achieving an efficient CO_2/N_2 and CO_2/CH_4 separation, crucial for the energy-efficient carbon capture, has remained a challenge [2]. ZIF-8 possesses a crystallographically-determined pore aperture of 0.34 nm, ideal for CO_2/N_2 and CO_2/CH_4 separations. However, the lattice flexibility and the gate-opening phenomenon in ZIF-8 is a bottleneck in realizing gas-sieving, critical for the membrane-based postcombustion capture, and selectivities have remained below 5 [3]. In this presentation, I will report a novel post-synthetic treatment, which drastically improves the capture performance of the ZIF-8 films [4]. Lattice stiffening is confirmed by the appearance of a temperature-activated transport, attributed to a stronger interaction of gas molecules with the pore aperture, with activation energy increasing with the molecular size ($CH_4 > CO_2 > H_2$). High CO_2/CH_4, CO_2/N_2, and H_2/CH_4 selectivities of up to 32.5, 29.3, and 175, respectively, and complete blockage of C3H6 is achieved. Spectroscopic and X-ray diffraction studies confirm that while the coordination environment and crystallinity is unaffected, lattice distortion and strain are incorporated in the ZIF-8 lattice, increasing the lattice stiffness. Overall, this method can improve the separation performance of several MOF membranes. References: [1] N. Rangnekar, N. Mittal, B. Elyassi, J. Caro, M. Tsapatsis, Chem. Soc. Rev. 2015, 44, 7128; [2] H. Yin, J. Wang, Z. Xie, J. Yang, J. Bai, J. Lu, Y. Zhang, D. Yina, J. Y. S. Lin, Chem. Commun. 2014, 50, 3699. 3. C. Zhang, R. P. Lively, K. Zhang, J. R. Johnson, O. Karvan, W. J. Koros, J. Phys. Chem. Lett. 2012, 3, 2130. 4. D. J. Babu, G. He, M. T. Vahdat, P. A. Schouwink, M. Mensi, K. V. Agrawal, Submitted.

MONDAY 10:30 am - Kings Garden 1

5c - New Asymmetric and Permselective Carbon Molecular Sieve Membranes

Jay (Junqiang) Liu (The Dow Chemical Company)*, Ted Calverley (The Dow Chemical Company), Dean Millar (The Dow Chemical Company), Mark Brayden (The Dow Chemical Company), Marcos Martinez (The Dow Chemical Company)

Microporous Carbon Molecular Sieve (CMS) membranes have shown excellent intrinsic permselectivities. However, industrially relevant defect-free asymmetric CMS membranes has been a challenge due to the melting and softening of most of the precursors during pyrolysis (decomposition temperature higher than glass transition temperature). In this talk, we will present a novel asymmetric CMS membrane derived from low cost polyvinylidene chloride copolymers (PVDC). PVDC dehydrochlorinate and crosslink before melting, which "locks" the morphology from polymer to carbon. We made for the first time asymmetric PVDC membranes using a phase inversion process and turned them into highly asymmetric Carbon Molecular Sieve (CMS) membranes using a two-step pyrolysis process. The PVDC copolymer composition, solution formulation, and casting conditions were optimized. The asymmetric CMS membranes showed very high permeances of 320 GPU for CO_2 and 8 GPU for C3H6, while maintaining good selectivities of 21 for CO_2/CH_4 and 8 for C3H6/C3H8. The asymmetric CMS membranes were repeatedly synthesized without statistically significant variation. The CMS membranes also have limited change of separation performance during the six months of N_2 storage.

MONDAY 11:00 am - Kings Garden 1

5d - Hitz zeolite membrane separation system (HDS) for gas separation

Masaya Itakura (Hitachi Zosen Corporation)*, Satoshi Imasaka (Hitachi Zosen Corporation), Kentaro Shinoya (Hitachi Zosen Corporation), Masashi Okada (Hitachi Zosen Corporation)

Hitachi Zosen Corporation (Hitz) has actively developed and commercialized our zeolite membrane separation systems for several kinds of industry use in liquid- and gas-separation. Hitz membrane elements are 16 mm in diameter and 1130 mm in length. The element is composed of a Al_2O_3 porous support tube with a closed end and a zeolite thin film deposited on outer surface of the tube by a hydrothermal synthesis process. The element constitution is only inorganic material, which enables the membrane to be used under high temperature and high-pressure conditions. We have developed six varieties of zeolite membranes (HZM-1 to HZM-6) with different composition, pore size and chemical affinity and the optimized zeolite membrane is selected depending on the separation use and the customer requirement. Regarding the gas separation application, for example, in the CO_2 separation performances of HZM-4 membrane for CH_4 upgrading the CO_2 permeance and the selectivity were 1.5×10^{-6} mol m^{-2} s^{-1} Pa^{-1} and 115 at 313 K, respectively. These permeance and separation performances were much higher than previous reports. Moreover, the permeance and separation performances retained under the coexistence water of 0.35 vol% in CH_4/CO_2 gas mixture. Hitz zeolite membrane system (HDS) for the CH_4 upgrading produce CH_4 at the yield of 99%. In addition, HZM-2 and HZM-6 membrane also has high potential for purification of hydrocarbons. For example, these membranes exhibit the good performance of olefin/paraffin separation and n/i-paraffin separation driven by the synergy effect of selective adsorption of the feed gas and molecular sieving function.

MONDAY 11:30 am - Kings Garden 1

5e - Insights on the Carbon Molecular Sieve Structure from Stability and Rejuvenation Tendencies

Samuel Hays (Georgia Institute of Technology)*, Oishi Sanyal (Georgia Institute of Technology), Nicholas Leon (Georgia Institute of Technology), Chen Zhang (University of Maryland), Pezhman Arab (Georgia Institute of Technology), William Koros (Georgia Institute of Technology)

Carbon molecular sieve (CMS) hollow fiber membranes provide high surface area-to-volume ratios, scalable manufactability and excellent separation performance for many key gas-pairs, thereby making CMS a frontrunner for next-generation membranes. Physical aging phenomena in CMS cause reduced permeability and increased permselectivity, analogous that seen in glassy polymer membranes. Unlike polymer membranes, however, physical aging in CMS has been thought to be irreversible, due to the lack of a clearly observable glass transition in CMS. This presentation provides a feasible method to restore permeability loss, thereby demonstrating that CMS physical aging can be significantly reversed. Such physical aging and regeneration results also provide new insights into fundamental features of CMS and provide a deeper understanding of the structure and morphology of this complex material. This presentation will address these features and their implications for practical largescale CMS membranes for several gas pairs.

MONDAY 12:00 pm - Kings Garden 1

5f - Fabrication of zeolitic imidazolate framework ZIF-8 membranes by ligand-induced permselectivation in vapor phase

Xiaoli Ma (University of Wisconsin-Milwaukee)*, Prashant Kumar (University of Minnesota), Nitish Mittal (University of Minnesota), Alexandra Khlyustova (University of Minnesota), Prodromos Daoutidis (University of Minnesota), Andre Mkhoyan (University of Minnesota), Michael Tsapatsis (Johns Hopkins University)

ZIF-8 membranes have emerged as a promising energy-efficient technology for gas separations, particularly hydrogen purification and propylene/propane separation. High quality ZIF-8 membranes are typically prepared by solution-based methods such as in situ, seeded growth, counter diffusion and microfluidic processing. In addition to cost and environmental disadvantages, the use of solvents in these methods makes it difficult to control the membrane thickness and defects due to the homogeneous nucleation/growth in the bulk solution and the possible re-dissolution of membrane materials. Here, we developed a vapor-phase processing method to address these challenges. Our method is based on the deposition of ZnO in a porous ceramic support using atomic layer deposition technique, followed by ligand vapor treatment to transform the non-selective ZnO deposits to a selective ZIF-8 membrane. All the precursors are introduced from vapor phase without the use of any types of solvents. The ZIF-8 nanocomposite membranes synthesized exhibit superior separation properties including high selectivity of hydrogen over hydrocarbons, and high propylene/propane separation performance maintained at high feed pressure (> 7 atm). These properties can be attributed to the absence of defects and the unique membrane structure that is partly present on the support surface and partly confined in the mesopores of the support.

MONDAY 9:30 am - Kings Garden 5

6a - Polymers with Side Chain Porosity for Ultrapermeable and Plasticization Resistant Materials for Gas Separations

Sharon Lin (MIT)0, Francesco Benedetti (University of Bologna), Yuan He (MIT), Chao Liu (University of Chinese Academy of Sciences), Yanchuan Zhao (University of Chinese Academy of Sciences), Hong-Zhou Ye (MIT), Troy Van Voorhis (MIT), Maria Grazia De Angelis (University of Bologna), Timothy Swager (MIT), Zach Smith (MIT)

There are very few solution-processable, linear polymers currently reported that have both ultrahigh CO_2 permeability and high selectivity. All of these structures possess rigid polymer backbones that result in poor polymer chain packing, leading to ultramicroporosity in solid-state chain-entangled films. We report the gas transport properties of two porous polymers obtained via Ring Opening Metathesis Polymerization (ROMP), CF3-ROMP and OMe-ROMP, which both possess flexible polymer backbones with rigid side chains. Both OMe-ROMP and its fluorinated counterpart, CF3-ROMP, exhibit exceptional plasticization resistance (CO_2 plasticization pressure > 51 bar). CF3-ROMP also possesses ultrahigh CO_2 permeability (> 21000 Barrer). In addition, the rate of physical aging for both CF3-ROMP and OMe-ROMP is slower than that of traditional polymers of intrinsic microporosity (PIMs), especially for gases with small effective diameters (He, H_2, and O2). The strategy of using ROMP to generate new polymers that have flexible backbones and rigid, pore-forming side chains shows promise for future studies.

MONDAY 10:00 am - Kings Garden 5

6b - Bench scale testing of next generation hollow fiber membrane modules at subambient conditions at the National Carbon Capture Center

David Hasse (Air Liquide)*, Shilu Fu (Air Liquide), Sudhir Kulkarni (Air Liquide), Trapti Chaubey (Air Liquide), Alex Augustine (Air Liquide), Andrew Hamilton (Air Liquide), Monaca Mcnall (Air Liquide)

Air Liquide is developing a hybrid CO_2 capture process based on sub-ambient temperature membrane operation in combination with cryogenic distillation (Ref 1-3). Prior work has shown extremely attractive membrane performance. The current work seeks to demonstrate performance of a commercial-scale, high-permeance hollow-fiber membrane, with actual flue gas from a coal fired power plant. Bench scale testing at 1,000 NM3/hr scale at the National Carbon Capture Center (NCCC, Wilsonville AL) is being conducted in a field-test unit incorporating pre-treatment, compression, drying, and auto-cooling steps using commercial scale membrane bundles. The membranes can be tested at pressures up to 16 bar with temperatures as low as -65°C. Results from the ongoing long-term testing of the high-permeance PI-2 membrane at NCCC will be presented. An updated techno-economic analysis will be shown along with a discussion of the impact / reduction of impurities (SOX / NOX / metals) in the cold membrane process. In addition, the PI-2 data will be placed in context with additional polymers over an extended temperature range (as low as -65°C) with CO_2/N_2 selectivity >200. Acknowledgement: Current development work is supported by the Department of Energy, National Energy Technology Laboratory under Award Number DE-FE0026422 (A. O'Palko, J. Figeroa, Program Managers). Prior work presented here was done under Award Number DE-FE0013163. References: 1. CO_2 Capture by Cold Membrane Operation with Actual Power Plant Flue Gas, Final Scientific Report, https://www.osti.gov/servlets/purl/1373105 2. 2017 NETL CO_2 Capture Technology Project Review Meeting https://www.netl.doe.gov/sites/default/files/2017-12/A-Augustine-Air-Liquide-Hollow-Fiber-Modules.pdf 3. Energy Procedia Volume 63, Pages 186-193 (2014) http://www.sciencedirect.com/science/article/pii/S1876610214018347

MONDAY 10:30 am - Kings Garden 5

6c - A Carbon Molecular Sieve Membrane-Based Reactive Separation Process for Pre-Combustion CO_2 Capture

Mingyuan Cao (Unversity of Southern California), Linghao Zhao (University of Southern California), Dongwan Xu (University of Southern California), Secgin Karagoz (University of California, Los Angeles), Patricia Pichardo (University of California, Los Angeles), Richard J. Ciora, Jr. (Media and Process Technology, Inc.), Paul Kt Liu (Media and Process Technology, Inc.), Vasilios I. Manousiouthakis (University of California, Los Angeles), Theodore T Tsotsis (University of Southern California)*

Reactive separation processes, such as membrane reactors (MRs) and adsorptive reactors (ARs), are attracting increased interest for industrial applications. We study here a hybrid system combining a MR and an AR in tandem (with the AR following the MR, and the MR's reject stream serving as the AR's feed). We apply this system to the water gas shift reaction (WGS) in the context of its application in the IGCC process for H_2 generation and simultaneous CO_2 capture from coal/biomass. The MR-AR hybrid system under study attains a high WGS reaction conversion exceeding equlibrium, produces a pure H_2 product for power generation, and delivers a high-pressure CO_2 stream ready for sequestration. In our research, specifically, we use carbon molecular sieve (CMS) membranes. Following the field-testing and performance-validation of these CMS membranes under coal/biomass gasification conditions, experiments have been carried-out in the laboratory to determine the membrane characteristics, and the MR performance under the IGCC-relevant conditions, i.e., for temperatures up to 300°C and pressures up to 25 bar, with different syngas compositions. As part of this work, a detailed model for the MR/AR system was also developed and validated using the experimental data. Using this model, a technical and economic analysis (TEA) was carried-out for process design/optimization and economic evaluation of the WGS-MR/AR system for a broad range of operating conditions and design parameters. The experimental and model findings manifest the ability of the WGS-MR/AR process to operate under the desired conditions and to improve the efficiency of the WGS reaction. They demonstrate, in addition, the potential of the system to carry out the in-situ separation of H_2 and CO_2 using the CMS membranes. We conclude from our study, that the CMS-membrane-based WGS-MR/AR system is a good candidate technology for incorporation into IGCC power plants for environmentally-benign power generation.

MONDAY 11:00 am - Kings Garden 5

6d - Synthesis of Polymers of Intrinsic Microporosity (PIM-1) with Poly(ethylene glycol) for Membrane Application

Volkan Filiz (Helmholtz-Zentrum Geesthacht)*, Gisela Bengtson (Helmholtz-Zentrum Geesthacht), Silvio Neumann (Helmholtz-Zentrum Geesthacht)

Polymers of intrinsic microporosity have been and continue to be the objects of many research papers since Budd and McKeown launched their first publication about PIM-1 in 2004. PIM-1 is a hydrophobic polymer formed as a molecular ladder; the polymer chains cannot pack tightly because of their contorted molecular structure, provided by spiro fused rings in the polymer chain. Thus, intrinsic nanopores are formed resulting in a large inner surface. Membranes prepared from this polymer show a large fractional free volume and therefore high permeability for many gases, but especially for CO_2 (4000 to 10,000 Barrer) because of an also very high solubility. Until now, the leading polymer of intrinsic microporosity PIM-1 has become quite famous for its high membrane permeability for many gases in gas separation, linked, however, to a rather moderate selectivity. The combination with the hydrophilic and low permeable poly(ethylene glycol) (PEG) and poly(ethylene oxides) (PEO) should on the one hand reduce permeability, while on the other hand enhance selectivity, especially for the polar gas CO_2 by improving the hydrophilicity of the membranes. Four different paths to combine PIM-1 with PEG or poly(ethylene oxide) and poly(propylene oxide) (PPO) were studied: physically blending, quenching of polycondensation, synthesis of multiblock copolymers and synthesis of copolymers with PEO/PPO side chain. Blends and new, chemically linked polymers were successfully formed into free standing dense membranes and measured in single gas permeation of N_2, O_2, CO_2 and CH_4 by time lag method. The aim of this study in polymeric membrane research is the combination of PIM-1, the rigid hydrophobic high flux polymer, with hydrophilic poly(ethylene glycol) (PEGs) of various polymer weights to form stable membranes of improved gas separation properties, especially for CO_2.

MONDAY 11:30 am - Kings Garden 5

6e - High-permeance polymer-functionalized graphene membranes that surpass the postcombustion carbon capture target

Guangwei He (L'Ecole polytechnique fédérale de Lausanne), Kumar Varoon Agrawal (L'Ecole polytechnique fédérale de Lausanne)*

Postcombustion carbon capture using high-performance CO_2-selective membranes has been identified as one of the most energy-efficient routes for reducing CO_2 emissions [1-3]. However, the capture performance, especially the permeance of the state-of-the-art membranes needs to be significantly increased to further improve the appeal of membranes for postconbustion capture, and importantingly, reduce the needed membrane area for separation because the capital cost of the membrane modules scales inversely with the permeance. Currently, with the exception of Polaris second generation membranes from Membrane Technology Research which yields a CO_2 permeance of 2000 GPU, there is no membrane technology that reaches the CO_2 permeance significantly higher than 1000 GPU, while maintaining a separation factor of 20 or above in the humid flue gas conditions. In this presentation, I will report a new class of sub-20-nm-thick organic-inorganic hybrid membranes, which we call as SPONG, comprising of nanoporous single-layer-graphene with a high porosity (up to 18.5%), functionalized with CO_2-phillic polymeric chains [4]. The hybrid membranes yield CO_2 permeance sixfold larger (6290 GPU) than the performance target while separating the CO_2/N_2 mixture mimicking flue gas streams with a separation factor well within the target. Overall, we report a number of membranes that yield CO_2 permeance up to 11790 GPU and CO_2/N_2 mixture separation factor up to 57.2. The hybrid membrane is conducive for scale-up because of its facile preparation involving simple coating steps. References [1] Merkel, T. C., Lin, H., Wei, X. & Baker, R. J. Memb. Sci. 359, 126-139 (2010); [2] Rao, A. B. & Rubin, E. S. Environ. Sci. Technol. 36, 4467-4475 (2002); [3] Bara, J. E., Camper, D. E., Gin, D. L. & Noble, R. D. Acc. Chem. Res. 43, 152-159 (2009); [4] He G., Huang S., Villalobos L. F., Zhao J., Mensi M., Oveisi E., Rezaei M., Agrawal K. V., Submitted.

MONDAY 12:00 pm - Kings Garden 5

6f - Fabrication of Thin Film Composite Membranes using Microporous Polymer Blends and Polybenzimidazole Nanoporous Supports for CO_2/N_2 Separation

Lingxiang Zhu (U.S. Department of Energy National Energy and Technology Laboratory)*, Surendar Venna (U.S. Department of Energy National Energy and Technology Laboratory), Ali Sekizkardes (Battelle/U.S. Department of Energy National Energy and Technology Laboratory), Victor A Kusuma (U.S. Department of Energy National Energy and Technology Laboratory), David Hopkinson (U.S. Department of Energy National Energy and Technology Laboratory)

Industrial gas separations usually use large-scale (~10,000 m^2 or more/plant) thin film composite (TFC) membranes that are comprised of a thin (100 nm - 1 μm) selective layer and a thick (> 20 μm) porous support layer. Fabrication and scale-up of the TFC membranes depend on selective layer material with good thin film forming ability and a porous membrane support with optimized pore structure (i.e., small pore size and high porosity) and thus minimal gas transport resistance. In this study, we fabricated TFC membranes based on a cost-effective polymer of intrinsic microporosity/ether side chain polyphosphazene (PIM-1/MEEP) blend that is a high-performance CO_2-selective membrane material developed in-house recently. PIM-1 and MEEP are soluble and miscible in chloroform, rendering PIM-1/MEEP blend exceptional coatability. To accommodate the thin film coating of PIM-1/MEEP, chloroform-resistant polybenzimidazole (PBI) nanoporous supports were fabricated via a phase-inversion method. These PBI nanoporous supports were characterized by the pore size of < 20 nm, surface porosity of ca. 15%, and CO_2 permeance of 250,000 GPU at 25°C. Defect-free sub-micron PIM-1 based TFC membranes were prepared on the PBI supports via a dip coating technique, and their pure- and mixed-gas CO_2/N_2 separation properties were systematically investigated. For example, 1-μm-thick PIM-1/MEEP TFC membranes showed pure-gas CO_2 permeance of ca. 800 GPU and CO_2/N_2 selectivity of 28 at 25°C. Their gas separation performance, membrane stability, and aging behavior in flue gas stream at the National Carbon Capture Center would be presented as well. Due to the great CO_2/N_2 separation performance, good processability, and the easiness in thin film coating of PIM-1/MEEP as exemplified on the flat-sheet PBI membrane supports, this presentation would also discuss the scale-up of PIM-1/MEEP based composite membranes on PBI hollow fiber membrane supports.

MONDAY 2:00 pm - Kings Garden 5

7a - Field trial of spiral-wound facilitated transport membrane module for CO_2 capture from flue gas

Yang Han (Ohio State University)*, Witopo Salim (Ohio State University), Kai Chen (Ohio State University), Dongzhu Wu (Ohio State University), Winston Ho (Ohio State University)

A field trial with 1.4 m^2 spiral-wound (SW) membrane modules fabricated from an amine-containing facilitated transport membrane was conducted with actual flue gas at the National Carbon Capture Center (NCCC) in Wilsonville, AL. Prior to the field test, the membrane modules were systematically evaluated with simulated flue gas to identify the optimal operating conditions. Significant responses to various operating conditions have been noted, particularly with respect to the feed flow rate, feed pressure, operating temperature, and CO_2 concentration. The concentration polarization behavior and feed side pressure drop were significantly affected by the choice of feed flow rate, which retroactively influences the stage cut, or the CO_2 recovery of the module. With the actual flue gas, the membrane modules showed ca. 1450 GPU CO_2 permeance and 185 CO_2/N_2 selectivity at 67°C with feed and permeate pressures of 4 and 0.3 atm, respectively. The module performance results were essentially identical to those obtained from both membrane modules and flat-sheet membrane samples tested at OSU using simulated flue gas. A 500-h stability was demonstrated despite the interference of system upsets and flue gas outages at NCCC. During the field trial, carbon capture rates of better than 40% were achieved by a single SW module with coal-derived flue gas. Except for a few flue gas upsets, the CO_2 purities in the captured stream were 94.5% or better on dry basis. In addition, a short transient time of the membrane system was observed, which benefits the dynamic integration of the carbon capture technology into the power plant. Further analysis of the tested membrane revealed that no significant amounts of Cr, As and Se were deposited onto the membrane. The data obtained from this project validate the membrane material and provide basis for the design and fabrication of full-scale SW module with a membrane area larger than 50 m^2.

MONDAY 2:30 pm - Kings Garden 5

7b - Gen-2 Proteus™ membrane and module development for CO_2 capture from gasification streams

Witopo Salim (Membrane Technology & Research Inc.)*, Karl Amo (Membrane Technology & Research Inc.), Trevor Carlisle (Oregon State University), Richard Baker (Membrane Technology & Research Inc.), Jenny He (Membrane Technology & Research Inc.), Jay Kniep (Membrane Technology & Research Inc.), Tim Merkel (Membrane Technology & Research Inc.), Vincent Nguyen (Membrane Technology & Research Inc.), Zhen Sun (Membrane Technology & Research Inc.), Jonathan Tan (Membrane Technology & Research Inc.), Bob Watson (Membrane Technology & Research Inc.), Erik Westling (Membrane Technology & Research Inc.)

Gasification of coal, biomass or other feedstocks is widely used in industry for the production of chemicals and power. In a carbon-constrained world, gas separation is an increasing important step to maintain the economic viability of these processes with CO_2 capture. Membrane Technology and Research Inc. has previously proposed a dual membrane process for CO_2 capture from gasification processes, including Proteus™ membranes for H_2 removal or purification and Polaris™ membranes for CO_2 recovery. This dual-membrane process can achieve a H_2 recovery of 99% with>90% CO_2 capture while producing high purity liquid CO_2 at an energy cost lower than conventional technology. This presentation focuses on the development of the H_2-selective Proteus membranes and modules. Field test results of Gen-1 Proteus bench-scale modules with a syngas slipstream at the National Carbon Capture Center (NCCC) in Wilsonville, AL showed promising results over 9,000 hours of testing. Membranes and modules showed H_2/CO_2 selectivities of 15 or more when treating real syngas containing up to 800 ppm H_2S. Based on these results, a preliminary techno-economic analysis of the MTR dual-membrane process resulted in a CO_2 capture cost of $49/tonne of CO_2, which is approximately 20% lower than the cost for the conventional absorption technology, Selexol™. New Gen-2 Proteus membrane under development at MTR will further reduce costs with an improved upper temperature limit and selectivity compared to Gen-1 Proteus. Initial field test results of the Gen-2 Proteus membrane at NCCC will be presented and future plans of Gen-2 Proteus membrane and module development will be discussed.

MONDAY 3:00 pm - Kings Garden 5

7c - One-pot surface modification of nanocellulose fibrils and effect of manipulated fibril surfaces in Hybrid Facilitated Transport Membranes for CO_2 capture

Saravanan Janakiram (Norwegian University of Science and Technology)*, Xinyi Yu (Norwegian University of Science and Technology), Luca Ansaloni (SINTEF Industry), Zhongde Dai (Norwegian University of Science and Technology), Liyuan Deng (Norwegian University of Science and Technology)

The successful use of membranes in CO_2 capture processes is governed by selection of appropriate membrane materials that has preeminent permeation, chemical and mechanical properties. Facilitated transport membranes have been proven to have superior permeation properties under humid conditions. These membranes transport CO_2 through reactional pathway in addition to solution diffusion. The transport properties of these materials can further be enhanced by incorporation of various fillers to result in hybrid membranes. The fabricated hybrid membranes synergistically inherit individual properties of both polymer and filler phase with properties that arise due to their interaction. Nanocellulose fibrils (NFCs) represent a class of renewable 1D bio nanomaterial. Thanks to their high aspect ratio, good mechanical properties, ability to form hydrogen bonding and bio-based origin, NFCs are considered one of the most promising candidate for the processing of polymer nanocomposites. Surface chemistry plays an important role determining the compatibility of nanocellulose fibrils with the polymer matrix. Different surface interactions lead to varying interfaces, which dominate the transport properties of the hybrid matrix. To study the effect of varying interfaces, NFC fibrils have been grafted with different small polymeric moieties using a rapid one-pot functionalization procedure mediated by EDC/NHS. The modified fillers were dispersed in sterically hindered polyallyl amine blend with polyvinyl alcohol and thin film composite membranes with selective layer thickness less than 300 nm were fabricated and tested. NFC functionalized with 4-arm polyethylene glycol resulted in good surface compatibility with matrix polymer leading to a superior gas permeation performance with CO_2 permeance as high as 650 GPU and highest CO_2/N_2 selectivity of 51 at varying compositions.

MONDAY 3:30 pm - Kings Garden 5

7d - DDR-type zeolite membrane: The first opportunity in CO_2-EOR industry

Junya Okazaki (JGC Corporation)*, Hiroaki Hasegawa (JGC Corporation), Nobuyasu Chikamatsu (JGC Corporation), Kenji Yajima (NGK Insulators), Katsuya Shimizu (NGK Insulators), Makiko Niino (NGK Insulators)

JGC Corporation (JGC) has announced the launch of a joint demonstration test with Japan Oil, Gas and Metals National Corporation (JOGMEC) on CO_2 recovery using a DDR-type zeolite membrane system, which has been co-developed by NGK Insulators, Ltd. (NGK) and JGC, to recover CO_2 from the associated gas of a CO_2-EOR (Enhanced Oil Recovery) oil field in Texas. To the best of our knowledge, our test is the first pilot scale experience of the use of zeolite membrane for the CO_2-EOR industry. The membrane has been fabricated by NGK as the world's largest zeolite membrane element to meet the needs of CO_2 separation in natural gas and CO_2-EOR industries. A thin and defect-free zeolite membrane is prepared onto the inner surface of a cylindrical monolithic porous alumina substrate having 1600 channels with a 180 mm in diameter and 1000 mm in length. Its surface area reaches 12 m^2 which is 200-300 times larger than conventional tubular type zeolite membranes of the same length. Recently, the interest in the practical application of CO_2-EOR has been growing rapidly. Since CO_2 purchase costs often account for most of the operating expense of a CO_2-EOR project, CO_2 recovery from associated gas to re-use is essential. For this purpose, polymer membranes are generally applied; however, their insufficient durability often increases the membrane replacement cost. On the other hand, the DDR-type zeolite membrane is not damaged by CO_2 even in very high CO_2 concentration conditions with elevated pressure. Moreover, it shows superior CO_2 permselectivity over hydrocarbons compared to polymer membranes. In this presentation, the advantage of the DDR-type zeolite membrane for CO_2 recovery from CO_2-EOR associated gas will be discussed. In addition, our recent activity for the demonstration test of the DDR-type zeolite membrane will also be introduced.

MONDAY 4:00 pm - Kings Garden 5

7e - Coating of hollow fiber membrane for carbon molecular sieve membrane preparation

Yuhe Cao (Georgia Institute of Technology)*, Kuang Zhang (Georgia Institute of Technology), Oishi Sanyal (Georgia Institute of Technology), William Koros (Georgia Institute of Technology)

This article provides a process for preparing an economical dual-layer asymmetric carbon molecular sieve (CMS) hollow fiber membrane having improved gas separation performance properties. The economically favored process comprises coating engineered support layer fibers to form a dense skin layer in a CMS precursor. Vinyltrimethoxysilane (VTMS) treatment was then employed to the coated precursor fibers, which suppressed the collapse during aggressive pyrolysis processing. Coated precursor fiber membranes were pyrolyzed, and the resultant CMS hollow fiber membranes were investigated for CO_2/CH_4 separation. CMS permeation results using CO_2/CH_4 (50:50) mixed gas CMS showed high CO_2/CH_4 selectivity and high CO_2 permeance.

MONDAY 4:30 pm - Kings Garden 5

7f - Ultrapermeable benzotriptycene-based PIMs that redefine the upper bounds for CO_2 separations

Alessio Fuoco (CNR)*, Bibiana Comesaña-Gándara (University of Edinburgh), Jie Chen (University of Edinburgh), Grazia Bezzu (University of Edinburgh), Mariolino Carta (Swansea University), Ian Rose (University of Edinburgh), Maria-Chiara Ferrari (University of Edinburgh), Johannes Carolus (John) Jansen (CNR), Elisa Esposito (CNR), Marcello Monteleone (CNR), Neil McKeown (University of Edinburgh)

Gas separation by polymeric membranes is an expanding technology for a number of industrial applications such as oxygen or nitrogen separation from air (i.e. separation O_2/N_2) or natural gas treatment and biogas upgrading (i.e. separation CO_2/CH_4). However, a further enhancement of the polymer transport properties is desirable to be cost effective in comparison with the traditional separation technology currently applied for CO_2 capture or removal. Up to now, only poly(trimethylsilylpropyne) (PTMSP) and few other polyacetylenes demonstrate ultrapermeability (P>20000 Barrer), but their industrial application is prohibited by their very poor gas selectivity. In this presentation, we report a series of ultrapermeable PIMs, based on benzotriptycene units, showing substantially higher selectivity than PTMSP [1]. Their extraordinary transport properties are attributable to the inefficient packing of their 2D polymeric chains which results in large interconnected pores that enhance gas permeability. We will show that the outstanding CO_2/CH_4 and CO_2/N_2 permselectivity of the whole family of benzotriptycene PIMs allows the introduction of a new upper bound to define the current state-of-the-art. The gas transport properties will be discussed in terms of permeability, diffusivity and solubility. For selected samples, a novel instrument will be used to determine not only mixed gas permeability, but also the diffusion coefficient of gas mixtures [2]. Ageing data and the effect of the temperature on the transport properties will be also discussed. Finally, we will anticipate the potential exploitation of these materials for large scale gas separation application. References [1] I. Rose et al., Polymer ultrapermeability from the inefficient packing of 2D chains, Nat. Mater. 16 2017. [2] S.C. Fraga et al., A Novel Time Lag Method for the Analysis of Mixed Gas Diffusion in Polymeric Membranes by On-Line Mass Spectrometry: Method Development and Validation. J. Memb. Sci. 561 2018.

MONDAY 2:00 pm - Ballroom 3

8a - CFD simulation of crimped hollow fiber membranes for liquid separation processes

Mohammad Younas (University of Engineering and Technology, Peshawar), Amir Muhammad (University of Engineering and Technology, Peshawar), Glenn Lipscomb (University of Toledo)*

Hollow fiber membrane contactors are considered a promising technology for intensification of separation processes ranging from haemodialysis to dehydration. Module performance depends on membrane materials, transport properties and module design parameters that control fluid flow through the module and the resulting development of permeation limiting boundary layers. If fluid boundary layer resistances are significant, module designs that introduce a cross-flow component to the shell flow can be beneficial. This can be done through the use of baffles and/or knitted fibers. Alternatively, crimped fibers can be used to introduce a cross-flow velocity component [1, 2]. Computational fluid dynamics (CFD) is used widely to simulate fluid flow in various areas of research. Commercials codes include COMSOL Multiphysics and Ansys Fluent which use the Finite Element Method (FEM) and Finite Volume Method (FVM) for discretization of the governing conservation equations, respectively. Commercially available HFMC module was taken for modelling and simulation of liquid-liquid extraction in this work. Simulations were performed for a flow-cell containing a single fiber based on symmetry considerations. The simulation domain was limited further to a short repeating segment of the fiber in the cell. The repeat unit of the flow-cell was created by translating the center of a circle along a sine wave with varying values of fiber diameter, crimp amplitude and crimp wavelength. Numerical approximations to the equations were obtained using COMSOL Multiphysics and Ansys. The geometry and mesh were created within each software and the same geometrical parameters for the flow-cell and boundary conditions were used. The model was validated with prior experimental results [3]. Velocity and concentration profiles were evaluated throughout the flow-cell. The study revealed that CFD is a promising tool for the simulation of coupled hydrodynamics and mass transfer processes.

MONDAY 2:30 pm - Ballroom 3

8b - Computational Fluid Dynamics Modeling for the Investigation of Multi-Layer Spacer Effects

Zachary Binger (The University of Arizona), Andrea Achilli (The University of Arizona)*

Forward osmosis (FO) and pressure retarded osmosis (PRO) are processes of interest for applications in water purification and energy recovery. By reducing hydrodynamic losses within membranes and maximizing the experienced osmotic pressure difference between streams, FO and PRO can be used to recover potable water from wastewater more efficiently and/or reduce the cost of desalination. The development of computational fluid dynamics (CFD) models, can be used to investigate the effectiveness of spacer geometries in reducing pressure loss and increasing water flux in FO and PRO. CFD work easily interfaces with 3D printing to create rapid prototyping to support experimental work. The geometries created with computer-aided design software for CFD simulation can be used by additive manufacturing machines such as 3D printers to get rapid prototyping for the physical testing of any CFD experiment. 3D printing techniques have the ability of printing resolutions down to 25 microns resulting in high resolution and smooth surfaces. This allows for creation of spacer shapes that would not be possible with traditional manufacturing techniques in addition to instant results for improving ensuing spacer designs. This research focuses on the use of multi-layer spacers to promote vortex shedding and improve water flux by reducing or eliminating stagnant eddies that form along membrane walls. CFD models have been created at the bench scale, approximately 100 cm^2, to investigate the effectiveness of multi-layer spacer geometries on vortex shedding. Implementing smaller wing-shaped filaments between larger mid-channel spacers can promote vortex shedding while still achieving redirection of low concentration bulk towards membrane walls. The wing-shape spacer will magnify horizontal velocities directed toward regions that stagnant eddies occupy, effectively breaking them up and increasing water flux while minimizing the increased pressure losses that the larger number of spacers cause.

MONDAY 3:00 pm - Ballroom 3

8c - Direct numerical simulation of unsteady mixing in direct contact membrane distillation systems with membrane spacers

Jincheng Lou (Colorado School of Mines)*, Jacob Johnston (Colorado School of Mines), Nils Tilton (Colorado School of Mines)

Direct contact membrane distillation (DCMD) is a thermal process in which warm feed and cool distilled water flow on opposite sides of a hydrophobic membrane. The temperature difference causes water to evaporate from the feed, travel through the membrane, and condense in the distillate. To date, CFD studies of DCMD focus primarily on the challenge of temperature polarization. For high concentration brines, however, concentration polarization is another major challenge that reduces system efficiency and leads to mineral scaling. In addition, unsteady mixing triggered by flow over membrane spacers is also not well understood. To study coupled temperature polarization, concentration polarization, and unsteady mixing in the DCMD treatment of high concentration brines, we develop an in-house CFD method for simulating 2-D heat and mass transport in plate-and-frame DCMD systems with cylindrical spacers. The coupled momentum, energy, and mass transport equations are solved using a finite-volume method with an efficient non-iterative unsteady scheme. The membrane spacers are simulated using recent advances in immersed boundary methods. We investigate temperature and concentration polarization under various operating conditions and spacer placement. At low Reynolds numbers, we find the flow is steady, with recirculation zones behind the spacers. At high Reynolds numbers, we find that the flow transitions to an unsteady regime with vortex shedding. For both regimes, the recirculation zones and vortices interact with the membrane surfaces, forming regions of high and low polarization, with corresponding variations in transmembrane heat and vapor flux.

MONDAY 3:30 pm - Ballroom 3

8d - Concentration polarization modeling for high-pressure membranes with engineered surface features

Zuo Zhou (Clemson University)*, David Ladner (Clemson University), Sapna Sarupria (Clemson University), Steven Weinman (Clemson University), Scott Husson (Clemson University), Ilenia Battiato (Stanford University), Negin Kananizadeh (Clemson University)

The goal of this research is to develop a computational framework to predict concentration polarization (CP) of patterned membranes designed for purification of saline or impaired waters. Extensive literature has studied the use of surface coatings on membranes to reduce fouling. More recently it has been proposed that the surface morphology can be modified (surface patterning) to give membranes fouling resistance, similar to what is seen in nature (i.e., shark skin). CP, a mass transfer related phenomenon, is an important driver for fouling. Our research hypothesis is that certain pattern geometries can significantly reduce membrane fouling by reducing CP. The work presented here focuses on computational fluid dynamics (CFD) modeling to simulate different surface features and evaluate how fluid flows in the space above the membrane. The computational models were built to simulate membranes patterned in the lab with a silicon stamp. The stamp produced a variety of surface features including line and grooves, pillars, and pyramids. The surface features varied in size from 2 µm to 512 µ-m with a pattern depth of 1 µ-m to 256 µ-m. Simulations evaluated the effects of velocity, concentration distribution, and shear stress on the fluid flow around the features. Velocity profiles showed that the surface features alter the fluid path, resulting in a lower velocity in between features, but a faster velocity on the elevated regions. Lower shear stress was observed in the valley region of the patterns, corresponding to higher concentration results. In general, patterns always increase CP, and this is due to the increased boundary layer thickness caused by higher roughness. Mass transfer coefficient was the highest among flat membranes. Analytical solutions were used to compare with the numerical results among flat membranes, confirming that the models predicted the results very well.

MONDAY 4:00 pm - Ballroom 3

8e - Analysis of Micromixers to Minimize Scaling Effects on Reverse Osmosis Membranes

Jeremy Walker (U.S. Army)*, Shawn McElmurry (Wayne State University), James Dusenbury (U.S. Army)

The standard diamond-shaped mesh spacers commonly used in spiral wound membrane elements results in non-ideal hydrodynamics that can lead to fouling, which ultimately reduces the flux of water through the membrane and the operational life of the unit. To address this shortcoming, a novel method for separating RO membrane leaves in spiral wound elements was developed and evaluated. Three-dimensional (3-D) printing was utilized to manufacture micromixers directly on membrane swatches. To enhance performance, a two-dimensional computational fluid dynamic model was used to select the optimal geometry and pattern of 3-D printed micromixers. The optimal geometry selected created unhindered flow between 0.2 m/s and 0.3 m/s, using an inlet flow velocity of 0.104 m/s, across 40% of the membrane surface. Laboratory experiments were conducted to evaluate the performance of micromixers and compared to unmodified membranes with a standard 20 mil (0.508 mm) mesh feed spacer. Pure water flux and salt rejection were found to be similar to standard membranes, indicating the 3-D printing process did not damage intrinsic membrane properties. Calcium sulfate scaling experiments were conducted. Scale initially began to form within 2 hours of treatment resulting in a flux decline ofapproximately 10% for both modified and unmodified membranes. Over 14 hours, an average flux decrease of 24% was observed for modified membranes compared to an average flux decrease of 78% for the unmodified membranes. This demonstrated the improved resistance to fouling created by the open channel design with optimal flow conditions. Micromixers printed directly to the membrane surface offers the ability to optimize feed channel hydrodynamics, reduce scale formation and minimize flux decline representing a significant advancement in membrane technology.

MONDAY 4:30 pm - Ballroom 3

8f - Do Hydrodynamic Instabilities Cause Roughness in Thin Polyamide Films of Reverse Osmosis Membranes?

Akshay Deshmukh (Yale University)*, Puskar Mondal (Yale University), Subhajyoti Chaudhuri (Yale University), Menachem Elimelech (Yale University)

Polyamide films form the all-important selective layer of thin-film composite (TFC) membranes, currently the gold-standard for reverse osmosis (RO) desalination. The physical structure of these thin films plays a critical role in determining key membrane properties, including permeability and selectivity. Recently, several spectroscopic studies have sought to elucidate the structure of polyamide films, which are formed by interfacial polymerization. However, fundamental aspects of polyamide film formation, including the origins of the chaotic ridge and valley morphology, remain unknown. In this study, we examine the potential for hydrodynamic instabilities to cause the extraordinary ridge and valley structure exhibited by thin polyamide films by destabilizing the reaction interface during polymerization. We begin by developing a transient moving-boundary reaction model to determine the rate of reaction between the diamine and tri-acid chloride monomers. We then determine the transient temperature profiles in the aqueous and organic layers as a function of their thermal properties. Finally, these temperature profiles are incorporated into a linear stability analysis that accounts for buoyancy (Rayleigh) and surface tension (Marangoni) forces. The two-layer Rayleigh-Bénard-Marangoni problem is solved using novel semi-analytical methods to account for transient temperature coupling. Our analysis demonstrates the rapid destabilization of the reaction interface by Marangoni forces emphasizing the need to account for surface tension gradients when investigating polyamide film formation. We show that buoyancy forces coupled with surface tension forces can readily lead to instabilities in both the aqueous and organic layers. While further experimental work is required to determine key parameters, including the enthalpy of reaction, our analysis provides a robust platform for assessing the impact of hydrodynamic instabilities on the roughness of thin polyamide films.

MONDAY 2:00 pm - Ballroom 4

9a - Structural and Transport Properties of Membranes in High-Salinity Desalination using Cascading Osmotically Mediated Reverse Osmosis

Xi Chen (Columbia University)*, Chanhee Boo (Columbia University), Ngai Yin Yip (Columbia University)

Management and treatment and high-salinity brines (>≈70,000 ppm TDS) have rapidly risen to be global environmental challenges. We recently proposed a novel cascading osmotically mediated reverse osmosis (COMRO) membrane technology for energy-efficient desalination of hypersaline brines. COMRO utilizes the innovative design of bilateral countercurrent reverse osmosis stages to lessen the transmembrane osmotic pressure difference, thus depressing the hydraulic pressure needed. In COMRO operation, membranes are challenged by the atypical working environments, including high pressures and high salinities. In this study, we experimentally investigate impacts of hydraulic pressures and salinities on structural and transport properties of thin-film composite polyamide membranes in high-salinity desalination using COMRO. Our study shows that raising the hydraulic pressure detrimentally affects the membrane water permeability, A. After the membrane is completely compacted, A remains constant under different pressures, and will gradually recover when the membrane is relaxed from pressurization. Our study also indicates that these changes in water permeability are governed by morphological changes in the active-support interlayer of the membrane. Membrane salt permeability, B, presents similar tendencies to A as hydraulic pressure changes. The membrane structural parameter, S, is demonstrated to be consistent at different salinities (0-293,000 ppm TDS). We found that salinities of up to 117,000 ppm TDS does not exhibit significant influence on membrane water permeability, whereas the salt permeability is substantially elevated when the salinity increases (up to 228,000 ppm TDS). This research systematically studies fundamental structural and transport properties of membrane under the high pressure and salinities in COMRO operations to advance the technology, and also provides insights into principal membrane transport mechanisms applicable for all Osmotically-Driven processes.

MONDAY 2:30 pm - Ballroom 4

9b - Utilization of reverse solute diffusion to enhance membrane performance during Osmotically-Driven processes - improved boron retention and silica scaling mitigation

Yining Wang (Nanyang Technological University)*, Weiyi Li (Southern University of Science and Technology), Chuyang Tang (University of New South Wales), Rong Wang (Nanyang Technological University)

This study presents the positive effects of reverse solute diffusion (RSD) during Osmotically-Driven processes. In one case, we utilize the reverse solute diffusion (RSD) of ions from alkaline draw solutions (DSs) and the concentration polarization of the hydroxyl ions to create a highly alkaline environment near the membrane active surface. The results show that boron (in feed) rejection can be enhanced by 40% by increasing the pH of NaCl DS to 12.5 in the active-layer-facing-feed-solution (AL-FS) orientation. The effect of RSD enhanced boron rejection was further promoted in the active-layer-facing-draw-solution (AL-DS) orientation, evidenced by the significantly increased boron rejection (from 40% to 90%) with a DS of pH 12. This observation can be explained by the enhanced RSD of OH- as compared with the AL-FS orientation and the presence of concentrative internal concentration polarization (ICP) in the AL-DS, which further increase the local pH at the membrane active surface. Similarly, in another case, silica scaling can be effectively controlled by lowering the pH of the DS in the AL-DS orientation, which can be attributed to the low local pH in the membrane support layer as a result of the ICP of the reversely diffused H+. The current study opens a new dimension for enhancing Osmotically-Driven membrane performance by using tailored DS chemistry, where the RSD-induced localized water chemistry change is taken advantage in contrast to the conventional method of chemical dosing to the bulk feed water.

MONDAY 3:00 pm - Ballroom 4

9c - Osmotic Membrane Desalination Performance Governed by Molecular Reflection at the Liquid-Vapor Interface

Akshay Deshmukh (Yale University), Jongho Lee (University of British Columbia)*

Hydrophobic nanoporous membranes that employ osmotically-driven vapor-phase water transport have great potential to simultaneously achieve high rejection of solutes and a high water permeability. While the permeability can be increased by reducing the membrane thickness, recent studies reported that considerable water transport resistances exist at the liquid-vapor interface associated with liquid-vapor phase change, which may limit the permeability enhancement. We develop a new theoretical framework to analyze osmotically-driven water transport across porous hydrophobic membranes by accounting for the transport resistances associated with both diffusion through the membrane and phase-changes at the liquid-vapor interfaces. We systematically derive the functional dependence of the membrane transport resistance on membrane pore size, pore aspect ratio, temperature, and condensation coefficient. A thermodynamic criterion is also derived for minimum membrane thicknesses above which membrane wetting becomes energetically unfavorable. The application of the developed framework to osmotic distillation (OD) reveals that for thick membranes (>> 1 μm) water flux is governed by a combination of molecular and Knudsen diffusion, showing an inverse proportionality to the membrane thickness. However, the resistance associated with phase-changes, arising from molecular reflection at liquid-vapor interfaces, becomes the transport-limiting factor as membrane thickness decreases (<~1 μm). The framework presented provides key design principles for membrane that utilize vapor transport for a range of applications, including desalination, water treatment, chemical concentration, and energy conversion.

MONDAY 3:30 pm - Ballroom 4

9d - Characterization of External and Internal Fouling of Forward Osmosis (FO) Process Through Ultrasonic Time Domain Reflectometry (UTDR) Method

Li Lai (Nanyang Technological University)*, Tzyy Haur Chong (Singapore Membrane Technology Centre)

Forward osmosis (FO) process is playing an increasingly important role in the desalination and wastewater treatment. Recent studies proposed that FO actually has higher fouling propensity, which is masked by self-mitigation effects when water flux decline. Moreover, water flux alone is insufficient to be a sensitive fouling indicator since it can be influenced by several factors apart from fouling. In this regard, the ultrasonic time domain reflectometry (UTDR) method was for the first time applied in fouling detection of FO in real-time manner. The signals are compared, transformed and analysed quantitatively. There are main conclusions: 1. The fluctuation of ultrasonic signal under foulants-free and hydraulic pressure-free channel were tested as 0.15% and 2.10% in 10 mins with respect to the arrival time and amplitude, respectively, and the values served as the threshold of reliable fouling indicators. 2. The amplitude and arrival time shift of selected peaks were adopted to differentiate the formation of cake layer and pore-clogging process, respectively. The amplitude of the first main peak was increased by 59.82±8.78% and by 0.72±6.05% at the conclusion of fouling process that lasted for 2 hrs on AL-DS mode and AL-FS mode, respectively. The trend of arrival time of the peak is also opposite in different membrane orientations. 3. On AL-FS mode, the cake layer formed on CTA membrane calculated through UTDR was 32.0±4.5 μm, which was in agreement with the thickness of 27.4±0.5 μm obtained via mass balance at end of fouling process. On AL-DS mode, the delay of amplitude enhancement on peak amplitude revealed a transition process of silica from fluid to glassy state when colloidal destabilization was prompted by severe ICP due to pore-clogging process. 4. Compared to CTA membrane, less symmetric and more porous TFC membrane imposed challenges on quantitative analysis of ultrasonic signal due to scattering and attenuation effect.

MONDAY 4:00 pm - Ballroom 4

9e - Incorporating membrane deformation into the boundary layer equation to model water and reverse salt flux in osmotic processes

Jaime A Idarraga-Mora (Clemson University)*, Alton O'Neal (Clemson University), Morgan Pfeiler (Clemson University), David Ladner (Clemson University), Scott Husson (Clemson University)

Osmotic processes (OP) involve the contact of two solutions with different solute concentration across a membrane interface to generate a water flux. OP utilize spacers with large openings to reduce detrimental concentration polarization. When hydrostatic pressure (HP) is applied in OP, such as in pressure-retarded osmosis (PRO), membranes undergo a mechanical deformation as they are compressed against the spacer. This deformation causes an increased reverse salt flux (RSF) as the HP increases. We developed a model that incorporates membrane deformation into boundary layer equations to describe water and RSF in OP. We deformed three commercial thin-film composite membranes using an Instron machine in creep test mode with the goal of correlating linear membrane strain with transport properties obtained via direct-flow permeation. We performed cross-flow PRO flux measurements and used literature data to train the model for predicting changes in RSF with increased HP. Using laser microscopy we were able to pinpoint the area on the membrane with increased deformation, and to measure the deformation profile. Finally, we correlated our PRO data with the opening size of feed spacer used in each experiment. Results showed that membrane deformation increases with feed spacer opening size at constant HP. The linear membrane strain reaches values above 15%, which are greater than the reported strain-to-fracture values for these membranes. RSF was not statistically different when measured using increasing and decreasing HP profiles, suggesting that membrane deformation is partially reversible. Best fits were obtained using a model that incorporated diffusion through an intact polymer, deformed polymer, and fracture sites. These results indicate that deformations in more than 2% of the membrane area lead to exponential increments in the RSF as HP increases. This work introduces membrane deformation in the design equations of membranes and modules for osmotic processes.

MONDAY 4:30 pm - Ballroom 4

9f - A Thin-Film Forward-Osmosis Membrane Made from a Graphene Oxide-Silver Metal-Organic Framework Nanocomposite: Antifouling and Antibiofouling Assessment

Mostafa Dadashi Firouzjaei (University of Alabama), Ahmad Arabi Shamsabadi (Drexel University), Mohammad Sharifian Gh (University of Virginia), Milad Rabbani Esfahani (University of Alabama)*, Ahmad Rahimpour (Babol Noshirvani University of Technology), Masoud Soroush (Drexel University)

Fouling and biofouling are still issues in thin-film nanocomposite (TFN) membranes. This paper describes a TFN membrane fabricated from a novel nanocomposite based on graphene oxide (GO) and a silver metal-organic framework (Ag-MOF). The TFN membrane has a more negative surface charge, higher hydrophilicity, and higher water permeability compared with its thin film composite (TFC) counterpart. Fluorescence imaging revealed that the GO-Ag-MOF TFN membrane kills E. coli more than the Ag-MOF TFN, GO TFN, and pristine TFC membranes by 16, 30, and 92%, respectively. Forward osmosis experiments with E. coli and sodium alginate solutions showed that the GO-Ag-MOF TFN membrane by far has the lowest water flux reduction among the four membranes, proving the exceptional anti-biofouling and antifouling properties of the GO-Ag-MOF TFN membrane.

MONDAY 2:00 pm - Kings Garden 4

10a - Tannic acid-iron network based green technique for enhanced membrane performance in water reuse

Hao Guo (The University of Hong Kong)*

Membrane-based water reuse technologies has arisen as a promising solution to address water scarcity. Currently, reverse osmosis (RO) and nanofiltration (NF) membranes are often used to remove solutes including dissolved ions from water for potential reuse. However, RO and NF are still not adequate enough to retain some organic micropollutants such as endocrine disrupting compounds (EDCs) (e.g., rejection rate as low as 50%), which is significantly jeopardizing the safety and quality associated with water reuse. We first reported a rapid and green techniques using tannic acid (TA)-iron (Fe) coordination networks to improve membrane rejection of organic micropollutants. Various molar ratios of TA-Fe were investigated to form a coating layer on a polyamide-based NF270 membrane. The coating process could be completed within 2 minutes as confirmed by quartz crystal microbalance. XPS analysis indicated the successful coating of TA-Fe on the membrane surface. The coated membrane with TA-Fe molar ratio of 1:3 (i.e., NF270-Fe3) showed the greatest enhanced rejection of NaCl. The rejection of neutral and hydrophilic molecular probes was used to evaluate membrane pore size change. The results showed that NF270-Fe3 had a reduced average pore radius of 0.40 nm compared to that of NF270 membrane (0.44 nm), suggesting the enhanced effect of size exclusion. The TA-Fe coating significantly improved membrane rejection of EDCs and antibiotics, while the highest rejection was achieved by NF270-Fe3. Compared to base NF270 membrane, NF270-Fe3 gave 2-3 times enhancement of rejection for EDCs (i.e., 4-5 times reduction for EDCs permeability) and one order of magnitude reduction of antibiotics permeability. Such a rapid and effective coating technique using green polyphenol-metal network assembly enable TA-Fe being a promising material for large-scale applications.

MONDAY 2:30 pm - Kings Garden 4

10b - Membrane Distillation (MD) and Reverse Osmosis (RO) processes for water recovery from pre-treated high strength brewery wastewater

Nawrin Anwar (Concordia University)*, Mahbuboor Choudhury (Concordia University), Saifur Rahaman (Concordia University)

Water recovery and reuse in the industrial sectors through advanced wastewater treatment can be effective means to address the water shortage issue as well as to reduce the wastewater load to the environment. Membrane-based separation processes can be efficient technology for producing freshwater for reuse within the industry thus combating the major challenge of freshwater scarcity. Beer production at brewery industries results in the discharge of a high quantity of wastewater (~3 to 10 L wastewater for 1 L of beer production) containing sugars, ethanol, soluble starch, total suspended solids, and volatile fatty acids. This study aims at investigating the potential of membrane-based separation processes (MD and RO distinctly) for pre-treated brewery wastewater for water recovery and possible reuse. Pre-treated brewery wastewater effluent with a reduction of the substantial amount of organics and total suspended solids was directed to MD and RO separately for enhanced water qualities for possible reuse within the brewery industry. The commercial membranes, implemented in industrial wastewater treatment, were used to evaluate the water recovery in MD, and RO processes through assessing the permeate flux decline, and contaminant rejections. The fouling characterization was observed by the surface morphology, and chemical composition of the membrane surfaces before and after the water recovery operation. Results revealed that both the processes could achieve high removal of organics (>94% COD removal), and nutrients (>99% TN and ~100% TP removal). Lower water flux drop, and higher water recovery were obtained by MD process compared to the RO process for 80 hours operation indicating the potential application of MD as an advanced treatment step for water recovery in food industries with high organic loadings.

MONDAY 3:00 pm - Kings Garden 4

10c - Field Demonstration of Real-Time Colloidal Particle Monitoring to Improve MF and UF Membrane Performance

Jana Safarik (Orange County Water District)*, Ganesh Rajagopalan (Kennedy/Jenks Consultants), Helia Safaee (Kennedy/Jenks Consultants), Megan Plumlee (Orange County Water District), Zita Yu (West Basin Municipal Water District)

Low-pressure membrane filtration processes provide reliable water quality for water reuse but are susceptible to fouling. Membrane fouling is attributed to deposition of colloidal particles (<200 nm) inside the membrane pores. Currently, no online techniques are available to directly measure colloidal particles in the feed water, which would facilitate appropriate pretreatment and real-time control. In this study, a real-time monitoring technology was implemented to track colloidal particles concentration and size distribution in the feed water to facilitate their targeted removal via coagulation. The mitigation strategy used Nanoparticles Tracking Analysis (NTA) (NanoSight NS500 by Malvern Instruments) and was tested at two sites: Orange County Water District (OCWD) and West Basin Municipal Water District (WBMWD). The fouling potential for the OCWD microfiltration (MF) pilot receiving targeted polyaluminum (PACl) coagulation was mitigated, with TMP values reduced by 60% after six-weeks compared to the control (no coagulant). Economic evaluation for OCWD's full scale advanced purification plant (100-MGD) indicated that the proposed approach can reduce energy consumption due to microfiltration by 28-35%, resulting in 3 GWh/year of projected energy savings, as well as $519,700/year of potential savings in membrane cleaning, replacement and operating costs. The ultrafiltration (UF) pilot demonstration at WBMWD targeted PACl coagulation slowed the rate of TMP increase by 43% compared to control and achieved a 63% reduction in energy consumption during pilot operation. This could translate to over 600 MWh/year of energy savings for a 14.4-MGD UF facility as well as potential savings of over $110,000/year in overall operating and maintenance costs. However, coagulant choice plays a critical role in the success of the targeted pretreatment strategy, as targeted FeCl3 dosing was not equally successful.

MONDAY 3:30 pm - Kings Garden 4

10d - Selective removal of phosphorus from wastewater using electromembrane process

Sanhita Chaudhury (Ben-Gurion University of the Negev)*, Oded Nir (Ben-Gurion University of the Negev)

With increasing demand for wastewater recycling, stricter regulations for phosphorus (P) concentration in the wastewater effluents are being placed, to ensure sustainable discharge and reuse. This requires removing P from effluent as a polishing step, which is mostly done by chemical precipitation, adsorption or membrane technology. However, precipitation and adsorption processes generate high amounts of waste and require substantial amount of chemicals to regenerate sorbent. In membrane-based processes, selectivity for P is low and liquid waste production can be substantial. However, electromembrane-based processes synergistically integrates both adsorption and membrane separation in a manner that can overcome the inherent disadvantages of both methods. With this in view, the present work aims to develop and test membrane-sorption processes for highly selective and efficient removal of P from effluents having low P concentrations. For this purpose, iron oxyhydroxides (hence refereed as FeO) nanoparticles modified membranes were prepared using a two-compartment permeation method, in which FeO nanoparticles were grown in the pores of commercially available polycarbonate track-etched membrane (0.2 μm dia). SEM and XRD methods confirmed the presence of amorphous FeO nanoparticles within the pores of the composite membrane (FeOm). The SEM also indicated that the cylindrical pores of the membrane were completely filled by the FeO nanoparticles (~50-60 nm), which was reflected in the extreme reduction in the water flux through the FeOm as compared to the corresponding pristine membrane. Due to the amorphous structure and nanoscale size of the FeO particles, high P uptake capacity (0.5 mmol/g) and large distribution coefficient (logKd = 5.5) was obtained. These promising results open interesting routes for chemical free electro-assisted regeneration of the sorbent and for electrodriven selective transport of phosphorus based anionic species.

MONDAY 4:00 pm - Kings Garden 4

10e - Water Recovery from Produced Water via Robust Membrane Distillation

Rong Wang (Nanyang Technological University)*

With many production activities, a substantial volume of produced water is generated annually in the oil and gas industry, making it one of the most water-intensive industries. Managing produced water has become a consequential part of the sustainable development of the oil and gas industry due to stringent environmental legislations, severe strain on water resources, and escalating costs of wastewater disposal. Direct-contact membrane distillation (DCMD) holds potential for the recovery of high quality water from produced water. However, membrane fouling and pore wetting are challenging issues for long-term DCMD operations due to the presence of low surface tension species in these waters. Thus, developing robust membranes with anti-fouling and anti-wetting properties for effective produced water treatment is very important. In this presentation, we report our efforts in making novel composite hollow fiber membranes based on co-deposition of polydopamine (PDA)/polyethylenimine (PEI) onto the outer surface of a commercial hydrophobic polyvinylidene fluoride (PVDF) substrate. The successful co-deposition was verified using different characterization techniques. The composite membrane is in-air hydrophilic and underwater superoleophobic that can prevent organics adhesion. The performance of the membrane was investigated via DCMD experiments by feeding a series of low surface tension solutions. The results suggest that the newly developed composite PVDF membrane could potentially be used for long-term water recovery from produced water via DCMD.

MONDAY 4:30 pm - Kings Garden 4

10f - Potable-quality water recovery from primary effluent through an integrated algal-osmosis membrane system

Lu Lin (New Mexico State University)*, Wenbin Jiang (New Mexico State University), Pei Xu (New Mexico State University)

A dual algal-osmosis membrane treatment system was studied for the recovery of potable-quality water from primary effluent of a municipal wastewater treatment plant. The core components of the integrated system include a mixotrophic algal process for removal of biochemical oxygen demand (BOD) and nutrients from the primary effluent, followed by a hybrid forward osmosis (FO)-reverse osmosis (RO) system for separation of biomass from the algal effluent and production of potable-quality water. In contrast to traditional wastewater treatment systems, mixotrophic metabolism of the algal system removes BOD and nutrients from the primary effluent through a single-step treatment and eliminates the input of aeration energy and supplemental carbon for denitrification. The hybrid FO-RO system was operated in batch mode and reached water productivity of 1.57 L/m^2-h in FO using seawater as draw solution; 3.50 L/m^2-h in brackish water RO (BWRO) and 2.07 L/m^2-h in seawater RO (SWRO) at 300 psi. Overall, the integrated algal-osmosis membrane system achieved complete removal of ammonia, fluoride, and phosphate; over 90% removal for calcium, sulfate, and organic carbon; 86-89% removal for potassium and magnesium. High resolution mass spectrometry and fluorescence excitation-emission matrix (FEEM) spectrophotometry were used to study the transport and fate of the organic contaminants in the algal and membrane processes. Broadband characterization using the high resolution mass spectrometry revealed broad removal of organic compounds, particularly wastewater surfactants upon algal treatment. Product water generated by the integrated system met both primary and secondary drinking water standards. Cleaning by NaOH solution was found to be effective in removing organic- and bio-fouling of the FO membrane and restoring membrane performance during longer-term algal wastewater filtration (700 hours) and at a water recovery of 90%.

MONDAY 2:00 pm - Kings Garden 1

11a - Ion Transport Through Perforated Nanoporous Graphene

Mandakranta Ghosh (University of Twente)*, Koen F. A. Jorissen (University of Twente), Jeffery A. Wood (University of Twente), Rob G. H. Lammertink (University of Twente)

In this work, we investigate the ion transport through a monolayer perforated graphene membrane. We use ion-beam bombardment to create nano pores in graphene supported on a polymer foil in order to study ion-transport across a single-atom thick charge-selective interface (Madauß et al., 2017). The ion selectivity of these 2D membranes are studied with respect to both monovalent and multivalent salts, by measuring the induced potential between two different concentration electrolyte compartments separated by the nanoporous graphene. Our experimental findings remarkably showed that the total electrical potential across a single-layer graphene membrane can be described as a cation-selective layer using a modified version of the Teorell-Meyer-Sievers (TMS) relationship (Ghosh et al., 2018). At low electrolyte concentrations, the potential enters into a Donnan regime where the potential shows a plateau vs. concentration at a fixed concentration ratio between compartments. At high salt concentrations, the potential is related to the induced diffusion potential based on different ionic mobilities of cation and anion, where for single-layer graphene the diffusion coefficients are consistent with the values in bulk solution which is atypical for ion-exchange membranes. TMS is derived for dense, finite width charge-selective interfaces, however the basic features of Donnan and diffusion-dominated regimes are observed for our single-atom thick membrane. This investigation reveals some of the fascinating features of ion-transport across 2D membranes and opens up exciting opportunities for further understanding of transport in such atomistically thin interfaces.

MONDAY 2:30 pm - Kings Garden 1

11b - Developing a new approach to describe ion sorption and transport in Nafion membranes

Rahul Sujanani (The University of Texas at Austin)*, Jovan Kamcev (University of California, Berkeley), Eui-Soung Jang (The University of Texas at Austin), Benny Freeman (The University of Texas at Austin), Donald Paul (The University of Texas at Austin)

Ion exchange membranes (IEMs) are widely used in water purification applications, such as reverse osmosis and electrodialysis and are also applied and explored for energy generation applications, such as fuel cells or reverse electrodialysis. IEMs are composed of charged polymers that contain charged moieties bound to the polymer backbone. The presence of these charged groups has significant impact on ion and water transport through these materials and allows for selective permeation of ionic species, making them attractive for the separations required for water and energy applications. However, a detailed understanding of the impact of polymer structure and morphology on ion and water transport properties in IEMs remains largely incomplete. We recently developed a theoretical framework, consistent with thermodynamics, to calculate, and in some cases predict, fundamental ion transport properties in charged polymers. We validated this framework via careful, systematic measurements of water and ion sorption, permeability, and ionic conductivity. We are now extending this framework to Nafion, a perfluorinated sulfonic acid polymer. Nafion has a significantly different chemical structure and morphology than the previous IEMs we have studied. Due to Nafion's exceptional ionic conductivity and robust mechanical and chemical stability, it is considered the benchmark material for fuel cells. The fundamental origins of these unique properties are often debated in the literature. Results for NaCl and $MgCl_2$ sorption in Nafion show that our framework describes the experimental data remarkably well. Transport measurements were also performed, showing similar behavior reported in the literature, and efforts are underway to model these properties. A key goal of these studies is to identify and elucidate the molecular features of IEMs that can be used to rationally design materials with rapid and selective ion transport.

MONDAY 3:00 pm - Kings Garden 1

11c - Co-transport of neutral solutes in Nafion 117 using In-situ ATR FTIR spectroscopy for multicomponent solution speciation

Breanna Dobyns (Auburn University)*, Bryan S Beckingham (Auburn University)

Membranes are widely utilized in many different applications: biological, water treatment, energy production, etc. Transport of species in dense polymeric membranes is described by the solution-diffusion model. For multiple solutes transport behavior has typically been quantified via diffusion cell experiments with ex-situ quantification of downstream concentrations which requires time-consuming aliquot analysis and introduces error through constantly decreasing downstream volume which increases calculation difficulty and introduces hydrostatic pressure issues. Therefore, single component permeability values are generally used to select a membrane for a particular application. Herein, we employ an in-situ attenuated total reflectance Fourier-transform infrared (ATR FTIR) spectrophotometer probe that enables characterization of downstream concentrations for multiple species every minute without the need for aliquotic sampling. We leverage this methodology to investigate emergent transport phenomena for methanol, ethanol, propanol, acetone and combinations thereof through Nafion 117. We find differences between the generally used theoretical selectivity (using single-component permeability values) and the true selectivity (using the multiple component permeability values) of up to 22%. The solution-diffusion model ($Pi = Ki \times Di$) quantifies changes in permeability, Pi, to be either thermodynamically driven (solubility, Ki) or kinetically driven (diffusivity, Di). Nuclear magnetic resonance (NMR) spectroscopy was used to determine the solubility of the solutes within the membrane via a commonly used adsorption/desorption technique. Once these values are known, the changes in permeabilities/selectivities are shown to be either kinetically or thermodynamically driven. This enables better understanding of transport behavior for multiple components within hydrated, dense polymer membranes.

MONDAY 3:30 pm - Kings Garden 1

11d - Functionalized Nanoporous Ceramic Membranes Towards Low-Cost Electrodialysis

Gregory M Newbloom (Membrion, Inc.)*, Ryan Kingsbury (Membrion, Inc.), Rachel Malone (Membrion, Inc.)

The State of California has experienced annual snow pack depletion and persistent drought that is impacting agriculture, industry and day-to-day life. They are not alone as many parts of the world face a fresh water crisis. Interestingly, California, for example, has 38 times more brackish water than fresh water. Electrodialysis is known to be one of the most efficient technologies for purification of brackish water and yet it is often not considered as a leading solution to the fresh water crisis. This is in part due to the cost and performance of the ion exchange membranes (IEM). IEMs suffer from high cost, high resistance, poor durability and other problems. In this presentation, we discuss the design and scale-up of novel, functionalized ceramic IEMs that are ion selective, experience near-zero swelling and are inexpensive at 30% of the production cost of current IEMs. In this work, we investigate the use of a surface-modified inorganic membrane created using sol-gel processing of siloxanes without calcination or sintering. Silica is an advantageous material for membranes because of its excellent durability, extremely low cost and lack of swelling in water. Membrane performance can be tuned by size exclusion (i.e., sub-5nm nanometer pores) and surface functionalization (i.e., silane chemistry) to achieve promising anion and cation-selective performance. Permselectivity and water permeance of these membranes are shown to be superior to an industry standard Neosepta, while area specific resistance has room for improvement. Area-normalized ion exchange capacities and accelerated degradation tests show nearly identical performance between the functionalized ceramic and Neosepta membranes. Design strategies for continued optimization will also be discussed. These promising preliminary results highlight the commercial potential for a new, inexpensive IEM with the potential to elevate the status of electrodialysis as a leading solution to the fresh water crisis.

MONDAY 4:00 pm - Kings Garden 1

11e - Polymerization of counterions in self-assembled, 1 nm pores of lyotropic liquid crystal anion exchange membrane to tune nanopore and ion transport properties

Michael J. McGrath (University of Colorado at Boulder)*, Samantha Hardy (University of Colorado at Boulder), Andrew Basalla (University of Colorado at Boulder), Hans Funke (University of Colorado at Boulder), Bryce Manubay (University of Colorado at Boulder), Zhangxing Shi (University of Colorado - Boulder), Douglas Gin (University of Colorado - Boulder), Richard Noble (University of Colorado - Boulder)

The well-defined 1 nm pores of a phase-segregated anion exchange membrane (AEM) made from a cross-linked, self-assembled liquid crystal are functionalized by in-pore polymerization of acrylate-based counter-ions. The resulting anionic polymer is trapped inside the pore and provides a route towards discrete pore modification that can significantly change the material's ion exchange and ion transport properties. After polymer formation inside the pores, the membrane loses its ion exchange capacity but the pore environment remains ionic, as evidenced by water and salt sorption measurements. Single-salt diffusion dialysis chloride/sulfate selectivities are > 100 in both dilute (0.05M) and concentrated (0.5M) solutions for the pore-modified membranes, whereas the unmodified membrane selectivities are ≤ 1 at the same concentrations. This work represents an exciting example of an ion selective membrane that relies on mechanisms other than Donnan exclusion, such as size and hydrophobicity, to achieve selectivity. Also, this modification route holds promise in future work to discretely design the pore size of the membrane for targeted nanofiltration separations.

MONDAY 4:30 pm - Kings Garden 1

11f - Quantifying tortuosity and diffusion over multiple scales in ionic and non-ionic polymer membranes

Louis Madsen (Virginia Tech)*

Self-organized polymer membranes conduct mobile species (ions, water, alcohols, etc.) according to a hierarchy of structural motifs that can span sub-nm to > 10 microns in scale. In order to understand transport in such materials, our group combines multiple types of dynamics and transport measurements (NMR spectroscopy, diffusometry, relaxation, imaging) with X-ray and neutron scattering, molecular dynamics simulations, and microscopy. We gain perspective on these combined data assisted by theories of porous media, electrolytic transport, and oriented matter. In this presentation, I will focus on the quantification of tortuosity in polymer membranes, meaning the restriction of pathways for diffusion or flow. I will discuss the measurement of tortuosity separately averaged over two distinct length scale ranges - from nanometer to bulk and from micron to bulk. We follow these multi-scale tortuosity values as a function of water uptake and polymer composition. I will present NMR diffusometry measurements of these two quantities in representative systems of hydrophilic-hydrophobic copolymer membranes, including proton-exchange membranes, hydroxide-exchange membranes, and all-neutral membranes. Using such knowledge, our group seeks to give precise feedback on the design of materials for use in, e.g., proton and hydroxide fuel cells, lithium batteries, and water desalination.

MONDAY 2:30 pm - Kings Garden 3

12a - Expanding the chemical palette for reliable chemical separations using molecular simulations and machine learning

David Sholl (Georgia Institute of Technology)*, Farhad Gharagheizi (Georgia Institute of Technology), Dai Tang (Georgia Institute of Technology)

The majority of research literature on chemical separations focuses on a relatively small collection of chemicals. In many ways this is appropriate because of the dramatic economic and environmental impacts of these species. The space of possible chemicals, however is vast; Carl Sagan's famous "billions and billions" is many orders of magnitude too small for chemical space. I will discuss early steps towards methods that may eventually allow rapid development of separations using nanoporous materials for a diverse range of molecules drawn from a broad chemical space.

MONDAY 2:00 pm - Kings Garden 3

12b - Diffusion, Reaction, and Network Structure Modeling of Interfacial Polymerization of Polyamide Membranes

Jeffrey D Wilbur (DuPont Water Solutions)*, Toni Bechtel (DuPont Water Solutions), Matthew Jansma (DuPont Water Solutions), Steve Rosenberg (DuPont Water Solutions), Dan Arriola (DuPont Water Solutions)

Polyamide membrane chemistry and network structure in reverse osmosis and nanofiltration membranes are a product of simultaneous diffusion and reaction of monomers during interfacial polymerization and are dependent on both feed solution compositions and process conditions. This process has been modeled using various approaches in order to explore the relationship between process inputs, membrane chemical features, and ultimately performance. We present an approach to combine a diffusion-reaction model similar to that described previously by V Freger and a simple statistical model of network structure related to polymerization theory developed by DR Miller and CW Macosko. Predictions suggested by this model are compared with experimental membrane structure data.

MONDAY 3:00 pm - Kings Garden 3

12c - Optimization of Mixed Matrix Membrane Materials for Post-Combustion Carbon Capture

Janice A. Steckel (U.S. Department of Energy National Energy and Technology Laboratory)*, Samir Budhathoki (U.S. Department of Energy National Energy and Technology Laboratory), Miguel Zamarripa-Perez (U.S. Department of Energy National Energy and Technology Laboratory), Paul Boone (University of Pittsburgh), Christopher Wilmer (University of Pittsburgh)

Highly selective and permeable membranes have the potential to allow for economically viable, membrane-based post-combustion CO_2 capture. While pure polymer membranes have a long history of usefulness in gas separations, they are characterized by a trade-off between permeability and selectivity known as the Robeson upper limit. Mixed matrix membranes (MMMs) are formed by incorporating inorganic porous nanoparticles, such as metal organic frameworks (MOFs), into a polymer membrane. Some MMMs have been shown to have selectivity and CO_2 permeability that are dramatically improved with respect to the parent polymer. Recently, we demonstrated that computations can accurately describe the performance that is observed when a wide variety of MOFs and polymers are combined in MMMs. In the current work, we use high-throughput atomistic simulations to calculate gas permeability in a large number of MOF structures. Monte Carlo simulations are used to calculate gas solubility and molecular dynamics simulations are used to calculate gas diffusion. These data are combined with experimental data for several polymers to predict the gas permeability of a large number of hypothetical MMMs using a predictive permeation model for the composite material. A multi-stage membrane system for carbon capture was optimized together with the material chosen for the membrane units. A mixed integer nonlinear programming model was used to determine the optimal design and operation of the post-combustion carbon capture plant. The cost of electricity was minimized as a function of plant configuration, process conditions and material properties. This computational screening and optimization study allows us to describe the features of the MOFs and the polymers that contribute to the best membranes, and to optimize the materials together with the process configuration and conditions for membrane-based CO_2 capture.

MONDAY 3:30 pm - Kings Garden 3

12d - Diffusivity in anion exchange membranes under low hydration: A molecular dynamics study

Dario Dekel (Technion - Israel Institute of Technology), Israel Zadok (Technion - Israel Institute of Technology), Srdan Pusara (Technion - Israel Institute of Technology), Simcha Srebnik (Technion - Israel Institute of Technology)*

Chemical degradation of cationic groups in anion exchange membranes (AEMs) has been recently shown to be significantly influenced by the hydration level at which the membrane operates. When the AEM is poorly hydrated, the cationic functional sites are more susceptible to attack by the diffusing hydroxide anions. Recent experiments on AEMS functionalized with various quaternary ammonium derivatives have shown significant decomposition of the membranes at low hydration. Using molecular dynamics simulations, we explore the behavior of common quaternary ammonium cations at very low hydration, focusing on solvation of hydroxide and its relation to cation-anion interactions. Diffusivity under different levels of hydration is shown to follow two distinct regimes. Analysis of the hydrogen-bonded water-hydroxide network within the different regimes reveal a unique water-bridged double-hydroxide charged complex that predominates at low hydration. These stable structures are distinct from previously reported structures under hydrated conditions, and presumably disrupt the hydrogen bonded network required for structural diffusion of hydroxide ions through water.

MONDAY 4:00 pm - Kings Garden 3

12e - Modeling of polymer membrane formation via phase inversion by mesoscopic phase-field methods: Investigating the development of structure on multiple scales

Michael R Cervellere (University of Arkansas/EMD Millipore)*, Xianghong Qian (University of Arkansas), David Ford (University of Arkansas), Paul Millett (University of Arkansas)

Phase inversion, a process that occurs when a polymer collapses from solution into a thin film upon thermal quenching or exposure to another phase, is the most widely used process for producing porous polymer membranes. Modeling of this process has been mostly limited to the continuum level and empirically-derived parameters, with little theoretical basis to connect the process conditions and important characteristics of membrane microstructure. We have developed a large scale three-dimensional mesoscopic simulation model to examine the dynamics of the phase inversion process and predict the morphology of pore structures that develop within polymer membranes. The computational approach is a phase-field method based on the Cahn Hilliard equation, which incorporates the basic thermodynamic and transport parameters relevant to phase separation in polymer solutions. The thermodynamics of mixing is described by the Flory-Huggins free energy of mixing for binary solutions, and a gradient term is included to govern the formation of interfaces between phases. The transport processes are assumed to be diffusive in nature and are dependent on the local temperature and polymer concentration. We present results for thermally induced phase separation (TIPS) processes using material parameters, polymer volume fractions, and quench rates that are relevant to industrial systems. Integrally skinned membranes can be formed via TIPS and controlling the morphology of the skin layer is of great industrial interest. The model is able to capture the formation of a dense skin layer and changing relevant parameters such as coagulation bath temperature and polymer weight % of the dope has a significant effect on thickness and pore size. This talk will focus on initial results of the Cahn Hilliard model and compare simulated structures and morphologies to current manufactured polymer membrane filters.

MONDAY 4:30 pm - Kings Garden 3

12f - Multiscale Modeling of Gas Permeation Through Poly(Dimethyl Phenylene Oxide) (PPO) and Its Pre-Steady State Polymer Response

Marielle Soniat (Lawrence Berkeley National Laboratory)*, Meron Tesfaye (Lawrence Berkeley National Laboratory), Daniel Brooks (CalTech), Nicholas Humphrey (USC), Lien-Chun Weng (Lawrence Berkeley National Laboratory), Boris Merinov (CalTech), William Goddard (CalTech), Adam Weber (Lawrence Berkeley National Laboratory), Frances Houle (Lawrence Berkeley National Laboratory)

In gas separations, glassy polymers are often used because of their high selectivity. However, in applications involving carbon dioxide or other small organic molecules, the interaction between the permeant and the polymer can increase permeability of the membrane over time or at high pressure, known as plasticization. This work reports a combined experimental-theory-multiscale modeling investigation of the transport of two gases, CO_2 and N_2, through the glassy polymer poly(dimethyl phenylene oxide) (PPO) to better understand the relationship of glassy polymer structure and dynamics with gas transport for both steady and pre-steady state conditions. The multiscale model consists of physically-based coupled reaction-diffusion kinetics and is solved using a stochastic algorithm, which allows for dynamic swelling of the membrane in the presence of CO_2. The model is validated against experimental data on the pre-steady state and steady state downstream pressure increase over time. The simulations show that polymer response differs for CO_2 and N_2 even at very short times and therefore affects the time-dependent permeation process. The connection between these molecule-polymer interactions and plasticization of the polymer during use will be discussed.

TUESDAY 9:30 am - Kings Garden 4

13a - Scaling mitigation in membrane distillation through electrokinetic mixing

Unnati Rao (UCLA)*, Guy Z Ramon (Technion - Israel Institute of Technology), Eric Hoek (UCLA), David Jassby (UCLA)

A major challenge faced by membrane-based desalination processes is the growth of mineral crystals on the membrane surface (scaling). Scaling restricts the flow of water through the membrane and can cause physical destruction of the membrane material. Scaling occurs when supersaturation conditions develop along the membrane surface, a process known as concentration polarization. The depletion of water along the membrane due to its passage to the permeate side leads to the formation of a concentration polarization layer. To mitigate scaling, concentration polarization must be minimized by limiting water recovery, encouraging turbulent conditions and the use of anti-scalants. Here, we report on an energy efficient technology that uses electrokinetic mixing to nearly eliminate calcium sulfate and silicate scaling in membrane distillation. Electrically conductive membranes are prepared by depositing a layer of carbon nanotubes (CNT) on a polypropylene support and crosslinking with polyvinyl alcohol. The application of low potentials (2V cell potential) to the membrane had a dramatic impact on scaling, with the impact being dependent on the frequency of the potential applied. Theoretical calculations reveal that the concentrations of ions along the membrane surface respond rapidly to the applied electrical potentials. We hypothesize that frequent shifts in membrane polarization induce electrokinetic mixing in the stagnant boundary layer and reduce available time for crystal nucleation, shifting the system from rapid heterogeneous nucleation on the membrane surface to slow homogeneous nucleation in the bulk fluid, which nearly eliminated membrane scaling.

TUESDAY 10:00 am - Kings Garden 4

13b - Donnan dialysis desalination with thermolytic salts

Hanqing Fan (Columbia University)*, Ngai Yin Yip (Columbia University)

Donnan dialysis (DD) is an ion-exchange membrane (IEM) process that utilizes the Donnan equilibrium principle to drive ion transport. In DD, the IEM separates two solutions with different counterions. Unlike conventional electrodialysis, no external electric current is applied. The IEM allows the transport of counterions in both directions but suppresses co-ion permeation, resulting in the redistribution of counterions until the same equilibrium membrane potential is reached for all different counterions, i.e., co-ions are retained but counterions are exchanged. Here, we proposed a novel desalination technology that couples DD with thermolytic salts and utilizing low-grade heat as energy input. Brackish feed water simulated by NaCl solution and working solution of NH4HCO3 flow through a cation exchange membrane (CEM) and then an anion exchange membrane (AEM) sequentially. In the CEM cell, Na+ in the feed migrates to the working solution chamber while NH4+ permeates in the opposite direction. On both sides, anions are retained. Similarly in the AEM cell, Cl- and HCO3- are swapped while cations remain. The next result is the substitution of NaCl with NH4HCO3. Heating the two outlet streams volatize the NH3 and CO_2, thus removing NH4HCO3 to yield desalinated product water and concentrated brine. The separated NH3 and CO_2 are then recycled to reform the working solution. The concept is experimentally validated by the desalination of brackish water simulated by 100mM NaCl feed to 17mM (i.e., WHO drinking water standard). We further present an analytical model for DD desalination. The results indicate the leakage of co-ions and uncharged species as the major inefficiencies. The research demonstrates a new desalination strategy that utilizes Donnan dialysis with low-grade heat sources, such as waste heat. On a broader perspective beyond desalination, this approach opens up new opportunities to tap low-grade heat for general ion separation.

TUESDAY 10:30 am - Kings Garden 4

13c - Electrically Conductive Membranes as Sensors: Stability, Conductivity, and Sensitivity

Nan Zhang (McMaster University), Amin Halali (McMaster University), Charles-François de Lannoy (McMaster University)*

Polymeric membranes are susceptible to various forms of fouling including organic, colloidal, and biofouling, as well as mineral scaling. The onset of fouling is most often monitored by changes in transmembrane pressures drop. The type of fouling is elucidated from feed water composition and in destructive membrane autopsies. The ability to monitor membrane fouling in-situ would enable immediate and finely tuned anti-fouling responses. This work focuses on electrically conductive membranes as fouling sensors. The presentation will also provide an overview of our recent conductive membrane chemistries, their stability, electrical conductivity, and production of flat sheet and hollow fiber configurations. To develop fouling sensors on membranes, we used electrochemical impedance spectroscopy (EIS) on the surface of electrically conductive membranes to precisely measure the onset of surface fouling. We fabricated an electrochemical cell containing an electrically conductive membrane as the working electrode, a graphite electrode as the counter electrode, and a NaCl solution as electrolyte. Conductive membranes were fabricated by coating PES and PVDF flat sheet and hollow fiber membranes with single-walled/double-walled CNT (SW/DWCNT) solutions. Membrane fouling was simulated by pressure depositing layers of latex beads onto the surface of the membrane. Changes in membrane permeance were used to correlated the extent of latex fouling, i.e. different thicknesses of latex fouled layers caused reductions in membrane permeance. These small changes in latex coverage on membranes were resolved with EIS quantified as Nyquist plots. Direct mechanistic interpretations of impedance impacts are possible and we have developed relevant equivalent circuits to model these systems. This research demonstrates that electrically conductive membranes are an ideal platform to use EIS to sense the onset and development of surface fouling.

TUESDAY 11:00 am - Kings Garden 4

13d - Electrocatalytic Membrane Reactor-based Integrated Processes for Landfill Leachate Treatment

Jianxin Li (Tianjin Polytechnic University)*, Hong Wang (Tianjin Polytechnic University), Junhao Ding (Tianjin Polytechnic University)

Treatment of nanofiltration (NF) retentate from landfill leachate is a significant challenge due to its poor biodegradability and high toxicity. One attractive potential alternative is to apply a chemical oxidation processes for pretreatment to convert the initially persistent organic compounds into more biodegradable intermediates, which could then be treated with a biological oxidation process at a considerably lower cost. This study reports the development of a novel integrated process, i.e. Fenton-electrocatalytic membrane reactor (ECMR)-biological aerated filter (BAF) process for the treatment of NF retentate from landfill leachate. The Fenton and ECMR processes were applied prior to the biological process to enhance the biodegradability of the leachate, making it suitable for the subsequent biological process. The results showed that the chemical oxygen demand (COD) removal rate of the effluent obtained from the Fenton process under the conditions of pH=3, molar [H_2O2]/[Fe2+] ratio of 2:1 (H_2O2 dosage of 180 mmol/L) was up to 64.6%. Simultaneously, the biochemical oxygen demand (BOD5) increased from 62 to 138 mg/L. The effluent from the Fenton process was then treated with the ECMR process under the operating conditions: an operating voltage of 2.0 V and a residence time of 10 min. The COD removal rate reached 21.3% and the ratio of BOD5 to COD (B/C) increased from 0.05 to 0.38, indicating that is biodegradable. After 48 hours of BAF treatment, the COD of the effluent declined to 88 mg/L and the total of chroma removal rate reached 99.5%. All the contaminants in effluent can satisfy Discharge Standard of China. In sum, our research opens a new avenue for landfill leachate treatment without concentrates.

TUESDAY 11:30 am - Kings Garden 4

13e - Nitrogen and Phosphorus Recovery from Agricultural Wastewater Effluents - Role of Electrically Conductive Membranes

Avner Ronen (Temple University)*, Kartikeya Kekre (Temple University)

Nitrogen and phosphorus are essential elements needed for agricultural crop fertilization, used in increasing quantities in order to feed an ever-growing population. Production of both nitrogen and phosphorus fertilizers is energy intensive and their demand is expected to increase in the coming decades due to changes in human population, diets, and agricultural practices. Agricultural wastewater from livestock (AWL) farms contains high concentrations of nitrogen and phosphorus in addition to organic loads. Currently, conventional wastewater treatment plants are unable to remove such high concentrations, which require complex treatment stages and reduce the efficiency of the treatment. An important strategy to mitigate phosphorus and nitrogen scarcity is to recycle them from waste streams back into agricultural production. Our research shows the ability of P and N recovery from concentrated and diluted solutions (1-100 mg L^{-1}) of synthetic AWL. The recovered P and N are in the form of Struvite which is considered an "eco-friendly" fertilizer. The recovery is based on an electrochemical reaction near an electrically conducting ultrafiltration membranes (ECM). The electrochemical reduction of dissolved oxygen and water molecules generate a local pH gradient in the close vicinity of the electrode. The pH near the cathode increases, leading to nucleation and growth of Struvite and, importantly, high yield with no chemical additives. As the recovery process utilizes permeate to control mass transfer, diffusion limitations are eliminated, leading to a significant increase in reaction rates and removal ratio. In terms of collecting the recovered Struvite, our ECMs have pores in the range of 10-100 nm which are significantly smaller than the size of Struvite particles, therefore, Struvite solid particles which are filtered by the ECM accumulate on the membrane's surface, to be collected by a pressurized backwashing process or simply blocking the permeate flux.

TUESDAY 12:00 pm - Kings Garden 4

13f - Simulations of Highly Selective Separations Based on Electromigration

Merlin Bruening (University of Notre Dame)*, Andriy Yaroshchuk (Polytechnic University of Catalonia), Mykola Bondarenko (National Academy of Sciences of Ukraine), Chao Tang (University of Notre Dame), Muhammad Ahmad (University of Notre Dame)

In ion transport through membranes, spontaneous or applied electric fields give rise to electromigration that sometimes leads to extraordinary selectivities, particularly when electromigration opposes convective transport. This presentation will explore electric field-induced selectivities in both ion-exchange membranes and in membranes that contain highly charged nanopores. In diffusion dialysis through an anion-exchange membrane, the Cl-/SO42- selectivity increases from 1.3 to 13 on going from a source-phase concentration of 0.005 M to 0.1 M. Simulations suggest that this occurs because electromigration disproportionately decreases SO42- transport due to its 2-charge and high concentration in the membrane. With diffusion dialysis of 0.1 M KCl, 0.1 M $MgCl_2$ through a cation-exchange membrane, concentration profiles and electromigration can even lead to negative Mg2+ fluxes with low ion concentrations in the receiving phase. During application of an electrical current through a membrane containing micro and nanoporous regions, trace-ion accumulation in the microporous layer should vary greatly with the ion diffusion coefficient to allow selective preconcentration of specific ions. The charged nanopores exclude coions (ions with the same charge as the nanopore walls) to deplete the region next to the nanopores of the dominant (high-concentration) salt and give rise to high local electric fields. Trace ions accumulate in the depletion region because convection moves them toward the nanoporous interface, but the opposing electromigration is greatest near the interface. When the trace-ion diffusion coefficients is half of the corresponding value for the dominant salt, calculated preconcentration factors reach 150. Much larger enrichment factors appear for divalent ions due to their relatively low diffusion coefficients.

TUESDAY 9:30 am - Ballroom 3

14a - Oil drop behavior on model nanofiltration membrane surfaces under conditions of hydrodynamic shear

Emily Tummons (Michigan State University), Charifa Hejase (Michigan State University)*, Zhefei Yang (University of Massachusetts Amherst), Jia Wei Chew (Nanyang Technological University), Merlin Bruening (University of Notre Dame), Vlad Tarabara (Michigan State University)

Large volumes of oily wastewater are produced in various industrial operations. Membrane filtration is the most cost-effective technology capable of removing oil droplets smaller than 10 µm, often a prerequisite for meeting environmental regulations. However, membrane fouling by oil limits the broader acceptance of this technology. Despite extensive studies, mechanisms of membrane fouling by emulsified oil remain poorly understood. The blocking laws do not apply to deformable particles such as oil droplets. Conditions that lead to oil accumulation into compressible cakes or formation of contiguous films also remain unclear. In the present work, we investigated the impacts of anionic surfactant and salt on oil droplet adhesion to model nanofiltration membranes. The Direct Observation Through the Membrane (DOTM) technique was employed to visualize oil drop deposition and adhesion to polyelectrolyte multilayers (PEM) with opposite surface charges, supported by Anopore ultrafilters. Hexadecane-in-water emulsion was stabilized by sodium dodecyl sulfate (SDS) and had a controllably varied concentration of monovalent (Na^+) and divalent (Mg^{2+}) counterions. Quartz crystal microbalance and FTIR analyses showed that PEM coatings were stable when exposed to SDS indicating their suitability for nanofiltration of surfactant-stabilized emulsions. Visualization by DOTM was complemented by a force balance analysis to predict the critical size for a droplet that would be swept off the surface by the crossflow drag. The results indicate that interfacial tension, electrostatic droplet-droplet and droplet-membrane interactions, and membrane's affinity for oil together govern attachment of oil drops to membrane surfaces. Crossflow-controlled coalescence was promoted under conditions of moderate affinity between oil drops and the polyelectrolyte-coated surface when the drops could reach a critical size and be removed by the crossflow shear prior to forming a contiguous surface film.

TUESDAY 10:00 am - Ballroom 3

14b - A case study review of the application of reverse osmosis membranes for water treatment and reagent recovery at mining and mineral processing operations

Chris Biederman (Hatch)*, Jeffrey Cobbledick (Hatch)

Water is an integral part of mining and mineral processing. It is used in almost every major ore processing stage and without ready access to it, the recovery of valuable minerals becomes prohibitively expensive or altogether not possible. Treating mine water presents a unique challenge as no two mine sites have the same water chemistry. Water chemistry at a mine site can be influenced by variations in the geochemical make-up of the ore body, the volume and quality of ground water seepage that is collected, any chemicals that are added to improve mineral recovery, and seasonal precipitation and run-off. Mine waters often contain scale-forming ions - such as calcium, magnesium, manganese, iron, sulfate, silica, and alkalinity - at exceedingly high concentrations. Many of these ions can cause significant issues for a membrane water treatment plant if not thoroughly considered during the design stage. This presentation will show the variations - sometimes extreme - in water chemistry between mine sites in different geographic regions of the world. This will be followed by three case studies to demonstrate the wide variation in potential uses for membranes in a mining environment and some of the pitfalls to avoid. The first case study will review the use of spiral wound UF and RO membranes to recover calcium thiosulfate reagent at a gold mine. The importance of stable pre-treatment, membrane selection and membrane flow control will be discussed. The second case study will demonstrate the challenges with retrofitting a reverse osmosis system to a brownfield water treatment plant at a nickel-copper mine. Challenges associated biological and colloidal sulfur fouling downstream of an advanced oxidation process will be presented. The final case study will be the review of a conceptual design for a greenfield reverse osmosis mine water treatment plant at a lithium mine that is projected to have water with 1 to 2 g/L of boron.

TUESDAY 10:30 am - Ballroom 3

14c - Self-Cleaning Nanocomposite Membranes with Phosphorene-Based Pore Fillers for Water Treatment

Joyner Eke (University of Kentucky)*, Isabel Escobar (University of Kentucky)

Photocatalytic membrane processes generate oxygen-reactive radicals under irradiation by light that could degrade pollutants in feed solutions. Phosphorene is a two-dimensional material made by exfoliating black phosphorus, a monotypic van der Waals crystal. Unlike other common two-dimensional materials, under light irradiation, phosphorene can generate reactive oxygen species (singlet oxygen, hydroxyl radicals and superoxide radicals). In this study, we exfoliated black phosphorus to synthesize phosphorene. Then immobilized the phosphorene nanomaterial into the membrane (sulfonated poly ether ether ketone and polysulfone blend). Studies were done under ultraviolet light irradiation. The dye, methylene blue was used as feed solution. Membranes were comprehensively characterized before and after modification. Results obtained showed enhanced flux recovery for the phosphorene modified membranes. Fluorescent studies also revealed lower coverage of methylene blue on these membranes as compared to the unmodified membranes. Mechanism of photocatalysis will be investigated and reaction equations proposed

TUESDAY 11:00 am - Ballroom 3

14d - Optimization of nanofiltration process for treating industrial wastewater of time-varying composition

Salman Alizadeh Kordkandi (McMaster University); Ryan LaRue (McMaster University)*; Abhishek Premachandra (McMaster University); Jacob Sitko (McMaster University); David Latulippe (McMaster University)

Industrial wastewater (IWW) contains a complex mixture of components that depends on the facility where it is generated (e.g. slaughterhouse, brewery, pulp & paper plant). While many studies have focused on using membranes to treat IWW from a single facility, there have been no studies that addressed the significant challenge faced by specialized treatment plants that are designed to receive shipments of IWW from many different facilities. The time-varying composition of the influent to these treatment plants is quite considerable, yet the composition of the effluent must consistently meet the mandated discharge criteria. In this study, we have focused on using nanofiltration (NF) membranes to remove the organic components in IWW that contribute to the chemical oxygen demand (COD) parameter. In collaboration with a local company, we have collected over 80 samples of the influent IWW to their facility and analyzed them using a liquid-liquid extraction process and gas chromatography-mass spectroscopy (GC-MS) to identify the most common chemical compounds. We then used a subset of 15 of those samples, that had considerable variation in chemical composition, and evaluated the performance of various commercial flat-sheet NF membranes (five were polyamide and one was polypiperazine-amide) from three different companies. We have found a significant difference in the performance (both volumetric throughput and quality) for those NF membranes with the same IWW sample and across different IWW samples. For example, the total reduction in COD varied from 30% to 80% for one particular IWW sample. Using the same GC-MS based analysis of the filtrate samples, we saw significant variation in the rejection of individual compounds. For example, the rejection of butanoic acid varied from 0 to 90%. We are currently working to develop a fundamental understanding of the relationship between the physico-chemical properties of the membranes and the filtration performance.

TUESDAY 11:30 am - Ballroom 3

14e - Adsorptive and Reactive Membranes for Remediation of PFOA, Chloro-organics, and RCRA Metals from water

DB Bhattacharyya (University of Kentucky)*, Hongyi (Derek) Wan (University of Kentucky), Anthony Saad (University of Kentucky), Md. Saiful Islam (University of Kentucky), Ashish Aher (University of Kentucky), Rollie Mills (University of Kentucky), Lindell Ormsbee (University of Kentucky), Evan Hatakeyama (Chevron)

The contamination of aquifers by toxic metals and organic compounds is a widespread problem. Traditional membranes, such as, RO, NF, UF provide many successful water applications. But incorporation of nanostructured materials with responsive and tunable properties in microfiltration type membranes has added immense value in through adsorptive separations and water detoxification in the area of pollutant remediation. The use of macromolecules in microfiltration membrane pores, such as, poly-acrylic acid (PAA) for pH responsive behavior, and Poly-N-isopropylacrylamide (PNIPAm) for temperature responsive behavior, provides tunable water permeability, and a platform for pollutant sorption/desorption and immobilization of pollutant degradation catalyst. In addition, we have created high capacity adsorptive membranes with SH and other functional groups in membrane pores for RCRA metals (such as, Hg) removal. The use of temperature responsive membranes (such as, PNIPAm) allow adsorption/desorption of organics by using TSA (temperature swing sorption/desorption) approach. Perfluorinated compounds (PFCs) such as Perfluorooctanoic acid (PFOA) and Perfluorooctane sulfonate (PFOS) have gained much attention due to their environmentally persistent nature. PNIPAm exhibits a phase transition from a hydrophilic state to a hydrophobic state at its LCST of 32 C. With PNIPAm hydrogel, initial PFOA sorption rate of about 28 mg/g gel/h was observed above LCST, and desorption below LCST. The presentation will water applications in three areas: (1) PFOA adsorption/desorption kinetics below above LCST by PNIPAm hydrogels, and in-situ polymerized NIPAm in membrane pores, (2) Chloro-organic degradation in PAA funcionalized PVDF membranes containing Fe/Pd nanoparticles, (3) high capacity sorption of toxic metals (Hg) with membrane pores functionalized with SH groups. This research is funded by NIEHS-SRP, NSF-EPSCOR program, and Chevron Corp.

TUESDAY 12:00 pm - Ballroom 3

14f - Selective separation of mono- and di-valent cations in electrodialysis during brackish water desalination: bench and pilot-scale studies

Xuesong Xu (New Mexico State University)*, Pei Xu (New Mexico State University), Guanyu Ma (New Mexico State University)

Selective separation of mono- and multi-valent ions has important applications in water reuse, desalination, and salt production. Innovative monovalent permselective cation-exchange membrane CR671 was developed by modifying the normal grade membrane CR67 with polyethyleneimine coating. Bench- and pilot-scale electrodialysis experiments were conducted at a brackish groundwater desalination facility to investigate desalination performance and ion permselectivity using the CR671 and CR67 under different operating conditions. Both normal grade (CR67 and AR204) and selective (CR671 and AR112B) membranes achieved same desalting efficiency. Na-selectivity in terms of relative transport number using sodium as standard ion was affected significantly by current density and linear velocity for the normal grade membranes while the selective membranes exhibited relatively stable good Na-selectivity. The Na-selectivity of the CR671 was demonstrated up to 9 and 5 times better than the CR67 during pilot- and bench-scale electrodialysis, respectively. Hydraulic retention time and electrodialysis stack staging had minor impact on Na-selectivity at high current density. With the same hydraulic conditions, overall desalination behavior and ion permselectivity were highly comparable between bench- and pilot-scale electrodialysis. It infers that bench-scale testing results can be used to simulate and project desalination performance and ion selectivity of pilot- and potentially full-scale electrodialysis applications.

TUESDAY 9:30 am - Ballroom 4

15a - Membrane solvent recovery in edible oil industry

Petrus Cuperus (SolSep BV)*, Ingrid Wienk (SolSep BV), Pieter Vandezande (VITO), Marco Pipolo (Bunge), Jan Kolijn (TUSTI), Will Fuchten (MTSA)

OSN technology is very well fitted for partial recovery of organic solvents in existing distillation installations. The past four years Solsep has been heavily involved in trialing such membranes in the industrial environment of such solvent: in particular acetone. In a last term of 2 years polymeric and ceramic membranes are tested for their stability, performance and window of operation. Currently used distillation/evaporation can be expensive (energy) and sometimes introduces degradation of products (lowering quality). Less heat input in the products by using membranes should improve continuous quality at lower costs. But when solvent is recovered with membranes the throughput of the membrane is not constant-due to concentration polarization and fouling. This leads to a point where membranes are less effective. In a constellation of membrane and evaporation this means that there are optimal conditions -for membranes in technical sense and for the process for the overall economics. In this area membrane permeation and separation as well as ease of evaporation and energy costs play a role. Inside-industry work has been focused on recovery of oil-acetone mixtures (oil: palm, shea). For operation on the long run clean-in-place procedure were evaluated. More than 2 years of in-industry piloting - NF with evaporation- has been done and will be addressed in the presentation. For economic evaluation extensive modeling work-backed-up by experimental work- suggest PBTs for different scenarios for implementation of the technology. We will discuss work on the removal of solvents and possible use in industry. Acetone-vegetable oil processing is presented in the development from lab-scale to long term pilot-scale executed -in "real life" .

TUESDAY 10:00 am - Ballroom 4

15b - Molecular Design of Polymeric Membranes for Organic Solvent Recovery

Tai-Shung Chung (National University of Singapore)*

Organic solvent nanofiltration (OSN), also known as solvent resistant nanofiltration (SRNF), is an emerging liquid separation method with substantial applicability in pharmaceutical, food and petrochemical industries. OSN not only aims to separate solutes or compounds with a molecular weight (MW) of 200-1000 g/mole from organic solvents but also to recycle the costly organic solvents used in related industries. As a result, the development of OSN technology may potentially reduce medicine costs and CO_2 emission. The solvent resistant nanofiltration membrane is the core of OSN processes which ideally should possess a pore size of 0.5-2.0 nm for an effective separation and purification of valuable solutes and recycling of organic solvents. In this presentation, we will introduce our recent efforts on membrane development for OSN systems. Various material and fabrication strategies to enhance membrane performance will be discussed.

TUESDAY 10:30 am - Ballroom 4

15c - Epoxy-based solvent resistant nanofiltration membranes prepared through phase inversion

Maarten Bastin (KU Leuven)*, Jasper Raymenants (KU Leuven), Ivo Vankelecom (KU Leuven)

A novel method has been successfully developed for the synthesis of solvent resistant nanofiltration (SRNF) membranes. The bulk chemistry of epoxy curing via ring-opening reactions has been applied in a phase inversion process, resulting in membranes with excellent solvent stability and water permeability, making them particularly suitable for the treatment of solvent/water mixtures. Several epoxy monomers were selected based on their size and number of epoxide functional groups and different amines were chosen as curing agent, based on their chemical composition, length and reactivity. Both the epoxy monomer and amine were dissolved in DMSO and left to react which, due to the increased number of crosslinks in the matrix resulted in an increase in viscosity. The solutions viscosity was monitored as a function of time and related to the reaction kinetics of the epoxy-amine curing system. Integrally skinned asymmetric membranes were prepared via phase inversion. The membranes were characterized via SEM, ATR, contact angle measurements and filtration experiments with different solvent water mixtures and dyes of various charges and size. Depending on the type of epoxy monomers and concentration, a range of membranes could be prepared, with NF properties (rose bengal retention of 95% in DMF/water and permeance of 1 - 3 $L\ m^{-2}\ h^{-1}\ bar^{-1}$) or UF properties which in their turn can be used as support for full epoxy thin film composite (TFC) membranes. In this case, on top of the UF-epoxy supports, an epoxy-cured top-layer can be added via interfacial polymerization, resulting in an ultrathin top-layer on an UF-support layer, all in the same poly(epoxy ether) material, giving the overall benefit of having the same swelling behaviour in similar chemical environments which decreases the chance of defects.

TUESDAY 11:00 am - Ballroom 4

15d - Nanoporous Graphene Membranes for Organic Solvent Nanofiltration

David Cheng (MIT)*, Rohit Karnik (MIT)

The separation of fine chemicals and pharmaceuticals from organic solvents during chemical processing is currently extremely energy-intensive due to the use of solvent evaporation for separation, which also prevents re-use of solvent leading to significant costs. Membranes that can easily handle solvents and separate them from small molecules, e.g. fine chemicals/active pharmaceutical ingredients, have the potential to significantly reduce energy consumption and enable solvent recovery and re-use, thereby lowering the energy costs and making chemical manufacture more environment-friendly. However, versatile membranes that are compatible with a range of solvents and also have the required selectivity are not readily available. Based on our recent success in developing nanoporous graphene membranes for aqueous and gas separation based on size-exclusion, we show in this work the development of nanoporous graphene membranes for performing pharmaceutical/fine chemicals separations from solvents. With its atomic thickness, high mechanical and chemical stability, and the ability to sustain selective, sub-nanometer pores in its rigid lattice, nanoporous graphene has tremendous potential as a material for next-generation membranes, yet little is known about transport of solvent and small molecules across nanoporous graphene. We reveal peculiar solvent and small molecule transport behaviors through nanoporous graphene and demonstrate size-based separation in solvents, potentially opening a new arena for replacing distillation with more efficient membrane separations.

TUESDAY 11:30 am - Ballroom 4

15e - OSN Technology in Petrochemical Industry: Looking Back, Looking Forward

Dhaval Bhandari (Exxon Mobil), JR Johnson* (Exxon Mobil)

Organic mixture separations are considered to be significantly more challenging than aqueous and gas separations due to the combination of high chemical, physical and often thermal constraints put on the membranes. ExxonMobil and Grace successfully commercialized the first large scale OSN technology, Max-DewaxTM for lube oil dewaxing in 1998. Key technical features of the technology that led to its successful commercial operation are highlighted. Over the last 10 years, homogenous catalyst recovery and solvent purification applications have also been developed and deployed. Organic liquid separation membranes are still considered 'emerging' while its cousin - gas separation membranes have enjoyed a relatively broader success in the petrochemical industry. Critical challenges in OSN commercialization are discussed and identified to be somewhat contrary to commonly held beliefs. The advent of dual challenge* is especially relevant in petrochemicals manufacturing where thermal processes still account for ~45% of energy consumption. Key development needs are identified for the next generation of OSN/OSRO technology to make an impact in this space. *dual challenge - ensuring the world has access to affordable and reliable energy supplies to support prosperity while reducing environmental impacts - including the risks of climate change. Weblink - https://corporate.exxonmobil.com/en/energy/energy-outlook

TUESDAY 12:00 pm - Ballroom 4

15f - Enabling Organic Solvent Nanofiltration and Reverse Osmosis using Ceramic Supported TFC Membranes

Jeffrey R McCutcheon (University of Connecticut)*, Lingling Xia (University of Connecticut), Marcus Weyd (Fraunhofer Institute for Ceramic Technologies and Systems IKTS), Mi Zhang (University of Connecticut)

We report on the performance of a ceramic supported polyamide thin film composite (TFC) membrane for organic solvent nanofiltration (OSN) and organic solvent reverse osmosis (OSRO) applications. The membranes are made by conventional formation of an aromatic polyamide selective layer directly onto a ceramic ultrafiltration membrane. The polyamide has demonstrated resistance to solvents while retaining molecular selectivity and the ceramic support is stable in most any solvent. We explore the use of different solvents and different dyes to probe differences in polyamide selectivity performance between different solvents and dyes. We also explore the use of these membranes in separating miscible solvent pairs via organic solvent reverse osmosis. In all, we note that certain solvents can alter the polyamide permselective properties and that these effects seem to be linked to the solvent's Hansen solubility parameter of the solvent. We also note that while these polymers demonstrate some level of selectivity with a particular solvent (i.e. "solvent activation"), changing from one solvent to another may result in unpredictable selectivity and permeance performance.

TUESDAY 9:30 am - Kings Garden 3

16a - Responsive CNT Composite Membranes for Protection Against Chemical Warfare Agents

Melinda L Jue (Lawrence Livermore National Laboratory)*, Chiatai Chen (Cornell University), Yifan Li (MIT), Eric Meshot (Lawrence Livermore National Laboratory), Ngoc Bui (Lawrence Livermore National Laboratory), Rong Zhu (MIT), Myles Herbert (MIT), Sei Jin Park (Lawrence Livermore National Laboratory), Steven Buchsbaum (Lawrence Livermore National Laboratory), Kuang Jen Wu (Lawrence Livermore National Laboratory), Timothy Swager (MIT), Francesco Fornasiero (Lawrence Livermore National Laboratory)

Stimuli responsive materials have many diverse applications in areas such as drug delivery, sensors, dynamic surfaces, and advanced textiles. In the field of responsive fabrics, high breathability and fast response to external stimuli are desired to combine comfort with "smart" functions. Materials of this type could be utilized, for example, to create protective garments that selectively respond to chemical threats without adding undue physiological stress to the user, thereby prolonging the safe wear. In this work, vertically-aligned single-walled carbon nanotubes (SWCNT) are used to create exceptionally breathable and protective composite membranes. Autonomous, smart response to environmental chemical threats is achieved thanks to an engineered copolymer layer grafted to the surface of the SWCN membranes, which collapses in the presence of organophosphate chemical threats to effectively block their transport through the membrane. This response is local and can be reversed with a simple base treatment. Moisture vapor transport rates that rival commercial breathable fabrics coupled with the ability to block both biological (by size sieving) and chemical threats make these membranes promising materials for next-generation protective garments for military, first responders, and medical personnel.

TUESDAY 10:00 am - Kings Garden 3

16b - Highly Efficient Dimethyl Ether Production from CO_2 Hydrogenation in a Dehydration Membrane Reactor

Huazheng Li (Rensselaer Polytechnic Institute)*, Fanglei Zhou (Rensselaer Polytechnic Institute), Qiaobei Dong (Rensselaer Polytechnic Institute), Shoujie Ren (Missouri University of Science and Technology), Xinhua Liang (Missouri University of Science and Technology), Naomi Klinghoffer (Gas Technology Institute), Shiguang Li (Gas Technology Institute), Miao Yu (Rensselaer Polytechnic Institute)

As an important industrial intermediate, Dimethyl Ether (DME), can also be used as a promising clean fuel in various ways. Traditionally, DME is produced either through methanol dehydration or a couple of consecutive reactions directly from syngas. However, the yield of DME is low, limited by the equilibrium, in both processes. In order to increase the DME yield, a novel catalytic membrane reactor to remove the produced water in-situ from the reaction system will be beneficial. In this study, a highly water selective and permeable NaA membrane, prepared on ceramic hollow fibers by an improved seeding method, was applied in a membrane reactor for DME synthesis directly from CO_2 hydrogenation over bifunctional catalysts (CZA/HZSM-5) at challenging conditions (temperature from 220 to 260°C and pressure from 300 to 500 psig). Current results showed that much higher CO_2 conversion (up to 73.4%) and DME yield (up to 54.5%) than those obtained in traditional packed bed reactor were achieved. These results solidly indicate that limitation of equilibrium have been successfully overcame, resulting from the timely and effective water in-situ removal from the reaction system by NaA membrane. Effects of mass ratio of CZA and HZSM-5 catalysts, catalyst mass per membrane area, gas hourly space velocity, catalysts gradients, etc, on catalytic reaction performance and catalyst stability were also intensively investigated.

TUESDAY 10:30 am - Kings Garden 3

16c - Poly-methacrylic acid Functionalized Membranes with Incorporated Reactive Pd/Fe Nanoparticles: Lab Scale to Groundwater Remediation Applications

Hongyi (Derek) Wan (University of Kentucky)*, Md. Saiful Islam (University of Kentucky), Nicolas Briot (Unuversity of Kentucky), Anthony Saad (University of Kentucky), Lindell Ormsbee (University of Kentucky), DB Bhattacharyya (University of Kentucky)

The detoxification of chlorinated organics, such as trichloroethylene (TCE) and polychlorinated biphenyl (PCB), in groundwater is an important area. With incorporated catalytic Pd/Fe particles, the functionalized commercial MF membrane modules can be used in the dechlorination treatment of groundwater. Membrane functionalization was achieved in both commercial flat-sheet PVDF membranes and spiral membrane modules. The Pd/Fe nanoparticles were synthesized inside the pores of membrane through in situ reduction after ion-exchange with carboxylate groups (methacrylic acid (MAA) was polymerized inside the pores). The MAA prevents nanoparticle agglomeration and metal ion loss. It also entitles the commercial membranes a pH-responsive behavior in the permeability: a 30-fold decrease was found when the pH increase from 2.3 to 10.5 (542 LMH/bar at pH 2.3). The reactive membranes were tested in the treatment of lab synergic PCB and field groundwater (collected from a Superfund site, which includes TCE, carbon tetrachloride, etc.). At 2.2 s residence time, above 90% of chlorinated organic species (10-1000 ppb level) in the groundwater were dechlorinated in the convective flow mode. Advanced characterization methods, such as FIB, EELS and XPS were applied to quantify particle size and distribution at the entire depth of the membrane pores. The depth profile of the membrane indicates that particle size was uniform inside membrane pores (24±6 nm) but slightly smaller than those nanoparticles located on the surface (39±9 nm). Furthermore, the use of TEM and XRD enables the calculation of Pd monolayer coverage on the Fe surface. The effects of Pd surface coverage and particle compositions on the H_2 production (Fe corrosion in water) and corresponding dechlorination rate were evaluated to find the optimal Pd content for the dechlorination. This research is supported by the NIEHS-SRP grant P42ES007380. Partial support is also provided by NSF KY EPSCoR grant (Grant no: 1355438).

TUESDAY 11:00 am - Kings Garden 3

16d - Investigation of a smart electrically responsive ultrafiltration membrane

Chia Miang Khor (UCLA)*, David Jassby (UCLA)

Stimuli responsive membranes that enable the smart control of substances flowing through can be enticing as alternative materials for water purification. These materials are capable of conformational changes when exposed to external stimuli such as temperature, pH, electric fields, and electrolyte concentrations. Membrane rejection properties can thus be tailored and facilitate membrane cleaning. Among the different types of external stimuli, electrical signals would have the most plausible prospect, due to the rapid nature of their application and the fact that the stimulus can be delivered through the membrane itself without the need to change the feed stream. In this work, a relatively environmentally benign method, surface initiated electrochemically mediated atom transfer radical polymerization (SI-eATRP), was employed for the surface modification of a carbon nanotube-coated ultrafiltration membrane with an anionic polymer. The resulting membrane was electrically responsive with tuneable pore sizes. Complete release or retention of dextran molecules with molecular weight 12,000 could be achieved through voltage reversals. This was accompanied by considerable changes in water flux through the membrane. Electrostatic repulsion between the grafted polyanionic brushes and charged species in the water was exploited with the application of different voltages to induce conformational changes of the polymer brushes and thus rejection of certain contaminants.

TUESDAY 11:30 am - Kings Garden 3

16e - Catalytic Membranes for Groundwater Treatment

Alexander J Sutherland (McMaster University)*, Charles-François de Lannoy (McMaster University)

Zero Valent Iron (ZVI) has received extensive scientific interest for its ability to reduce highly toxic chlorinated organic contaminants (COCs) present in groundwater. The development of Zero Valent Iron nanoparticles (nZVI) has been proven to enhance reactivity as a result of increased surface area. Despite its remarkable potential to remediate COCs, nZVI faces challenges to its reactivity due to particle aggregation and an inability to selectively reduce desired contaminants over other compounds in solution. The goal of this project is to develop a permeable membrane surface composite that allows for the stabilization of nZVI nanoparticles to reduce chlorinated contaminants in the feed. Various conductive materials have been investigated as stabilizers and scaffolds for conducting current to regenerate nZVI reactivity. The results of this project have shown that Carbon Nanotubes (CNTs) can be used as a conductive network to stabilize nZVI to mitigate aggregation, channel electrons to support nZVI as a catalyst, and act as an electrode for electrochemically reducing COCs alongside a graphite counter electrode. By applying a -2V potential to a surface layer of CNT-stabilized nZVI on a Polyethersulfone microfiltration support membrane, model contaminant solutions of 0.25 mM Methyl Orange (MO) can be entirely degraded within three-hour batch tests. This is a marked improvement over membrane electrodes that utilize CNTs alone, which achieve only 70% degradation with the same experimental parameters. Using microfiltration membranes and facilitating the removal through the application of a low voltage presents significant advantages over conventional membrane separation. Firstly, eliminating the requirement of nano-scale pores possessing high trans membrane pressure gradients to accomplish the separation. Secondly, these membranes produce a waste stream of completely degraded contaminants, rather than a concentrated contaminant retentate.

TUESDAY 12:00 pm - Kings Garden 3

16f - Plasmonic membrane catalytic activation of peroxide for quantized oxidation via residence time control

Hao Tang (University of Washington), Guozheng Shao (University of Washington), Bruce Hinds (University of Washington)*

Many industrial oxidation processes based on peroxide, have a difficulty of over oxidation in homogeneous solution principally due to faster oxidation rates of already oxidized species and high statistical probability of secondary oxidation near the completion of reaction. Ideal for stepwise oxidation is to limit the residence time of target molecule in a reaction zone to allow for single (or quantized) reaction events. This can be achieved in a membrane geometry where catalyst along a pore length and set flow velocity can precisely control residence time for oxidation. It is also known that Au nanoparticles can catalytically activate peroxide under light irradiation due to the formation of concentrated surface plasmon electric fields and thus hypothesized to be present in nanoporous planes. A plasmonic membrane was synthesized by evaporation of 25 thick Au films onto pore entrances anodized aluminum oxide membranes (AAO) with pore diameters of 20-200nm. This allowed solutions of peroxide, benzene, methyl blue to flow through membrane and interact with Au surface plasmon upon exit of the membrane. Under light illumination of 10-100 mW/cm^2

quantum efficiencies (photon/peroxide radical) of 200-800% were seen, indicating a mechanism of field induced activation of peroxide compared to a hot electron injection mechanism. Dye and benzene to phenol oxidation are demonstrated.

TUESDAY 9:30 am - Kings Garden 5

17a - Mixed-matrix membranes formed from imide-functionalized UiO-66-NH_2 for improved interfacial compatibility

Qihui Qian (MIT)*, Zach Smith (MIT)

Mixed-matrix membranes (MMMs) formed by dispersing metal-organic framework (MOF) particles in polymers have attracted significant attention because these composite systems can potentially surpass the separation performance of pure polymers alone. However, performance improvements are often unrealized because of poor interfacial compatibility between the MOF and the polymer, which results in interfacial defects. From a practical perspective, strategies are needed to address these defects so that MMMs can be deployed in real-world separation processes. From a fundamental perspective, strategies are needed to reliably form defect-free MMMs so that transport models can be applied to estimate pure-MOF property sets, thereby enabling the development of robust structure-property relationships. To address these interfacial challenges, we have developed a method to surface functionalize a UiO-66-NH_2 MOF with a nanoscopic shell of covalently tethered 6FDA-Durene oligomers. When combined with a high molecular weight 6FDA-Durene polymer of identical chemical structure to that of the imide-functional MOF surface, defect-free MMMs with uniform particle dispersions can be formed. With this technique, both permeabilities and selectivities of select gases in the MMMs were improved at loadings ranging from 5 wt% to 40 wt%. At a 40 wt% loading, CO_2 permeability and CO_2/CH_4 selectivity were enhanced by 48% and 15%, respectively. Additionally, pure-MOF permeabilities for H_2, N_2, O2, CH_4, and CO_2 were predicted from the Maxwell Model due to zero defects in the MMMs formed. Experiments have shown that this technique can be generalized to other MOF/polymer pairs, enabling the prediction of pure MOF transport property sets, which is impossible when defect-free pure-MOF membranes cannot be formed.

TUESDAY 10:00 am - Kings Garden 5

17b - Janus polymers bearing tri(n-alkoxy)silyl side groups: glassy membrane materials with properties of rubbers

Yury P Yampolskii (A.V.Topchiev Institute of Petrochemical Synthesis, Russian Academy of Sciences)*, Ludmila Starannikova (A.V.Topchiev Institute of Petrochemical Synthesis, Russian Academy of Sciences), Dmitry Alentiev (A.V.Topchiev Institute of Petrochemical Synthesis, Russian Academy of Sciences), Maxim Bermeshev (A.V.Topchiev Institute of Petrochemical Synthesis, Russian Academy of Sciences), Eugene Finkelshtein (A.V.Topchiev Institute of Petrochemical Synthesis, Russian Academy of Sciences)

A series of polymers of a new type, which contain rigid main chains (glassy nature) and flexible substituents (rubbery nature), were prepared from norbornene derivatives with Si(OR)3 -groups where R= Me, Et, n-Pr, n-Bu. Two series of polymers, promising materials for membrane separation of gaseous hydrocarbons, were considered: metathesis and addition types substituted polytricyclononenes. The prepared addition polymers were glassy, while the glass transition temperature of metathesis polymers depended on the length of the alkyl-group R and varied in the range from -44 to 61oC. The dramatic tuning of polymer gas-transport properties was demonstrated by the change of the polymer main chain structure and the length of Si(OR)3 -group. For example, we observed for the first time that obtained metathesis polymers turned out to be more permeable than their addition isomers. The novel polymers are not large free volume materials (Permeability coefficients P(O2) are in the range 16-190 Barrer), so they are devoid of common disadvantage of highly permeable polymers, fast aging and reduction of gas permeability. Meanwhile, all the studied metathesis and addition polytricyclononenes exhibited solubility controlled (or reverse) selectivity of permeation of hydrocarbons: ideal selectivity P(C_4)/P(C1) for them was found to be in the range 13-49, higher than for many polymers with similar properties. The permeability coefficients of butane are also relatively great, up to 8100 Barrer. Doubtless, this new group of polymers can be considered as materials for membrane separation of hydrocarbons of natural and associated petroleum gases. * This work was supported by Russian Science Foundation (grant #17-19-01595)

TUESDAY 10:30 am - Kings Garden 5

17c - Solution processable metal organic frameworks for gas separations: from porous liquids to mixed matrix membranes

Anastasiya V Bavykina (KAUST)*, Alexander Knebel (Leibniz University Hannover Institute for Physical Chemistry and Electrochemistry), Shuvo Datta (KAUST), Magnus Rueping (KAUST), Mohamed Eddaoudi (KAUST), Juergen Caro (Leibniz University Hannover Institute for Physical Chemistry and Electrochemistry), Jorge Gascon (KAUST)

Conventionally the virtue of porosity is considered for porous solids, e.g. Metal Organic Frameworks (MOF), Porous Polymers or zeolites. Often their processability is not a focus of attention which hampers the application. We demonstrate that MOFs can be made solution processable via surface functionalization. We report on a solution processible ZIF-67. When placed in mesitylene, ZIF-67 predictably sediments. When it is modified and distributed within the same solvent, a clear solution formed with no sedimentation occurring for at least 24h. The morphology and textural properties of the modified MOF do not alter from the ones of its parent, which was confirmed by XRD, N_2 adsorption and SEM. Mesitylene has a kinetic diameter of 8.5Å and does not penetrate the pores, which are smaller in size. The system can be considered as a porous liquid - a novel class of porous materials. The modified MOF can either be directly employed as liquid membranes (LMs) or be co-processed with polymers to yield highly loaded mixed matrix membranes (MMMs). In a LM scenario, the permeation tests were carried out using a U-shaped permeation cell with equimolar C3H6/C3H8 and CO_2/CH_4 mixtures. While a pure solvent shows a separation factor for C3H8/C3H6 mixtures close to 1, addition of the solid pore generator increases the separation. The permeance is higher than in solid membranes. The formation of dispersions of a MOF in a liquid is not limited by formation of porous liquids and application as LMs. We studied the advantages that this processability offers. Dispersions of functionalized ZIF-67 were blended with 6FDA-DAM in the shape of MMMs. While MMMs based on pure MOF maintain good physical resistance at low loadings, increasing the concentration of MOF results in brittle composites. In contrast, MMMs made from functionalized ZIF keep good mechanical strength even at ca. 50 wt.% loadings. The results obtained for this MMM are among the best MMMs ever reported for C3H8/C3H6 separation.

TUESDAY 11:00 am - Kings Garden 5

17d - Enhancing CO_2/N_2 Selectivity and Elimination of Langmuir Sorption within High Tg, Glassy Polynorbornene Membranes

Christopher Maroon (University of Tennessee), Jacob Townsend (University of Tennessee), Kevin Gmernicki (University of Tennessee), Daniel Harrigan (Aramco Services Company), Ben Sundell (Aramco Services Company), John Lawrence (Aramco), Shannon Mahurin (Oak Ridge National Laboratory), Konstantinos Vogiatzis (University of Tennessee), Brian Long (University of Tennessee)*

We will describe how the CO_2/N_2 gas separation performance of alkoxysilyl-substituted vinyl-added polynorbornenes (VAPNBs) may be significantly enhanced via careful monomer design. More specifically, we will show that by increasing the incorporation of 5-tris(2-methoxyethoxy)silyl-2-norbornene in vinyl-added polynorbornenes, a substantial increase in CO_2/N_2 selectivity may be realized with minimal decrease in CO_2 permeability. This trend avoids the traditional permeability/selectivity "tradeoff" relationship, and produces an optimal membrane whose performance nears the 2008 upper bound for CO_2/N_2 separations. Though these polymers were initially designed to maximize CO_2 solubility, subsequent gas sorption studies revealed that these high Tg glassy membranes virtually lack any Langmuir sorption component, and indicate that their improved CO_2/N_2 selectivity is due to decreased N_2 solubility within the matrix. This is further supported via computational modelling which suggests that the source of this apparent "N_2-phobicity" is likely explained through comparative analyses of polymer-polymer and polymer-gas interactions. Finally, mixed-gas permeation tests were performed to provide a more realistic look at these membrane's real-world gas separation performance.

TUESDAY 11:30 am - Kings Garden 5

17e - Ultra-High Permeability Mixed-Matrix Membranes as a Next-Generation Carbon Capture Technology for Post-Combustion

Sameh K Elsaidi (U.S. Department of Energy National Energy Technology Laboratory)*, Surendar Venna (U.S. Department of Energy National Energy Technology Laboratory), Ali Sekizkardes (Battelle/U.S. Department of Energy National Energy and Technology Laboratory), Mona Mohamed (University of Pittsburgh), David Hopkinson (U.S. Department of Energy National Energy Technology Laboratory)

Mixed matrix membranes (MMMs) have great potential for separation applications including post-combustion carbon capture. To reduce the cost of capture and make them economically viable, new formulations with ultra-high permeability and moderate selectivity must be identified. Here, we present a bottom-up design of MMMs wherein the constituent materials have been designed for high gas separation performance and compatibility. MOF nanoparticles and polymers are decorated by desired functional groups that can chemically interact with each other to form defect free MMMs. The custom formulation of MMMs is important in scale-up production for practical applications. Successful formulation requires a blend of art and science. Here, two, three or four-components, were chosen for their compatibility, where each component has a different function and all are formulated with specific quantiles to fine-tune the physicochemical properties for defect free custom-formulated mixed-matrix membranes (CF-MMMs). PIM-1 and MEEP80 were combined with various MOF nanoparticles with different surface areas, pore sizes, and binding sites to conduct a parametric analysis of ten different membranes (CF-MMM-1 to CF-MMMs-10). The chemical interaction between the MOF nanoparticles and the polymer was the key facet for optimizing the interfacial compatibility between the polymer matrices and MOF fillers. Systematic variation of the compositions of these components and the type of MOF nanoparticles were performed to optimize the membrane separation performance affording enhanced CO_2 permeability (4200-25670 barrer) and CO_2/N_2 selectivity (17-34) compared with the neat polymers. The permeability/selectivity values surpass Robeson upper bound and suggest the potential of these membranes for practical CO_2 separations. The scaling-up of these membranes were performed by preparing their components in a kilogram-scale, while their precise formulation allows technical feasibility of CF-MMMs fabrication.

TUESDAY 12:00 pm - Kings Garden 5

17f - Composite gas separation membranes from metal-induced ordered polymeric frameworks

Zhihua Qiao (Tianjin University), Song Zhao (Tianjin University), Menglong Sheng (Tianjin University), Jixiao Wang (Tianjin University), Zhi Wang (Tianjin University), Chongli Zhong (Tianjin University), Michael D Guiver (Tianjin University)*

Metal-induced ordered microporous polymers (MMPs) are a new class of porous frameworks [1]. They are readily synthesized by polymer-directed chemical synthesis (PDCS) from functional polymers, small organic linkers and divalent metal ions. The small organic linkers coordinate with metal ions, and then the resulting unit cells self-assemble along the extension of the polymer chains, incorporating the polymer chains to construct three-dimensional frameworks as 50-100 nm nanoparticles. Variations in framework characteristics and pore sizes were achieved though choice of small organic linkers, metal ions, polymers and polymer molecular weight. Compared with conventional metal organic frameworks, MMPs show improvements in crystal size controllability and hydrolytic stability. MMP/mPSf membranes were fabricated by coating MMP dispersions onto modified polysulfone (mPSf) substrate, resulting in defect-free ultrathin gas-selective coated layers. Composite membrane MMP-3/mPSf had a CO_2 permeance of >3000 GPU and mixed gas CO_2/N_2 (15/85) selectivity of 78, stable under both humid and dry gas feed conditions. MMP-1/PSf membrane had a CO_2 permeance of >5500 GPU and mixed gas CO_2/N_2 selectivity of 35, placing them in the region of interest for CO_2 capture applications [2]. [1] Nature Materials 18 (2019) 163-168. [2] Journal of Membrane Science 359 (2010) 126-139.

TUESDAY 2:00 pm - Kings Garden 5

18a - High temperature gas separation properties of sub-micron polybenzimidazole membranes

Melanie M Merrick (The University of Texas at Austin)*, Benny Freeman (The University of Texas at Austin)

Polybenzimidazoles (PBIs) are a class of polymer materials that have the potential to improve and expand the applications of commercial gas separation membranes. While many commercial membranes can only perform gas separations up to 100°C, PBIs are both chemically and thermally stable at temperatures up to 400°C. A potential application of PBI membranes is the high temperature hydrogen-hydrocarbon separations proposed by the NSF Engineering Research Center for Innovative and Strategic Transformation of Alkane Resources (CISTAR). Two novel aspects of this work are the previously unstudied ethane and ethylene gas separation properties at elevated temperatures and the temperature dependence of gas transport in sub-micron films over an expansive temperature range. Hydrocarbon gas separation properties of Celazole® PBI membranes were characterized at temperatures up to 190°C. Gas permeabilities, gas selectivities, activation energies of permeation, and physical aging rates differed significantly between thin sub-micron and bulk 20-micron Celazole® films. In particular, the activation energies of permeation were 1.5 to 4 times higher in an aged 450-nm thin film than in a 27-micron bulk film, suggesting a denser membrane structure and greater size-sieving ability. The first known evidence of physical aging in PBIs showed an 80 percent decrease in methane permeability over a 1000-hour period at 190°C for a 450-nm Celazole® film. This study elucidates the effects of elevated operating temperature on thin films by encompassing a broader and higher temperature range than previous studies of flat, sub-micron polymer membranes.

TUESDAY 2:30 pm - Kings Garden 5

18b - Thermally Rearranged Semi-interpenetrating Polymer Network (TR-SIPN) Membranes for Olefin/Paraffin Separations

Young Moo Lee (Hanyang University)*, Won Hee Lee (Hanyang University), Jong Geun Seong (Hanyang University), Ho Hyun Wang (Hanyang University), Sun Ju Moon (Hanyang University)

Thermally rearranged (TR) polymers have received much attention in membrane gas separation applications due to their outstanding gas permeability [1]. Rigid chain structures constructed by TR processes can induce microporous morphology, resulting in improvement in gas transport characteristics [2]. Recent studies have aimed to use rigidity of TR polymers to enhance gas permeability by introducing cross-linkable agents [3] or preformed rigid moieties [4] into the TR polymer. In particular, TR polymers architected by network structures have been recently demonstrated to maximize rigidity and gas permeability [5]. Here, we report for the first time, TR-SIPN membranes composed of both linear and network TR polymers to attain polymer rigidity and to enhance olefin/paraffin separation performances. Facilely prepared TR-SIPN membranes exhibited noticeable rigid and microporous characteristics judging from BET surface area and gas sorption coefficients compared to conventional linear TR polymers. Moreover, the polymer matrices of TR-SIPNs, investigated by SAXS and gas permeation models, showed ideal structures in which large and small channels coexisted parallel to the gas transport direction. In olefin/paraffin separations, they retained more than 10 times higher gas permeabilities in ethylene and propylene gases without sacrifice in olefin/paraffin selectivities. As a result, they surpassed upper bound plots for ethylene/ethane and propylene/propane gas pairs. Furthermore, entirely rigid TR-SIPNs responded to outstanding plasticization resistances under aggressive pressures. References: [1]H.B. Park et al., Science 2007 318 (5848), 254-258; [2] M.D. Guiver and Y.M. Lee et al., Science 2013 339 (6117), 284-285; [3] M. Calle et al., Macromolecule, 2015 48 (8) 2603-2613; [4] Y.B. Zhuang J.G. Seong et al., Macromolecules 2015 48 (15), 5286-5299; [5] Y.S. Do and W.H. Lee et al., Chem. Commun., 2016 52 (93) 13556-13559.

TUESDAY 3:00 pm - Kings Garden 5

18c - Conformation-controlled molecular sieving effect for membrane-based propylene/propane separation

Yang Liu (Georgia Institute of Technology)*, William Koros (Georgia Institute of Technology), Mohamed Eddaoudi (KAUST), Gongping Liu (Georgia Institute of Technology), Youssef Belmabkhout (KAUST), Zhijie Chen (KAUST)

Membrane-based propylene/propane separation can reduce the operation cost in petroleum refining and petrochemical industries; however, identifying a suitable molecular sieve for membrane development is still challenging due to the close size (~0.15Å difference) between these two components. Our study addresses such a critical scientific question by a novel strategy. Specifically, with both C-C single bonds rotated (double eclipsed conformation), the C3H8 molecule exhibits its minimum end-on essentially triangular shape but with a rotation energy barrier. Our strategy is to design/screen a "triangular" pore-aperture to allow only the double eclipsed C3H8 conformer to pass through, effectively preventing all other C3H8 conformers from diffusion jumps. This combination of factors leads to a very low C3H8 diffusivity. On the other hand, the more compact C3H6 molecule can pass through the pore-aperture with relative ease. According to our theoretical calculations based on transition state theory, a membrane with such desired pore-aperture can offer a C3H6/C3H8 diffusion selectivity over a thousand. We further identified a MOF, Zr-fum-fcu-MOF, possessing the desired contracted triangular pore-aperture and fabricated mixed-matrix membranes using this MOF as an efficient filler. The fabricated hybrid membranes display very attractive C3H6/C3H8 separation performance, far beyond the current trade-off limit of polymer membranes. Back-calculations using composite theory suggests the MOF displays a diffusion selectivity, like the ideally targeted, above 1000. This is the highest value reported for the C3 pair to date, and we refer to this phenomenon as "conformation-controlled molecular sieving effect". Our work establishes a novel bottom-to-up strategy to develop membranes for gas separation and, meanwhile, refines the thinking on the molecular sieving effect to focus on the detailed geometrical characteristics comprising each gas molecule.

TUESDAY 3:30 pm - Kings Garden 5

18d - Mixed matrix materials containing soluble discrete metal-organic polyhedra in rubbery polyethers for CO_2/gas separation

Haiqing Lin (The State University of New York at Buffalo)*, Junyi Liu (The State University of New York at Buffalo), Cressa Fulong (The State University of New York at Buffalo), Tim Cook (The State University of New York at Buffalo)

Incorporating metal-organic frameworks (MOFs) in polymers is an effective strategy to improve membrane gas separation properties. For these mixed matrix materials (MMMs) to be fabricated into selective layers of 100 - 200 nm in industrial membranes, the MOFs particles should be smaller than 50 nm and have good compatibility with polymers to avoid defects at the interface of polymers and particles. In this work, soluble discrete metal-organic polyhedra (MOP) with particle sizes of ≈5 nm was synthesized and added into polyethers derived from poly(ethylene glycol) diacrylate (PEGDA) and poly(1,3 dioxolane) acrylate (PDXLA), which are leading materials for CO_2/N_2 and CO_2/H_2 separation. Interestingly, the introduction of the MOP-3 in the polyethers decreases glass transition temperature and even induces crystallization of polyPDXLA at higher loadings, indicating strong interactions between the polymer and particles. Increasing the MOP loading increases CO_2 permeability with a slight decrease in CO_2/N_2 and CO_2/H_2 selectivity. For example, as the MOP content increases from 0 to 30 wt% in polyPDXLA, CO_2 permeability increases from 190 Barrers to 580 Barrers, and CO_2/N_2 selectivity decreases from 70 to 62 at 35°C. These results can be satisfactorily modeled using the Lewis-Nielsen model with the MOP exhibiting much higher gas permeability than the polyethers. These MOP-based MMMs exhibit the separation properties above the Robeson's upper bounds for CO_2/N_2 and CO_2/H_2 separation. This talk will also compare these MMMs with those containing inherently insoluble ZIF-8 nanoparticles for the morphology and CO_2/gas separation properties. Despite a body of studies of MMMs comprising MOFs, this work represents one of the few on the MMMs based on amorphous rubbery polymers elucidating the effect of soluble discrete MOP loading on the polymer properties and demonstrating their impact in achieving superior CO_2/H_2 and CO_2/N_2 separation properties

TUESDAY 4:00 pm - Kings Garden 5

18e - Accessing novel microporous polymers to enhance yield, stereochemical control, and membrane performance in the recovery of light hydrocarbons from natural gas

John Lawrence (Aramco Services Company)*

Light hydrocarbons, i.e. propane and butane, provide additional value when recovered from natural gas due to their widespread use in energy and chemicals production. Current industrial approaches utilize energy-intensive refrigeration processes to recover light hydrocarbons, though the application of gas separation membranes can dramatically lower energy input and overall cost. Herein, we describe our efforts in developing novel microporous polymers to fabricate membranes capable of selectively removing light hydrocarbons from natural gas streams. Utilization of a new synthetic pathway enabled the synthesis of stereochemically pure monomers bearing bulky, rigid substituents with varying connectivity to the polymer backbone in order to study how increasing polymer rigidity through pendant group chemistry affects polymer synthesis, chain packing, and membrane performance. Immediate benefits of the enhanced stereochemical control gained using this synthetic pathway were evident in the polymerizations, where the one polymerization route proceeded in yields up to 100%. NMR analysis confirmed monomer and polymer structure, while XRD revealed increased interchain spacing when bulky pendant groups are present. More importantly, single and mixed gas permeation studies demonstrated a dramatic increase in C_4H_{10} permeability and C_4H_{10}/CH_4 selectivity when pendant group structure and connectivity were optimized. For example, at 800 psi in mixed gas feeds, increased cyclic content in one of the polymer structures improved C_4H_{10} permeability from 321 Barrer to 739 Barrer and C_4H_{10}/CH_4 selectivity from 9.8 to 12.8 when compared to the analogous polymer lacking this connectivity. Even more interesting, changing the substituent off the pendant group had a profound effect on permeation properties and further enhanced performance, achieving a C_4H_{10} permeability of 6520 Barrer and C_4H_{10}/CH_4 selectivity of 19.6.

TUESDAY 4:30 pm - Kings Garden 5

18f - Insight into the transport in polymeric and mixed matrix membranes via analysis of unique mixed gas diffusion coefficients

Johannes Carolus (John) Jansen (National Research Council of Italy)*, Alessio Fuoco (National Research Council of Italy), Marcello Monteleone (National Research Council of Italy), Elisa Esposito (National Research Council of Italy)

Gas separation by polymer membranes is an expanding technology in many application fields, such as oxygen or nitrogen separation from air (i.e. separation O_2/N_2) or biogas upgrading [1] and natural gas treatment (i.e. separation CO_2/CH_4), and others. Over the years, many new polymers have been developed, and their performance is typically expressed in terms of their position in the Robeson diagram [2,3]. Improvement of the performance requires a rational design of the polymer [4] and detailed knowledge of the transport properties. For a better understanding of the membrane performance, studies must not be limited to the more common single gas permeability alone, but they should be extended to gas mixtures and to diffusion as well as solubility. In this work we discuss a novel method that uses on-line mass spectrometric analysis of the permeate composition during the transient phase, enabling the unique determination of mixed gas diffusion coefficients, besides permeability coefficients [5,6]. Using some typical examples, we will discuss the difference between pure and mixed gas permeation and diffusion in standard membranes, the effect of porous fillers in mixed matrix membranes, and the effect of the gas-membrane affinity in membranes with anomalous transport. For instance, we will show how a MOF in low free volume polymers decreases the apparent diffusion coefficient via its increased sorption capacity, while functional groups in high free volumes effectively decrease the diffusion coefficient and promote selectivity by competitive sorption. [1] E. Esposito, J.C. Jansen et al., Energy Environ. Sci. 2019, 12, 281-289. [2] L.M. Robeson, The upper bound revisited, J. Membr. Sci., 2008, 320, 390-400. [3] R. Swaidan et al., ACS Macro Lett., 2015, 4, 947-951. [4] B. D. Freeman, Macromolecules, 1999, 32, 375. [5] S.C. Fraga, J.C. Jansen et al., J. Membr. Sci., 2018, 561, 39-58. [6] M. Monteleone, J.C. Jansen et al., Membranes, 2018, 8, 73.

TUESDAY 2:00 pm - Kings Garden 4

19a - Minimizing the cost of membrane distillation

Timothy V Bartholomew (Carnegie Mellon University)*, Alexander V Dudchenko (Carnegie Mellon University), Nicholas Siefert (U.S. Department of Energy National Energy Technology Laboratory), Meagan S Mauter (Carnegie Mellon University)

This work presents a modeling platform for assessing the energy consumption and cost of membrane distillation (MD). We leverage multi-objective nonlinear programming models to maximize the gained output ratio (GOR) and minimize the levelized cost of water (LCOW) of membrane distillation for multiple module designs (i.e. direct contact, air gap, conductive gap, sweeping gas, and vacuum) and system configurations (i.e. single or multi-stage, pass through or looping). This optimization approach identifies the energy or cost optimal design and operation for a specified water recovery, feed salinity and flowrate, and process and financial parameters. Our system-scale model includes a detailed finite difference module-scale model that explicitly considers the pressure drop along the module and non-constant solution properties (i.e. density, diffusivity, non-ideality, specific heat) and mass and heat transfer coefficients. We explore the influence of parameter uncertainty inherent in emerging technologies by simulating process performance across a range of parameter values to develop distributions for the expected GOR and LCOW and identify generalizable guidelines for energy efficient and low-cost membrane distillation.

TUESDAY 2:30 pm - Kings Garden 4

19b - Process Optimization Using Perturbation Expansion Solutions for Membrane Gas Separation Modules

Norfamila Che Mat (Universiti Malaysia Sarawak), Glenn Lipscomb (University of Toledo, Toledo)*

Simulation and optimization of hybrid membrane gas separation processes, such as those proposed for carbon capture, require high fidelity module performance simulations. Unfortunately, these simulations require iterative solution techniques which can be sensitive to the initial solution guess and thus require extensive iteration (i.e., computational time) or fail to converge. Perturbation solutions to the isothermal, constant transport property, multi-component cross-flow and counter-current flow performance equations are presented. Predictions from the direct, non-iterative solution are compared to solutions obtained using numerical approximations. Excellent agreement is found in all cases even for systems with broad ranges of recovery for feed gas components. The results also are excellent initial guesses for simulations that account for module non-idealities. The use of these solutions in optimizing staged membrane systems is discussed. The perturbation solution can dramatically speed up optimization and help identify unfeasible configurations.

TUESDAY 3:00 pm - Kings Garden 4

19c - Multi-physics Simulation of Hollow Fiber Vacuum Membrane Distillation Using OpenFoam

Albert Kim (University of Hawaii)*, Ho Ji (Korea Research Institute Of Ships And Ocean Engineering), Deok-Soo Moon (Korea Research Institute Of Ships And Ocean Engineering), Hyeon-Ju Kim (Korea Research Institute Of Ships And Ocean Engineering)

A VMD module often consists of three regions of feed solution, low-pressure gas, and porous membrane. At interfaces of the regions, at least two transfer phenomena are coupled. On the feed-membrane interface, liquid feed evaporates, and vapor molecules migrate through void spaces of the porous membrane. Momentum and heat transfer in the liquid feed can be described using coupled Navier-Stokes and Fourier equations. Migration of vapor molecules and heat transfer is at the level of molecular kinetic motion through the membrane pores, where molecular convection and diffusion are dominant. In the (low-quality) vacuum phase of the distillate, it is questionable to apply continuum-based flow models or rarefied gas models. Overall, modeling studies on VMD are fundamentally challenging because the phenomena are not only intrinsically coupled for transports, but also of multi-scales at the molecule, pore, and module levels. In industrial applications, a VMD module has a large number of fibers of an order of a few hundred or thousand. As each hollow fiber is treated as a mass source of vapor distillate, a continuum model for packed hollow fibers is fundamentally of flaw to predict the VMD performance. In this vein, the current paper investigates the VMD module of realistic sizes, containing hollow fibers of an order of 100, and simulate the coupled transport of momentum, heat, and mass in a seamlessly combined manner. For this study, OpenFOAM (an open-source CFD package) is used to solve for the thermal flow of liquid water and vapor molecules. We used a theoretical interpretation to enhance the computational efficiency and modified OpenFOAM solvers to rigorously investigate the multi-physics, multi-scale phenomena. Computational techniques and physical interpretations will be discussed in detail.

TUESDAY 3:30 pm - Kings Garden 4

19d - Scale-up of Facilitated Transport Membrane Module for CO_2 Capture from Flue Gas

Kai Chen (Ohio State University)*, Witopo Salim (Ohio State University), Yang Han (Ohio State University), Dongzhu Wu (Ohio State University), Winston Ho (Ohio State University)

In order to modularize a composite membrane with an ultrathin selective layer, it is a challenge to scale up the membrane fabrication procedure from lab to pilot scale. Herein, a knife-coating method was successfully integrated into a roll-to-roll fabrication machine, making a 165-nm thick facilitated transport membrane for CO_2 capture from flue gas with 14 inches in width and > 1200 feet in total length. The scale-up membranes were subsequently used to make spiral-wound (SW) elements and modules. To increase the membrane packing density in a single element, a detailed multi-leaf rolling procedure was developed to scale up the elements efficiently. Although the fabrication of a multi-leaf SW membrane module has been well recorded in the literature, the majority of the methods had all the membrane leaves directly attached to the central permeate tube, thus, the number of leaves was limited by the central tube surface area. In order to improve the fabrication, the glue lines were modified to allow an arrangement of a stack of staggered membrane leaves not attached to the central tube before rolling. Moreover, the spacer thickness was reduced to as low as 6 mil, which further increased the membrane area to 2.94 m^2 in the most recent modules. The modules also reproduced the transport performance of the flat-sheet membrane under various testing conditions. For instance, a CO_2 permeance of 1450 GPU and a CO_2/N_2 selectivity of 185 were showed with a simulated flue gas at 67°C. In addition, satisfactory pressure drops were also achieved. From the first successful SW element made in house to the most recent ones, a scale-up factor of >70 was achieved. A predictive model was developed to relate the operating conditions to the module separation performance, accounting for the limitations set by concentration polarization and pressure drop. The experience gained is helpful to design a full-sized element (40 inches in length and 8 inches in diameter) for industrial application.

TUESDAY 4:00 pm - Kings Garden 4

19e - Development of industrial scale polyvinylidene fluoride transfer membrane

Marta E Bojarska (GVS Filter Technology)*, Deb English (GVS Filter Technology), Gil Tavares (GVS Filter Technology)

One of many life science applications is blotting, which is a method of proteins, DNA or RNA transferring onto a polymeric material that is called a transfer membrane. Typically transfer membranes are manufactured from polyvinylidene fluoride (PVDF), nitrocellulose (NC) or Nylon (NY). In techniques such as Western Blotting or Dot Blot, which are used for protein transfer, mainly PVDF and NC membranes are used. Transfer of proteins onto the membranes mostly depends on material ability to bind the proteins. The binding might be based on a hydrophobic or electrostatic interaction between the membrane and the protein. Present work will mainly focus on the development of polyvinylidene fluoride membrane for protein transfer. Hydrophobic PVDF membrane can be manufactured via Non-solvent Induced Phase Separation (NIPS), where the cast film of polymeric solution (PVDF + solvent + additives (optional)) is immersed into the coagulation bath (non-solvent or mixture of solvent and non-solvent). This technique, depending on the process parameters, allow obtaining both asymmetric and symmetric membranes. One also need to have in mind that factors such as grade of polyvinylidene fluoride, solvent type, as well as process parameters of dope solution preparation, casting and coagulation can influence membrane surface characteristic and its affinity towards proteins. In this work, the main focus will be put into subtle differences between the membranes morphology and surface characteristic that allows obtaining better sensitivity during Western Blotting or Dot Blot. At the same time, standard membrane characterizations such as air flow rate, water breakthrough, capillary flow porometry, SEM, contact angle, and protein binding were performed. Present work provides an insight into an industrial membrane manufacturing with a scientific touch for details in reference to life science applications.

TUESDAY 4:30 pm - Kings Garden 4

19f - Low energy seawater desalination using multistage electrodialysis

Gijs Doornbusch (Eindhoven University of Technology)*, Michele Tedesco (Wetsus), Jan Post (Wetsus), Zandrie Borneman (Eindhoven University of Technology), Kitty Nijmeijer (Eindhoven University of Technology)

Due to the economic and population growth, unavoidable problems arise, such as fresh water scarcity. Electrodialysis (ED) is a well-established electrochemical desalination technology based on ion exchange membranes. In an ED system, ions migrate through anion and cation selective membranes due to the applied stack potential. Energy consumption of the system strongly depends on the salt concentration of the feed solution. Due to the high operational costs compared to seawater reverse osmosis (RO), ED is currently not used for seawater desalination. Although RO is the state-of-the-art seawater desalination technology, due to technological limitations further reduction of energy consumption (2 kWh/m3) is not apparent. A strong benefit of ED compared to RO not seriously considered so far, is the possibility of staging of different ED stages. In such an approach each ED stack is operated at its optimum conditions (different current densities) aligned with the governing feed water conditions at that stage, unlike in the case of single-stage operation. This research shows the potential of staging in ED as an effective approach for seawater desalination. We experimentally compared single-stage ED with multistage ED (four desalination stages) and investigated the effect of operational parameters for both systems. Different current regimes show that multistaging in ED can be an effective tool to desalinate seawater. Using one single stage with the same membrane and electrode area as the four stages of the multistage does not allow for full desalination. Oppositely, drinking water can be produced in the multistage when operating each stage close to 95% of the measured limiting current density. The lowest energy consumption obtained while reaching the lowest diluate concentration is 3.6 kWh/m3, confirming that ED can be considered as an energy-effective desalination technology, potentially able to desalinate seawater at a competitive energy consumption compared to RO.

TUESDAY 2:00 pm - Kings Garden 3

20a - Air Products Hybrid Membrane Adsorption

Cory E Sanderson (Air Products and Chemicals)*, Don Henry (Air Products and Chemicals)

Light noble gases, primarily Helium and Neon, are rare in Earth's atmosphere and upper mantle and difficult to produce. Yet the gases are highly valuable to the many diverse markets which depend on their specific chemical and physical properties. Air Products has developed a hybrid membrane and adsorption process to extract and purify a light noble gas from a feed gas stream which has low and variable concentrations of the light noble gas at both high recovery and low unit cost. The process combines the best features of both unit operations- high light noble gas selectivity compared to bulk gases of Methane and/or Nitrogen, with a unique recycle control loop to create a process which has superior performance over any membrane or adsorption only process. The process has flexibility in both scale and source of gas which gives it a wide application and potential implementation strategy.

TUESDAY 2:30 pm - Kings Garden 3

20b - Hybrid Distillation and Facilitated Transport Membrane Processes for C3 Splitter Debottlenecking

Kenneth Pennisi (Compact Membrane Systems), Christine Parrish (Compact Membrane Systems)*, Sudip Majumdar (Compact Membrane Systems)

Propylene and propane are very important hydrocarbons in the global economy. Propylene is used to manufacture polypropylene, a very widely used polymer, as well a variety of important chemicals such as, acrylonitrile, propylene oxide, cumene, acrylic acid, and isopropanol. Propane is a widely used fuel. Obtaining these two hydrocarbons in sufficient purity for end uses almost always involves distillation to separate them from each other. Due to their low relative volatility, the distillations columns have many trays and operate at high reflux ratios. Consequently, capital and energy costs of the separation are high. Many propylene producers have been forced to operate their distillation columns at maximum capacity in an effort to keep up with rising demand. Since demand generally continues increasing year over year, many producers currently have a need for a means to meet incremental demand increases that is not as capital intensive as replacement of the distillation column. Compact Membrane Systems, Inc. has developed a novel membrane technology which can efficiently separate propylene and propane, as well as other olefin/paraffin mixtures. When retrofitted to an existing distillation column, the membrane can debottleneck the separation step of the production process. Furthermore, economic analysis has shown the approach to be very cost effective. In this paper, we will examine some of the options for configuration of the retrofit and evaluate the process economics of each.

TUESDAY 3:00 pm - Kings Garden 3

20c - Water Recovery via Forward Osmosis - Freeze Concentration

Georgios Kolliopoulos (University of Toronto), Jeffrey Martin (University of Toronto), Chenbo Xu (University of Toronto), Vladimiros Papangelakis (University of Toronto)*

Contaminants accumulate in industrial process circuits before their disposal in tailings ponds. The unstable nature of the latter often leads to catastrophic accidents such as acid mine drainage and is a threat to the prosperous development of the local communities as well as a liability for industry and governments. Therefore, energy efficient water recovery technologies are of critical importance, especially in today's era of increased environmental awareness. Recently, at the University of Toronto, we developed a proprietary green technology system to recover water from such contaminated aqueous streams. Our technology is a climate-driven hybrid forward osmosis - freeze concentration, which relies on two-steps to achieve fresh water recovery. First, water spontaneously permeates from multi-component effluents into environmentally benign single-solute inorganic draw solutions with no external energy input via forward osmosis. The above is the result of the osmotic pressure gradient between a dilute feed solution (i.e., the effluent) and a more concentrated solution (i.e., the draw solution). Subsequently, the resulting inorganic diluted solution is exposed to a temperature below its freezing point to recover water as ice, while regenerating the draw solution. Water recovery by freezing is similar to distillation but is 7 times more energy efficient. The cold winter climate in northern territories, where many industrial activities (e.g., mining) took place, operate, or are scheduled to be in operation in the next few years, can be leveraged to produce fresh water and decontaminate the land upon stabilizing the aqueous wastes via forward osmosis - freeze concentration. Hereby, we report on experimental forward osmosis and freeze concentration results for a series of inorganic draw solutions, the estimated required energy to drive our technology system, as well as a comparison of our technology with industry standards.

TUESDAY 3:30 pm - Kings Garden 3

20d - Leverage membrane technology in process intensification and new application design - from early stage design to technical scale implementation

Peter Kreis (Evonik Technology & Infrastructure GmbH)*, Kah Peng Lee (Evonik Fibres GmbH), Lars Peters (Evonik Fibres GmbH), Erik Hoving (Evonik Cooperation), Steven Pedersen (Evonik Canada Inc.), Kumar Abhinava (Evonik Cooperation)

For many separation process applications, conventional technologies like distillation, extraction and absorption are still the first choice in a conceptual separation design. New separation technologies such as membrane or membrane based hybrid separations are often not considered although advanced membrane materials and simulation infrastructure has significantly developed in recent years. Step-changes in process intensification and product-by-process applications are possible, if these technologies are considered at early stage. Beyond their stand-alone operation, membrane units are particularly effective when combined with established, often thermal, separation processes in an intelligent configuration. In this way, membrane processes have maximum impact and exhibit their advantage in synergy with other basic unit operations, while the limitations of a stand-alone membrane process are mitigated. For example, early stage design with organic solvent nanofiltration have led to new industrial efficient and low energy consuming applications in homogeneous catalyst recovery but also in the recovery of THC and CBD from cannabis. Another emerging field of using membranes over classical separation processes is upgrading of biogas to renewable natural gas and in the field of natural gas sweetening. \In this talk, we give an overview about the application diversity of membranes. We show why it is critical to consider membrane processes early on during the conceptual design process and how membrane technology leads to success.

TUESDAY 4:00 pm - Kings Garden 3

20e - Carbon Capture utilizing Hybrid Membrane/Liquifaction

David Hasse (Air Liquide)*

Air Liquide has developed mulitple hybrid CO_2 capture processes based on hollow-fiber membrane operation in combination with cryogenic distillation. (ref 1, 2, 3) Membranes offer unique ability to achieve high recovery, while distillation allows unrivalled purity. CO_2 purification is limited by the triple point of CO_2. If it was not for this thermodynamic barrier, large scale distillation similar to air separations would be the preferred method for CO_2 capture at large scales. However, CO_2 liquifaction is limited to low recovery at low concentrations unless uneconomically large pressures are used. Membranes offer a solution to raise the recovery while maintaining purity. Three separation techniques will be discussed. CO_2 capture from air fired coal, where the CO_2 is at low concentration, CO_2 capture from oxy-fired coal, where the CO_2 is at high concentration, and CO_2 capture internal to an SMR, where the diluent gas is hydrogen. Each case provides unique issues, and a hybrid membrane solution is an optimal solution. This is a true hybridization, with a recycle loop between processes. References: Energy Procedia Hasse el al, volume 63, pages 186-193 (2014) Energy Procedia Lockwood et al, volume 63, pages 342-351 (2014) Energy Procedia Pichot et al, volume 114, pages 2682-2689 (2017)

TUESDAY 4:30 pm - Kings Garden 3

20f - Membrane distillation hybridized with a thermoelectric heat pump for energy-efficient water treatment and space cooling

Yong Zen Tan (Nanyang Technological University), Nick Guan Ping Chew (Nanyang Technological University), Wai Hoong Chow (Nanyang Technological University), Rong Wang (Nanyang Technological University), Jia Wei Chew (Nanyang Technological University)*

The current concept for cooling the indoors is far from ideal with respect to the total energy consumed and waste discharged. A novel concept for improving the energy efficiency is proposed via hybridizing the heat pump with a membrane distillation (MD) unit for simultaneous space cooling and water treatment. MD is well-acknowledged for utilizing low-quality waste heat for water treatment, which makes it feasible for coupling with a heat pump to make use of both the hot and cold reservoirs of the pump. Accordingly, the objective of the current effort was to investigate via experiments the efficacy of a thermoelectric heat pump coupled with a sweep-gas MD system (T-SGMD) by measuring the cooling capacity, condensate production and power consumption. The results from this study can be extended to other heat pumps. Three key highlights emanated from this study. Firstly, condensate production per unit energy consumed can be doubled with the T-SGMD system relative to thermoelectric dehumidification alone. Secondly, cool air recycle affected the condensate flux the most without a drastic loss of cooling compared to other tested parameters during the operation of the T-SGMD. Lastly, the T-SGMD system was able to provide an increase in condensate produced per unit energy without a loss in cooling capacity per unit energy input. These advantages of coupling heat pumps with MD, leveraging on the current advancements in MD, is promising for a hybridized system for decentralized water treatment, dehumidification and space cooling.

TUESDAY 2:00 pm - Ballroom 3

21a - Reduction of Brackish Water Desalination Brine Volume using Membrane Evaporation Coupled with Activated Sludge Aeration Waste Heat

Drew W Johnson (The University of Texas at San Antonio)*

Desalination processes produce highly concentrated brine effluent which is costly to dispose of for inland brackish water desalination facilities. The current study focuses on membrane evaporation processes to reduce brine disposal volume where brine concentrate is a cold fluid that enters inside hollow fiber lumens while exposing the outer surface to low grade heat obtained from compressed air at wastewater treatment facilities. Aeration is the most energy-intensive operation in wastewater treatment, amounting to 45-75% of plant energy costs. During aeration, the compression of the air causes an increase in air temperature (often modeled as adiabatic). This waste heat can be utilized to drive heat and mass transfer operations for evaporating brine concentrate from inside the membrane. As an example, the activated sludge water recycling center for the city of San Antonio TX treats 303 MLD of wastewater in a two-stage aeration mode by providing 1,400 MLD of air with a rated discharge pressure of 186 kPa and a temperature of 345 K. For these conditions, membrane evaporation was evaluated at the bench scale for a brackish water brine at different brine flow rates, airflow rates and air temperatures, extent of fouling and possible means to control fouling. The evaporation rate was found to increase with airflow rate and air temperature and decrease with brine flow rate. Model predictions, derived based upon literature values for heat and mass transfer correlations, agreed well with the measured evaporation rates. Fouling was observed in experiments for brine with and without ferrous iron. Cleaning of the membranes by flushing the fibers with brine at higher flow rates was not able to alleviate fouling. Fouling could be alleviated when the pH of the brine solution was lowered for the duration of the studies conducted. The economics of the proposed process appear favorable if energy demands are neglected when using waste heat.

TUESDAY 2:30 pm - Ballroom 3

21b - Smart High Recovery RO Boosts Recovery Rate to More Than 90% for Reuse Purposes

Ronit Erlitzki (AdEdge Water Technologies)*

Concentrate management is a key issue for the use of reverse osmosis (RO) technology. Reducing the amount of RO concentrate is critical for all inland RO applications. This paper will present a case history of a reverse-osmosis plant located in Southeast Asia that is using municipal treated tertiary effluent for industrial reuse purposes. The recovery rate (RR) of the system was limited to 75% due to scaling from relatively high feed water concentrations of phosphate, calcium, and iron, resulting in frequent CIP cycles. Attempts to increase RR by conventional methods to 85%, led to increase in CIP frequency from 1 - 3 months to about two weeks. To improve the performance in a cost-effective manner, the existing 2,000 gpm RO was retrofitted in early 2017 with Flow-Reversal (FR) technology. Currently, the retrofitted train is operating at 90% RR with the same CIP interval that was required to achieve the lower RR. The RR is expected to increase gradually to achieve higher than 90% while keeping the balance between RO recovery, CIP intervals, and water production optimization. The increase in water production has translated into 60% decrease in concentrate volume, a 20% growth in annual revenue, and a return on investment of 1.5 years. The Flow-Reversal RO technology employs conventional RO equipment with two unique functions to improve recovery. The first is allowing each array of RO vessels to periodically function as the third stage in the process. The second, which happens concurrently, is periodically reversing the direction of the feed stream through each one of the pressure vessels. This approach reduces the adverse effects of scaling by rotating to the next group of pressure vessels before scale is precipitated out, and by switching the direction of feed stream to flush out what small amount of scale that might have embedded on the membrane surface.

TUESDAY 3:00 pm - Ballroom 3

21c - Membrane Distillation for Enhanced Water Recovery by Inland Brackish Water Desalination Plants

Zhewei Zhang (University of Pittsburgh)*, Radisav Vidic (University of Pittsburgh)

Inland brackish water (BW) desalination plants typically use reverse osmosis (RO) for the production of drinking water. However, low water recovery (i.e., 50 - 75%) results in fairly large quantities of the reject (concentrate) stream that needs to be disposed at considerable cost and potentially significant adverse environmental impacts. Membrane distillation (MD) offers a unique opportunity to improve the performance of these BWRO plants by recovering additional clean permeate from the reject stream, thereby reducing the volume of concentrate that needs disposal. Unlike RO that uses pressure to drive water through the membrane, MD relies on the vapor pressures difference between the feed and permeate to reject all inorganic and non-volatile organic contaminants and recover almost pure water on the permeate side. Membrane fouling and wetting due to precipitation of common salts on the feed side (e.g., gypsum, silica) affect MD performance in terms of flux decline and deterioration in permeate quality. Therefore, appropriate feed pretreatment to remove potential scaling components and ensure high water recovery from the reject stream is essential for the application of this treatment technology. In this study, barite precipitation was utilized to remove sulfate from actual BWRO concentrate as the key precipitate that would interfere with MD operation. The effect of barium concentration and mixing condition was evaluated for its impact on the removal of sulfate and silica as the most common inorganic scales. The pre-treated BWRO concentrate was treated in direct contact membrane distillation (DCMD) system to achieve at least 80% water recovery. The performance of optimized pretreatment process and DCMD design and operating parameters will be discussed together with the impact of MD treatment on the overall performance of inland desalination plants in terms of water recovery and environmental impacts.

TUESDAY 3:30 pm - Ballroom 3

21d - Selective Separation of Industrial Relevant Metal Ions from High TDS Water using Wafer-Enhanced Electrodeionization (WE-EDI)

Humeyra B Ulusoy Erol (University of Arkansas)*, Christa Hestekin (University of Arkansas), Jamie Hestekin (University of Arkansas)

There is a need for an efficient and economical process of ion-selective separation. In hydraulic fracturing, for instance, several million gallons of water are used in a site, most of which are disposed of by deep well injection taking the water from the ecosystem and causing earthquakes. However, due to the high amount of ions in wells at fracking sites, pressure driven processes for water recovery are impossible to use. This project will demonstrate an emerging environmentally friendly Wafer Enhanced Electrodeionization (WE-EDI) technology which attempts to eliminate the shortcomings that come with normal EDI and separates effluent streams into concentrated ions and clean water. In WE-EDI, ion-exchange resins are replaced by wafers that consist of cation and anion exchange beads and a polymer binding agent. In this study, selective separation of these ion exchange wafers and the effect of electroosmotic drag on the membrane structure have been performed for various anions and cations ($Na+$, $K+$, $Ca2+$, $Cl-$) and improved selectivity and higher water recoveries have been observed while significant modeling has taken place to predict the selective separation as a strong function of the equilibrium constants between the ions and the beads.

TUESDAY 4:00 pm - Ballroom 3

21e - Pilot Evaluation of Closed Circuit Reverse Osmosis (CCRO) for Treatment of RO Concentrate for Potable Reuse

Han Gu (Orange County Water District)*, Megan Plumlee (Orange County Water District), Mike Boyd (Desalitech), Ran Nadav (Desalitech), Michael Hwang (Jacobs), Jim Lozier (Jacobs)

The Orange County Water District (OCWD) Groundwater Replenishment System (GWRS) purifies secondary wastewater for potable reuse and currently discharges 18 million gallons per day (MGD) of RO concentrate (ROC) to the ocean. This will increase to ~23 MGD after plant expansion in 2023. OCWD seeks to identify a cost-effective approach to minimize RO concentrate waste and generate new water supply for the region. A field pilot evaluation of Desalitech's Closed Circuit Reverse Osmosis (CCRO) technology was conducted at OCWD to recover water from ROC. CCRO recirculates its concentrate until the desired recovery is reached. The objective of the test was to evaluate the performance of CCRO technology in terms of recovery, clean-in-place (CIP) frequency and permeate quality. This presentation will focus on challenges experienced (organic fouling, mineral scaling, and ROC water quality fluctuations) and adaptive control strategies that improved performance at GWRS. When CCRO was operated with GWRS ROC as the feed, rapid permeability decline was observed and suspected to be caused by organic fouling based on membrane autopsy. Treating ROC while using GWRS RO feed (i.e., microfiltration effluent) to refill the CCRO side conduit at the end of each cycle extended run time by more than two fold. In addition, high cross flow velocity was shown to improve mixing and reduce fouling. ROC water quality fluctuations also created a major challenge for maintaining performance. To address these fluctuations in feed water quality, the maximum recovery, concentrate conductivity, and feed pressure were used as thresholds to to engage the end/re-start of each cycle. Instead of running at a preset maximum recovery, the pilot is operated in a variable recovery mode where the cycle-to-cycle recovery is controlled by both the maximum concentrate conductivity and recovery setpoint. Two long-term, continuous runs were successfully completed to confirm the optimal CIP frequency of the CCRO pilot.

TUESDAY 4:30 pm - Ballroom 3

21f - Brine Treatment for Minimal and ZLD

Malynda Cappelle (The University of Texas at El Paso)*, W. Shane Walker (The University of Texas at El Paso)

Globally, demand for water and energy are increasing with increasing population. High recovery desalination can mitigate environmental and economic concerns with brackish desalination by minimizing the amount of liquid waste. Zero Discharge Desalination (ZDD) has been demonstrated to achieve up to 98% recovery from brackish water while producing suitable water for drinking water, industrial, or agricultural purposes. ZDD combines a pressure-driven desalination process (reverse osmosis or nanofiltration) to produce high quality drinking and irrigation water and an electrically-driven process (electrodialysis metathesis) to reduce concentrate volume and improve system recovery. A study was performed using a mathematical ZDD model, which was expanded to include byproduct recovery, specifically sodium chloride (which is used for the EDM process in ZDD), as well as calcium sulfate and magnesium hydroxide. The updated ZDD model was used to evaluate the economic tradeoffs in ZDD design. The model results indicate that for a fixed water chemistry, the unit cost of water 1) decreases with decreasing salt rejection of the primary desalting membrane, 2) decreases with increasing RO/NF recovery, and 3) decreases with decreasing EDM membrane resistance. A sensitivity analysis was performed for a ZDD design with NF operating at 75% recovery (hybrid NF90-NF270 staging), and the results of this analysis suggest that the unit cost 1) increases by 13-18% if more efficient, more expensive EDM membranes are employed, 2) is substantially affected by EDM equipment and power unit cost assumptions, and 3) is only slightly impacted by RO/NF equipment and evaporation pond cost assumptions; however the membrane choice and RO/NF recovery do have a noticeable impact on product water quality, ZDD specific energy consumption, and ZDD system recovery. The incorporation of salt recovery decreased the cost of all ZDD designs by approximately 10-30%, even when revenue from salt sales was excluded.

TUESDAY 2:00 pm - Ballroom 4

22a - Novel Spirocyclic Polymers for Membrane-based Organic Solvent Separations

Ronita Mathias (Georgia Institute of Technology)*, Kirstie Thompson (Georgia Institute of Technology), Dhaval Bhandari (Exxon Mobil), JR Johnson (Exxon Mobil), Huaxing Zhou (Exxon Mobil), M.G. Finn (Georgia Institute of Technology), Ryan Lively (Georgia Institute of Technology)

Solution-processable microporous polymers have the potential to outperform traditional polymeric membranes in the area of organic solvent separations. In this work, novel spirocyclic polymers were incorporated into thin-film composite membranes with the aim of improving solute rejection when compared against current spirocyclic polymers such as PIM-1. The separation system of interest is a complex mixture of hydrocarbon solvents that are typically found in the light fractions of crude oil. These compounds were utilized not only to create a molecular weight cut-off (MWCO) curve but also to assess the performance of this class of membranes in the fractionation (as opposed to purification) of complex mixtures. We find that the novel spirocyclic polymers have higher rejections than PIM-1 membranes, although the permeances are lower using our non-optimized thin film composite fabrication techniques. A MWCO of ~ 250 Da can be obtained, which is accompanied by high selectivities amongst different classes of hydrocarbons of a similar size. Moreover, high selectivities are also observed amongst molecules within the same class with only a sub-nm difference in size. Such size- and shape-selective separation enables the use of these polymeric membranes in a wide range of applications.

TUESDAY 2:30 pm - Ballroom 4

22b - 3D Printed fouling resistant composite membranes

Davide Mattia (University of Bath)*

The formation and growth of fouling layers is a long-standing challenge for the continuous operation of membrane processes. Membrane patterning is considered as a potential route for fouling mitigation by promoting shear stress and creating localised turbulence near the membrane surface. Patterned membranes have been fabricated using a variety of techniques, e.g., moulding and stamping, which lack sufficient fidelity and flexibility, limiting their actual use. 3D printing allows overcoming these challenges by enabling the design and fabrication of complex patterns impossible to manufacture otherwise.1 In this paper we report the fabrication of 3D printed composite membranes with a double-sinusoidal, i.e. wavy, pattern, resulting in enhanced fouling resistance whilst reducing water, chemical cleaning and energy consumption.2 The 3D composite membrane was obtained by conformal deposition of a thin polyethersulfone (PES) selective layer onto a 3D printed wavy support. Flat (non-patterned) and wavy (patterned) 3D composite membranes were characterised and tested in terms of permeance, rejection, and anti-fouling property using oil-in-water emulsions and by filtering Bovine Serum Albumin (BSA) through a cross-flow ultrafiltration set-up. CFD simulations in terms of velocity profile, local surface shear rate, and transmission probability were used to optimise the design of the wavy pattern and the experimental observations. Results revealed that the wavy composite membrane had significantly higher permeance recovery compared to the flat one, with negligible irreversible fouling and rejection >96% for both oil and BSA. The wavy 3D composite membranes maintained the same level of permeation after 10 complete filtration cycles, using only pure water as cleaning agent. 1. Low, Z.-X.; et al., J Mem Sci 2017, 523, 596-613. 2. Al-Shimmery, A.; et al., J Mem Sci 2019, 574, 76-85.

TUESDAY 3:00 pm - Ballroom 4

22c - Next-Generation Membranes Using Thin-Film Lift Off

Mackenzie Anderson (UCLA)*, Brian McVerry (UCLA), Na He (UCLA), Eric Hoek (UCLA), Richard Kaner (UCLA)

For the past 30 years, thin-film membrane composites have been the state-of-the-art technology for reverse osmosis, nano- and ultrafiltration, and gas separation. However, traditional membrane casting techniques, such as phase inversion and interfacial polymerization, limit the types of material that are used for the membrane separation layer. Here we describe a novel thin-film lift off (T-FLO) technique that enables the fabrication of thin-film composite membranes with new materials for desalination, organic solvent nanofiltration, and gas separation. The active layer is cast separately from the porous support layer, allowing for the tuning of the thickness and chemistry of the active layer. A fiber-reinforced, epoxy-based porous support layer is then cured on top of the active layer that forms covalent bonds between the layers. Upon submersion in water, the cured membrane lifts off from the substrate to produce a robust, free-standing asymmetric membrane composite. We demonstrate the fabrication of three novel T-FLO membranes for chlorine-tolerant reverse osmosis, organic solvent nanofiltration, and gas separation. The isolable nature of support and active layer formation paves the way for the discovery of the transport and selectivity properties of new polymeric materials. This work introduces the foundation of T-FLO membranes, which enables exciting new materials to be implemented as the active layers of thin-film membranes, such as high performance polymers, 2-D materials, and metal-organic frameworks.

TUESDAY 3:30 pm - Ballroom 4

22d - Interfacial Junctions Control Electrolyte Transport Through Charge-Patterned Membranes

Feng Gao (University of Notre Dame), Bill Phillip (University of Notre Dame)*

Unique transport mechanisms emerge when nanostructured substrates are patterned with multiple chemistries. For example, due to their charge-patterned structure, charge mosaic membranes are capable of transporting whole salts more rapidly than neutral molecules of comparable or smaller sizes. Here, we elucidate the molecular interactions that promote the transport of whole salts through patterned membranes by using ink jet printing to create a family of membranes with systematically varying the pattern geometry and size. For patterns that consist of equivalent areal coverages of positively-charged and negatively-charged domains, the effects of these geometric parameters were encapsulated in a single variable, the interfacial packing density, that quantified the fraction of the membrane surface covered by junctions between oppositely-charged domains. Experimentally, the transport of symmetric electrolytes (i.e., KCl and $MgSO_4$) increased with the value of the interfacial packing density, while the interfacial packing density did not affect the transport of asymmetric electrolytes (i.e., K_2SO_4 and $MgCl_2$). Simulations demonstrate that for symmetric electrolytes, the structural charge heterogeneity reduces the barrier to ion partitioning thereby promoting salt transport. For asymmetric electrolytes, the charge heterogeneity skews the local availability of ions from the stoichiometric ratio of the salt thus hindering salt transport. These findings demonstrate the promise of accessing new transport mechanisms that could revolutionize chemical separations and sensing applications through chemical-patterning of membranes.

TUESDAY 4:00 pm - Ballroom 4

22e - Fabrication of Inside-Out Isoporous Hollow Fiber Membranes via Spinning and Coating Methods

Kirti Sankhala (Helmholtz-Zentrum Geesthacht, Geesthacht)*, Joachim Koll (Helmholtz-Zentrum Geesthacht, Geesthacht), Maryam Radjabian (Helmholtz-Zentrum Geesthacht, Geesthacht), Clarissa Abetz (Helmholtz-Zentrum Geesthacht, Geesthacht), Volker Abetz (Helmholtz-Zentrum Geesthacht, Geesthacht)

Isoporous block copolymer membranes enable to tailor the selective surface with promising energy-efficient separations.1-3 The membranes are fabricated via evaporation-induced self-assembly of block copolymers and non-solvent induced phase separation (SNIPS).2 The isoporous block copolymer membranes were first reported in flat sheet geometry and afterwards the self-assembled structures were achieved on the outer surface of hollow fibers.2,3 Self-supported hollow fiber membrane geometry provides a higher active surface area for lesser volume.3,4 So far, evaporation under normal environmental conditions has been a common approach to achieve microphase separation of block copolymers resulting in regular self-assembled structure. Our recent works introduced a novel approach of controlling the kinetics of microphase separation of block copolymers by using gas flow. PS-b-P4VP diblock copolymer solutions were used for fabrication of technologically-relevant inside-out isoporous hollow fiber membranes. Integral asymmetric and composite inside-out isoporous hollow fiber membranes of inner diameter ca. 1 mm were fabricated by spinning and coating processes, respectively.5,6 This novel approach of controlled evaporation via gas flow seems the only feasible way to achieve evaporation-induced self-assembly in compact geometries. [1] V. Abetz. Macromolecular Rapid Communication 2015, 36:10. [2] K.-V. Peinemann, V. Abetz, P.F.W. Simon. Nature Materials 2007, 6:992. [3] M. Radjabian, J. Koll, K. Buhr, U. Vainio, C. Abetz, U.A. Handge, V. Abetz. Polymer 2014, 55:2986. [4] K. Sankhala, D. C. F. Wieland, J. Koll, M. Radjabian, C. Abetz, V. Abetz. Nanoscale 2019, DOI: 10.1039/C8NR06892E [5] K. Sankhala, J. Koll, M. Radjabian, U. A. Handge, V. Abetz. Advanced Materials Interfaces 2017, 4:1600991. [6] K. Sankhala, J. Koll, V. Abetz. ACS Macro Letters 2018, 7:840.

TUESDAY 4:30 pm - Ballroom 4

22f - Empirical evidence suggests that polyamide film formation during interfacial polymerization in thin film composite (TFC) membranes is not self-limiting

Kasia Grzebyk (University of North Carolina at Chapel Hill)*, Orlando Coronell (University of North Carolina at Chapel Hill)

Thin-film composite (TFC) membranes used in reverse osmosis typically have a polyamide active layer formed by interfacial polymerization (IP) between monomers m-phenylenediamine (MPD) and trimesoyl chloride (TMC). In IP, a support layer containing an aqueous MPD solution makes contact with TMC in an organic solvent. The monomers meet at the aqueous/organic interface, where MPD diffuses into the organic solvent and reacts with TMC to form the polyamide. It is commonly thought that MPD and TMC react to form a film so dense that MPD passage into the organic phase is restricted, thereby terminating polyamide growth. Variations can be employed in IP to alter the polyamide, but regardless of the variation, TFC polyamide layers exhibit a typical maximum thickness of ~200 nm. It is, therefore, believed that IP is self-limiting (Chai and Krantz, 1994, J Membr Sci, 93, 175-192; Cahill et al, 2008, MRS Bull, 33, 27-32). However, in recent work, Ukrainsky and Ramon (J Membr Sci, 2018, 566, 329-335) promoted IP in a microfluidic flow cell and found that IP proceeded so long as monomers were supplied; films grew orders of magnitude thicker compared to conventional TFCs. Accordingly, we investigated the self-limiting quality of IP in TFC membrane fabrication. Our results show that MPD can permeate the active layer of the four TFCs tested. In cross-flow filtration, MPD passage was 22.3%, 4.1%, 3.5%, and 2.8% for NF90, ESPA3, SW C_4+ and hand-cast TFC membranes, respectively. Furthermore, a method for additional polyamide formation atop TFC membranes was established, where membranes are exposed to a continuous source of MPD during IP. Resulting polyamide thicknesses were ~10x greater than that of conventional TFCs. Our findings indicate that IP terminates due to MPD depletion during IP, not due to its inability to penetrate the polyamide. Our work therefore suggests that novel IP protocols may be developed to produce TFCs with enhanced performance.

WEDNESDAY 9:30 am - Kings Garden 4

23a - Creation of well-defined "mid-sized" micropores in carbon molecular sieve membranes for organic solvent separations

Yao Ma (Georgia Institute of Technology)*, Dhaval Bhandari (Exxon Mobil), JR Johnson (Exxon Mobil), Ryan Lively (Georgia Institute of Technology)

Carbon molecular sieve (CMS) membranes are promising candidates for the separation of organic molecules due to their stability, ability to be scaled to practical form factors, and the avoidance of expensive supports or complex multi-step fabrication processes. A critical challenge is the creation of "mid range" (e.g., 5-9 Å) microstructures that allow for both facile permeation of organic solvents and selection between similarly-sized guest molecules. Here, we create these microstructures via the pyrolysis of a microporous polymer (PIM-1) under low concentrations of H_2 gas. The introduction of hydrogen inhibits aromatization of the decomposing polymer network and ultimately results in the creation of a well-defined bimodal pore network with the smallest micropores existing at 5.1 Å. The "H_2-assisted" CMS membranes show a dramatic increase (~15 times) in p-xylene ideal permeability, while still maintaining commercially attractive p-xylene/o-xylene selectivities when compared to CMS membrane pyrolyzed under a pure argon atmosphere. This technique has the potential to be a platform technique for the creation of CMS microstructures capable of separating a wide variety of organic solvent molecules.

WEDNESDAY 10:00 am - Kings Garden 4

23b - PoreDesigner for computational optimization of channel proteins and porous organic cages for highly selective membrane-based separations

Ratul Chowdhury (Penn State University)*

Monodispersed angstrom-size pores embedded in a suitable matrix are promising for highly selective membrane-based separations. They can contribute to energy savings in water treatment and small molecule bioseparations. Aquaporin-based membranes are commonplace in biological membranes to this end, but their low structural stability, modest selectivity and non-tunable selectivity thwarts their applicability as synthetic industrial membranes. We have developed PoreDesigner, a fully automated workflow to redesign the robust beta-barrel Outer Membrane Protein F as a scaffold to access three specific pore designs that exclude solutes larger than sucrose (>360 Da), glucose (>180 Da), and salt (>58 Da) respectively. PoreDesigner further enables us to design any specified pore size (spanning 3-10Å), engineer its pore profile, and chemistry. These redesigned pores may be ideal for conducting sub-nm aqueous separations with permeabilities exceeding those of classical aquaporin-based biological water channels, by more than an order of magnitude at over 10 billion water molecules per channel per second. The versatility of PoreDesigner is not limited to designing water channel proteins and is being used to design channel proteins for DNA sequencing. In other efforts, tools have been developed to simulate and explain the biophysics of water permeation and complete salt rejection through porous pillar-appended pillar[4]arene (PAP4) cages immobilized in biomimetic membranes. The in silico approach was used to assess the energetics, stability, and lowest energy structure of a bouquet of such channels whose unique interlaced organization offered a low-activation salt-rejecting continuous water path with water permeabilities of billion molecules per second.

WEDNESDAY 10:30 am - Kings Garden 4

23c - Crack-free, large-area, single-layer graphene membranes with a record performance in gas mixture separation

Shiqi Huang (L'Ecole polytechnique fédérale de Lausanne)*, Kumar Varoon Agrawal (L'Ecole polytechnique fédérale de Lausanne), Jing Zhao (L'Ecole polytechnique fédérale de Lausanne)

Among the emerging membrane platforms, single-layer graphene based membranes are highly promising attributing the atomic thickness. As a result, they have been acclaimed as the ultimate gas separation membrane, capable of yielding ultrahigh gas permeance and an attractive molecular selectivity [1]. However, the development of these membranes face two major bottlenecks: 1) crack-free fabrication of large-area membranes; 2) incorporation of high-density of size-selective nanopores. Herein, we report novel approaches to address these bottlenecks achieving a record separation performance from a relatively large-area graphene membrane [2,3]. The mechanical strength of single-layer graphene was increased by reinforcing it with 100-nm-thick nanoporous carbon film, which eliminated crack formation in the suspended graphene domains even on macroporous supports with 5-μm-sized support pores. A high pore-density (2.1×10^{12} pores/cm^2) with a tight pore-size-distribution was incorporated by a novel decoupled pore-nucleation and pore-expansion strategy using a combination of oxygen plasma and ozone that reduced the concentration of pores larger than 0.38 nm to 22 ppm, resulting in a record gas mixture separation performance (H_2 permeance of 1340 to 6045 GPU; H_2/CH_4 separation factor of 15.6 to 25.1; H_2/C3H8 separation factor of 38.0 to 57.8) [3]. Implementation of these strategies in a homemade novel millisecond reactor led to attractive CO_2/CH_4 separation factor of 18. The membranes were stable at high temperature (150 °C) and could withstand transmembrane pressure difference of 7 bar. Overall, we report the first successful implementation of single-layer graphene film in gas mixture separation and demonstrate that graphene based membranes are indeed capable of reaching the predicted high performance in gas separation. References 1. Wang et al., Nat. Nanotechnol., 2017, 12 (6), 509. 2. Huang et al., Nat. Commun., 2018, 9, 2632. 3. Zhao et al., Sci. Adv., 2019, 5, eaav1851.

WEDNESDAY 11:00 am - Kings Garden 4

23d - Unprecedented CO_2/H_2 Selectivity and CO_2 Permeance Demonstrated in Facilitated Transport Membranes with Tunable Amine-CO_2 Chemistry

Yang Han (Ohio State University)*, Winston Ho (Ohio State University)

CO_2-selective, amine-containing facilitated transport membranes are of great interest for syngas purification since high-pressure H_2 can be retained upon CO_2 removal. A variety of amine-containing polymers have shown decent chemical and thermal stability at aggressive conditions, but their CO_2/H_2 separation properties are largely limited by the hydrostatic compaction and severe carrier saturation associated with the high syngas pressure. Herein, we report a new approach to enhance the CO_2 permeability by manipulating the steric hindrance of the amine carriers. A series of α-aminoacids with different alkyl or hydroxyethyl substituents are deprotonated by 2-(1-piperazinyl)ethylamine, leading to involatile amine carriers with different degrees of steric hindrance. For hosting the low MW amine carriers, a crosslinked and water-swellable polymer network is synthesized from polyvinylalcohol and polyvinylamine. In order to avoid membrane compaction, perforated graphene oxide mono-sheets are dispersed as reinforcement fillers. In the presence of moisture, a bulkier alkyl substituent to the amino site (increasing steric hindrance) destabilizes the carbamate adduct and thus drastically increases the chemisorption of CO_2, while the incorporation of ethylene oxide groups provides additional physisorption of CO_2. The enhanced CO_2 solubility significantly mitigates the carrier saturation behavior, and an unprecedented CO_2/H_2 selectivity greater than 100 is demonstrated at 107°C and 12.5 atm of CO_2 partial pressure. As the CO_2 partial pressure reduces to 0.4 atm, a less hindered amine yields a higher reactive diffusivity of CO_2, resulting in a CO_2 permeance of 435 GPU with a selectivity greater than 500. The performances of these reaction-mediated polymeric membranes are well above the theoretical upper bound, and they open up a new avenue for designing a highly-selective membrane process for syngas purification.

WEDNESDAY 11:30 am - Kings Garden 4

23e - Engineering the Nanochannels in Reduced Graphene Oxide Membrane for Dye Desalination

Liang Huang (The State University of New York at Buffalo)*, Haiqing Lin (The State University of New York at Buffalo)

Graphene and its derivatives, graphene oxide (GO) and reduced graphene oxide (rGO), are excellent platforms for developing size-selective membranes because of their atomic thickness, high mechanical strength, and chemical inertness. However, they usually show moderate rejections for salt ions and high rejections for various dye molecules. Therefore, they are not applicable for dye desalination, which is a big challenge in textile wastewater treatment and involves the separation of dyes from salts (NaCl, Na_2SO_4). On the other hand, commercial nanofiltration (NF) and ultrafiltration (UF) membranes are also not efficient in dye desalination, because NF membranes will reject both divalent salts and dyes, while UF membranes have a low rejection for dye molecules. Here we prepared solvated rGO membranes by filtrating rGO dispersion through Nylon microfiltration films. The filtration process was stopped immediately as the rGO dispersion in filter disappeared, while the membrane was still wet, which is the critical step to retain high water flux and strong size-sieving ability. The nanochannel size in solvated rGO membranes is smaller than the sizes of most dye molecules (such as Direct Red 80) and larger than those of divalent salts. So the solvated rGO membranes can reject about 99.0% of Direct Red 80 while showing almost 0% rejection for Na_2SO_4. Interestingly, we found that the nanochannels in solvated rGO membranes can be narrowed in salt solutions, which was proven by measuring their molecular weight cutoff (MWCO) curves in 60 g/L Na2SO4 solution. As a result, the solvated rGO membranes exhibited higher dye rejection (99.9% for Direct Red 80) in 60 g/L Na_2SO_4 solution with almost zero rejection for Na_2SO_4. More importantly, their pure water permeance is as high as 80 $L/m^2 \cdot h \cdot bar$, which is about 8 times that of NF membranes and 2 times that of UF membranes (MWCO: 2000), suggesting their promise for practical applications in dye desalination.

WEDNESDAY 12:00 pm - Kings Garden 4

23f - Impact of module design on heat transfer and Nusselt correlation selection in membrane distillation

Alexander V Dudchenko (Carnegie Mellon University)*, Meagan S Mauter (Carnegie Mellon University)

Membrane distillation (MD) is a thermally driven process that can treat high salinity waters utilizing low grade heat or renewable sources, making it an attractive low cost treatment process for heavily impaired waters (e.g. waste brines, produced water, saline ground waters). The driving force in MD process (flux, water recovery) is governed by the heat transfer to/away from the membrane surface, which if poorly estimated can lead to an inaccurate process performance prediction. Although, the effect of flow channel design, fluid flow rates, and thermophysical properties has been well understood in heat exchanger design, leading to development of over a dozen empirical Nusselt correlations, these relationships have not been validated in MD. Here, we present a novel approach that leverages our fundamental understanding of vapor transport and energy transfer through the membrane, removing need for high accuracy equipment for estimation of heat transfer rates. Using this method, we demonstrate how module design (e.g. length, width) impacts heat transfer in experimental modules and apparent membrane performance. We further extend the method to allow quantitative validation of existing Nusselt correlations. Utilizing a rich experimental data set for open and spacer filled channels we demonstrate how validity of these correlations changes across different module designs, providing guidance for accurate correlation selection.

WEDNESDAY 9:30 am - Kings Garden 3

24a - Experimental and Theoretical Investigations of the Loss of Protein Sieving Due to Fouling of ATF Microfilters

Alex Apostolidis (Amgen)*, Glen Bolton (Amgen)

A prevalent technology to enable perfusion in cell culture processes is alternating tangential flow (ATF) filtration. ATF operations can be limited by sieving loss due to membrane fouling. In this study, ATF filters were fouled using a range of levels of a model cell lysate stream and antibody sieving was measured as a function of processing time. The sieving of antibody through filters with different fiber lengths or diameters were studied using the model stream. The effects of feed and permeate flux were investigated. The results suggest that sieving can be improved using conditions that lead to higher membrane sweeping and backflow: longer fibers, smaller diameter fibers, higher feed flow rate, and lower permeate flow rates. The results were well described by theoretical models that were recently developed to capture the physics behind microfilter fouling. The results were then confirmed in bioreactor experiments with model solution and cell culture. These observations can lead to operational improvements that can improve the yield and robustness of ATF filtration.

WEDNESDAY 10:00 am - Kings Garden 3

24b - pH and Excipient Partitioning in Ultrafiltration / Diafiltration Processes for Formulation of High Concentration Monoclonal Antibody (mAb) Products

Parinaz Emami (Penn State University)*, Youngbin Baek (Penn State University), Nripen Singh (Bristol-Myers Squibb), Andrew Zydney (Penn State University)

Ultrafiltration (UF) and diafiltration (DF) are used for the final formulation of monoclonal antibody therapeutics, placing the product into the desired buffer at the target concentration. A number of studies have demonstrated that the pH and excipient / salt concentration at the end of the DF can be significantly different than that in the DF buffer, but there are still significant uncertainties over the origin of these phenomena. The objective of this study was to develop a mass balance model to describe the pH and excipient profiles throughout the UF/DF process accounting for the effects of Donnan equilibrium, buffer -protein and protein -protein interactions. Experimental studies were performed with two mAbs provided by Bristol-Myers Squibb. UF and DF were conducted using Pellicon® 3 TFF Cassettes with 30 kDa Ultracel membranes (MilliporeSigma). The DF was performed from a phosphate buffer into a low ionic strength histidine buffer, in some cases followed by concentration up to 200 g/L by UF, with the solution pH and histidine concentration evaluated over the course of the process. There was a small increase in pH during the initial stage of the diafiltration even though the feed and DF buffer were both at pH 7. This increase in pH was due to the shift in effective buffer pKa due to the reduction in ionic strength. The pH over the latter portion of the DF was determined by the mAb charge via Donnan equilibrium effects, with the final pH being below that in the DF buffer for a negatively-charged mAb and above that for a positive mAb. The pH continued to change during the final UF due to a combination of Donnan effects and the change in mAb charge with increasing mAb concentration associated with intermolecular electrostatic interactions between the proteins. Model calculations were in good agreement with the experimental data during both DF and UF, providing a framework for the design and optimization of UF/DF processes to achieve the desired mAb formulation.

WEDNESDAY 10:30 am - Kings Garden 3

24c - Effect of membranes molecular weight cut-off on peptides migration and selectivity during electrodialysis with filtration membranes

Sabita Kadel (Laval University)*, Jacinthe Thibodeau (Laval University), Carole Laine (Amer-Sil), Laurent Bazinet (Laval University)

Filtration membranes (FM) are an integral part of electrodialysis with filtration membranes (EDFM); therefore, their properties have significant impact on peptide separation and overall performance. Thus, the objective of this study was to understand the effect of FM properties (MWCO, porosity, pore size distribution, surface charge, contact angle, etc.) on global and selective peptides migration to the anionic (A-RC) and cationic (C+RC) peptides recovery compartments. In this study, six polyether sulfone (PES) FM with different molecular weight cut-offs (MWCO) of 5, 10, 20, 50, 100 and 300 kDa were used to separate peptides from complex whey protein hydrolysate. As expected, global migration rate of peptides to both recovery compartments was the lowest for 5 kDa (A-RC= $2.41\pm0.34 g/m^2\cdot h$, C+RC=$3.19\pm0.15 g/m^2\cdot h$) and the highest for 300 kDa (A-RC= $11.28\pm1.40 g/m^2\cdot h$, C+RC=$10.92\pm1.65 g/m^2\cdot h$): the migration rate as a function of MWCO increased in linear way. Surprisingly, the relative abundance of identified cationic peptides migrating to C+RC through studied FMs were found to be similar. However, in A-RC, the relative abundance of negatively charged peptides, notably the lower molecular mass peptides IDALNENK (MW=915.4661 Da) & VLVLDTDYK (MW=1064.5753 Da) was the highest for 5 and 10 kDa. On the other hand, the relative abundance of highly negatively charged peptides with higher molecular mass TPEVDDEALEK (MW=1244.5772 Da) & VYVEELKPTPEGDLEILLQK (MW=2312.2515 Da) increased with an increase in MWCO. These results indicate that MWCO as well as FM properties such as surface charge, porosity, and pore size distribution have significant impact on the selective migration of anionic peptides but no effect on cationic peptides when negatively charged FM (PES) were used.

WEDNESDAY 11:00 am - Kings Garden 3

24d - Scale down depth filtration case study using Pall Supracap™ 50 to predict large scale STAX™ capacity and product quality

Patricia Rose (Merck Pharmaceuticals)*, Jennifer Pollard (Merck Pharmaceuticals)

Depth filtration is commonly used as either the primary or secondary step in mammalian cell culture harvest. Early key data to predict filter capacity, processing time and product quality could be generated from small scale cell culture batches if a robust filtration scale down model is available. However there are inherent issues with scale down filtration. While process parameters such as flux can be maintained across scales, there are limitations to aligning equipment and membrane module geometry. Also the cell culture; cell density, cell viability, debris and media, can have different responses at different scales. Slight differences in flow path and membrane configuration can change how the cells interact with each other and the filtration media. The cells can settle and/or become damaged differently compared to large scale. These inconsistencies can impact the small scale depth filter capacity to either significantly over or under forecast the filtration requirements for large scale. This paper will present a scale down model case study using the Pall Supracap™ 50 depth filter to predict performance of the Pall STAX™. Scale down strategy qualification and using direct translation of operating parameters or matching pressure profiles will be discussed.

WEDNESDAY 11:30 am - Kings Garden 3

24e - Unraveling the Plugging Mechanisms during a Combined Tangential-flow and Depth Filtration Process

Xianghong Qian (University of Arkansas)*, Da Zhang (University of Arkansas), Daniel Strauss (Asahi Kasei), Parag Patel (Asahi Kasei), Ranil Wickramasinghe (University of Arkansas)

The rapid advancement in upstream cell culture operations has led to a significant increase in product titers. However, this high-level of productivity is accompanied by a rather high cell density which places a significant burden on the traditional downstream clarification and purification operations, especially during the initial clarification and capture steps. Depth filtration, centrifugation and tangential-flow filtration (TFF) have been used to remove cells, cell debris and other impurities. Depth filter consists of a thick porous bed that can trap particles within the filter in contrast to screen type filters that largely reject particulate matter based on surface filtration. On the other hand, the tangential flow of the feed to the membrane surface leads to the suppression of cake growth during TFF operation. BioOptimal MF-SL from Asahi KASEI operates as a hybrid TFF-depth filter. However, the plugging mechanism(s) for BioOptimal MF-SL and other hybrid filters are poorly understood. Here a combined pore blockage and cake filtration model has been able to describe well the fouling mechanism of this hybrid filter in the normal flow mode whereas a resistance in series model has been able to predict performance of the filter operated in the TFF mode. Significant insights have been obtained in understanding the fouling mechanism and optimal operation condition of this hybrid filter.

WEDNESDAY 12:00 pm - Kings Garden 3

24f - Functionalized Microporous Membranes for Protein Capture and Analysis

Joshua Berwanger (University of Notre Dame)*, Merlin Bruening (University of Notre Dame), Liu Yang (University of Notre Dame), Hui Yin Tan (University of Notre Dame), Weijing Liu (University of Notre Dame)

Microporous membranes are an attractive alternative to resins for protein affinity capture, especially for protein identification and quantification. Convective flow through micron-sized pores in membranes minimizes diffusion limitations to allow protein binding in short residence times. Derivatization of micropores with poly(acrylic acid) introduces -COOH groups for simple functionalization with a range of affinity molecules. Moreover, the poly(acrylic acid) films show low nonspecific adsorption. This talk highlights the use poly-electrolyte modified membranes for purification of His-tagged proteins, binding and analysis of therapeutic monoclonal antibodies, and immobilization of small-molecule drugs for determination of their protein targets. Functionalization of membranes with Ni-NTA complexes enable purification of polyhistidine-tagged proteins in minutes when using a spin membrane format. Immobilized mimotopes (epitope-mimicking peptides) selectively capture therapeutic monoclonal antibodies out of serum enabling quantitation and analysis. Recent work is exploring the immobilization of small-molecule drugs to identify intended and unintended protein targets. Initial reflectance FTIR studies allow rapid screening to develop the immobilization conditions that lead to minimal nonspecific adsorption. In summary, functionalized microporous membranes can accomplish affinity capture of various proteins more quickly and efficiently than bead-based methods.

WEDNESDAY 9:30 am - Ballroom 3

25a - A systematic approach to remove bottlenecks in high concentration UF/DF for subcutaneous application of biological drug substances

Hasin M Feroz (Bristol-Myers Squib)*, Dongyoun Jang (Bristol-Myers Squibb), Jessica Hung (Bristol-Myers Squibb), Yan Chen (Bristol-Myers Squibb), Erinc Sahin (Bristol-Myers Squibb), Andrew Zydney (Penn State University), Melissa Holstein (Bristol-Myers Squibb), Sanchayita Ghose (Bristol-Myers Squibb)

The current industry focus on subcutaneous administration of biological drug products necessitates purified mAb at concentrations of ~100 g/L, more than an order of magnitude higher than those used for most intravenous applications. However, UF/DF-based (ultrafiltration / diafiltration) approaches to buffer exchange and concentrate to high concentrations poses specific challenges from an operational, formulation, and molecular stability perspective. We identify these challenges and propose strategies to remove the bottlenecks for high concentration UF/DF applications. Most notable challenges of high concentration operations are identified as 1) high pressures exceeding safety limits in current UF/DF devices due to exponential increase in viscosity at high mAb concentrations, 2) recovering high yields of the concentrated drug substance from the system with minimal dilution, and 3) throughput and material loss during post UF/DF sterile filtration. Mitigation strategies include the use of viscosity-modifying excipients at high concentrations, investigating UF/DF devices with modified membranes and/ or improved flow channel design to address pressure concerns at higher concentrations, as well as implementing a low flux plug-flow flush strategy to recover the concentrated drug substance from the hold-up volume of the UF/DF skid at minimal dilution. From a formulation perspective, we are challenged with higher aggregation rates at high protein concentrations, due to higher frequency of intermolecular interactions. We also investigated larger differences in pH and excipient concentrations (diafiltration buffer vs retentate composition) during both diafiltration and ultrafiltration as a function of isoelectric point and concentration of proteins, buffer, pH and excipient concentrations. Our findings suggest that protein concentration plays an important role in protein stability as well as the accuracy of chemical composition of the final formulation.

WEDNESDAY 10:00 am - Ballroom 3

25b - Continuous Diafiltration for cGMP Biomanufacturing

Akshat Gupta (MilliporeSigma)*, Beth Goodrich (MilliporeSigma), Herb Lutz (MilliporeSigma)

Diafiltration is a critical unit operation which is part of practically every downstream purification train of monoclonal antibody or bio therapeutic modality production. It is typically employed to provide an efficient and robust method for exchanging buffer matrix for the therapeutic modality and offers simultaneous clearance of small molecule and elemental impurities. Continuous manufacturing and process intensification approaches has been an area of keen interest and various technologies are being investigated to make biotherapeutic manufacturing process more efficient in terms of facility utilization and capital cost reduction by reducing reliance on large scale facilities. Technologies like multi-column cycling, inline viral inactivation and single pass tangential flow filtration (SPTFF) for protein concentration have been developed in past few years, offering significant improvement over their existing batch equivalents. Ability to develop a similar technology for diafiltration suitable for GMP manufacturing has remain rather elusive. Our approach here is based on a shift in perspective wherein the process design is based on membrane utilization and system footprint as opposed to flux and process time in a typical batch diafiltration process. This rather flexible technology offers 6-8-fold increased membrane utilization, up to 3-fold reduction in pump passes and system footprint can be substantially decreased. Buffer usage, extent of buffer exchange, and product yield are consistent with a constant volume diafiltration process. In this presentation, we will present a system design for this concept and a 24-hour experimental data demonstrating operational robustness. Implementation into a fully continuous, or intensified process train and flexibility of the system to accommodate complex diafiltation schemes for challenging molecules will also be discussed.

WEDNESDAY 10:30 am - Ballroom 3

25c - Understanding the role that patterning plays on membrane biofouling by visualization using light imaging and simulation

Anna Malakian (Clemson University)*, Bowen Ling (Stanford University), Ilenia Battiato (Stanford University), Scott Husson (Clemson University)

The goal of this project is to develop the basic science needed to design new fouling-resistant membranes. We present findings on the relationship between membrane surface patterning and protein fouling that support two hypotheses (i) geometric patterns will reduce membrane fouling compared to the control, and (ii) computational models will accelerate the discovery of new membrane surface morphologies and methods to improve membrane fouling resistance. Experimental efforts combined flux decline measurements with light microscopy to study the fouling of as-received and patterned polyethersulfone membranes during ultrafiltration of bovine serum albumin (BSA, model foulant). Patterned membranes were prepared by embossing a "tire-tread" micropattern (TTP) on the surface. Both membrane types were labeled with 5-DTAF. BSA labeled with Alexa Flour 488 was used to visualize the fouling profiles with confocal laser scanning microscopy (CLSM), which provided three-dimensional images of membrane surfaces patterns and co-localized protein foulant. In parallel, numerical simulations were used to compare fouling patterns with visual analysis of the CLSM images and to explore the full potential of the TTP. We utilized the customized simulator SUMs (Stanford University Membrane solver) within the OpenFOAM framework. The solver uses a finite volume method to study the dynamic couplings among flow, solute transport, and surface fouling. From the numerical study, we found that the TTP acts as a fouling-focusing mechanism. When the flow alternates, the foulant shifts its preferential accumulation zone and self-cleans the previously fouled region. This innovative design provides a possibility to consecutively alternate flow direction during operation to periodically clean zones of the membrane surface. The overall approach is expected to aid in the design of new membranes with tailored surface structures that prevent the irreversible deposition of foulants in prone-to-foul regions.

WEDNESDAY 11:00 am - Ballroom 3

25d - Treating Poultry Processing Wastewaters by Combined Electrocoagulation and Ultrafiltration

Ranil Wickramasinghe (University of Arkansas)*, Kamyar Sardari (University of Arkansas)

Ultrafiltration (UF) is an emerging technology of interest for the treatment of highly impaired industrial wastewaters. The US poultry industry produces over 60 billion gallons of poultry processing wastewater (PPW) requiring treatment. In this work, nine different commercially available ultrafiltration membranes, having different pore sizes (10 to 300 μm) and made of different polymeric materials (polyethersulfone and regenerated cellulose), were screened for treating PPW streams obtained from bird washer and chiller operations. Bird washer wastewater was found to be more fouling as it contained higher biochemical oxygen demand (BOD), chemical oxygen demand (COD), oil and grease (O & G) and total suspended solid (TSS) compared to the chiller wastewater. The presence of suspended particles can lead to plugging of the membrane pores. Thus, it is of significant importance to select the most appropriate membrane (pore size, polymeric material, flux, etc.) that minimizes the fouling and maximizes contaminate rejection. For the feed streams considered here, membranes with 30 kDa nominal molecular weight cut-off provided the most stable performance in laboratory scale tangential flow filtration. Development of a practical ultrafiltration process will involve pretreatment of the feed stream. Here electrocoagulation (EC) was used to pretreat PPW. EC is shown to be effective in removing suspended solids and organic compounds which foul the membrane. Higher EC reaction times result in higher contaminate removal. We show that EC can remove over 85% of fats, oil and grease and total suspended solids from the PPW. The EC-UF process results in lower flux decline and enhanced contaminate rejection. Finally, long-term EC-UF experiments, were conducted with membrane cleaning, to study the feasibility of the combined process.

WEDNESDAY 11:30 am - Ballroom 3

25e - Mass transfer assisted TIPS process to precisely tailor the hollow fiber membrane surface and sublayer structures

Hideto Matsuyama (Kobe University)*, Chuanjie Fang (Kobe University), Saeid Rajabzadeh (Kobe University)

A novel method was used to tailor the surface and sub-layer structure of the poly (vinylidene fluoride) (PVDF) hollow fiber membranes in the TIPS process by the co-extrusion of different types of solvents at the outer layer of the extruded polymeric solution. By changing solvent type, membrane surface structure was controlled in an extensive range, e.g. pore size varied from 24 to 600 nm, and water permeability changed from 3 to more than 4000 L m^{-2} h^{-1} bar^{-1}, with a slight change in the membrane mechanical strength. A slight mass transfer that happened during the air gap distance, before starting phase separation, changed the membrane sublayer structure considerably from a bicontinuous to a composite-like structure, spherulites embedded in the bicontinuous network. The segregation of the diluent molecules or PVDF chains to the interface between the extruded solvent and polymer solution was assessed to tailor the membrane surface and sub-layer structures. The diffusion of extruded solvents with good compatibility to PVDF into the polymer solutions altered the phase separation mechanism and produced a novel composite-like structure in the sub-layer of the membrane. This membrane structure enhanced permeation stability drastically based on its unique structure to dissipate the applied hydrodynamic pressure and resist against the membrane compaction. Mass transfer phenomena at the interface were assessed by using molecular dynamics simulation (MD) to support our proposed explanations. Finally, membrane structures and characteristics are summarized based on the change in the competitive ternary interactions between the polymer, diluent, and extruded solvent. This kind of summarization potentially can be used as a guideline for selecting appropriate extruded solvents to tailor membranes with desired structures and characteristics.

WEDNESDAY 2:00 pm - Ballroom 4

25f - Organic solvent resistant membranes obtain by using non-toxic solvents

Stefan Chisca (King Abdullah University of Science and Technology)*, Gheorghe Falca (King Abdullah University of Science and Technology), Tommaso Marchesi (University of Bologna), Valentina Elena Musteata (King Abdullah University of Science and Technology, Suzana Nunes (King Abdullah University of Science and Technology)

Membrane technology is a sustainable process which can replace the conventional separation methods in chemical, pharmaceutical and petrochemical industries [1]. However, the membrane fabrication process involve in general the use of toxic solvents such as DMF and NMP [2]. Therefore, another main challenge of this technology beside to design membranes which are easy to process and are stable in wide range of pH and organic solvents, is to improve the sustainability of membrane fabrication. One way to achieve this is to replace the common toxic solvents with greener ones, such as ionic liquids, which are chemically stable, non-flammable and has relative low toxicity and volatility [3]. In this study we report the synthesis of organic solvent resistant polytriazole membranes by dissolving the polymer in ionic liquid at 85°C. The membranes were obtain by phase inversion method. The resulting solution was cast on a glass plate and then it was immersed after different time in water. To obtain crosslinked polytriazole membranes we exposed them to two different diepoxy crosslinkers. The performance of the membranes were evaluated in DMF at 20°C, 65°C, 85°C and 105°C. The permeances were in the range of 4 - 10 L m-2 h-1 bar-1 with the MWCO in the range of 1000 g/mol. We used dynamic mechanical analyzer to correlate the membranes performance with the mobility of the crosslinkers and polytriazole chains and we investigated the mechanical properties by measuring the creep recovery. Moreover we used scanning electron microscopy to investigate the morphology of the membranes before and after filtration at 105°C. Acknowledgment This work was sponsored by King Abdullah University of Science and Technology (KAUST), CRG6 URF/1/3441-01-01. References 1. D.S., Sholl, R.P., Lively, Nature 532, 435-437 2016. 2. G. Szekely, M. F. Jimenez-Solomon, P. Marchetti, J. F. Kim. A. G. Livingston, Green Chem., 16, 4440-4473, 2014. 3. P. C. Marr and A. C. Marr, Green Chem. 18, 105-128, 2016.

WEDNESDAY 9:30 am - Ballroom 4

26a - Artificial Water Channels - toward biomimetic membranes for desalination

Mihail Barboiu (Institut Europeen des Membranes)*

This lecture discusses the incipient development of the first artificial water channels systems. We include only systems that integrate synthetic elements in their water selective translocation unit. Therefore, we exclude peptide channels because their sequences derive from the proteins in natural channels. We review many of the natural systems involved in water and related proton transport processes. We describe how these systems can fit within our primary goal of maintaining natural function within bio-assisted artificial systems. In the last part, we present several inspiring breakthroughs from the last decade in the field of biomimetic artificial water channels. All these examples demonstrate how the novel interactive water-channels can parallel biomolecular systems. At the same time these simpler artificial water channels offer a means of understanding water structures useful to understanding many biological scenarios. Moreover they can be used for the preparation of highly selective membranes for desalination. [1] M. Barboiu and A. Gilles, Acc. Chem. Res. 2013, 46, 2814-2823. [2] M. Barboiu, Angew. Chem. Int. Ed. 2012, 51, 11674-11676. [3] Y. Le Duc, et al., Angew. Chem. Int. Ed. 2011, 50(48), 11366-11372. [4] E. Licsandru, et al., J. Am. Chem. Soc., 2016, 138, 5403-5409. [5] M. Barboiu, Chem. Commun., 2016, 52, 5657- 5665. [6] I. Kocsis, et al. Science Adv. 2018, 4, eaao5603.

WEDNESDAY 10:00 am - Ballroom 4

26b - Self-Assembly of Long-Lasting Lipid Bilayers and the Effect of Temperature of their Ion Rejection

Yair Kaufman (Ben Gurion University of the Negev)*, Shiju Abraham (Ben Gurion University of the Negev), Tabea Heckenthaler (Ben Gurion University of the Negev), Yacov Morgenstern (Ben Gurion University of the Negev)

Supported lipid bilayers have promising potential for diverse separation processes, such as water desalination and small-molecule/ion separation, as well as for biosensors. However, the preparation of long lasting (thermodynamically stable) defectless (no holes) lipid bilayer is challenging, and for some applications, these are crucial requirments. During the talk, I will shed-light on the vesicle fusion mechanism, which is the commonly used mechanism to prepare supported lipid bilayers with or without membranes proteins. Then, based on theoretical analysis and experimental results, I will describe the conditions that render these bilayers thermodynamically stable. Lastly, I will highlight the profound effects of temperature on the resistance of lipid bilayers to ion transport (i.e., ion rejection).

WEDNESDAY 10:30 am - Ballroom 4

26c - Peptoid-based membrane-mimetic 2D nanomaterials with incorporated natural and synthetic channels

Chunlong Chen (Pacific Northwest National Laboratory)*

From a materials perspective, cell membranes, which consist of 3.0 - 5.0 nm thick lipid bilayers, provide a great source of inspiration for developing functional 2D nanomaterials because they exhibit sequence-specific water and ion transport and the ability to self-repair. Inspired by that, we recently developed sequence-defined and lipid-like peptoids that self-assembled into highly stable and self-repairable membrane-mimetic 2D materials. These membranes were formed through a solvent-induced crystallization process, in which inter-peptoid hydrophobic interactions drove the anisotropic packing of peptoids to form a bilayer-like membrane structure. They exhibit a number of properties associated with cell membranes, including spontaneous assembly at interfaces and the ability to self-repair. We further demonstrated that these membranes are superior to lipid bilayers and other assembled 2D materials because: 1) they are free-standing, atomically ordered, and highly stable in pure organic solvents, biological environments, as well as high temperature; and 2) a broad range of active functional groups, including chromophores, can be incorporated and patterned within membranes as peptoid side chains. To take advantages of the unique feature of these peptoid membranes, we further demonstrated the incorporation of both natural and synthetic channels into these membranes using the co-crystallization approach. Because peptoids are easy to synthesize, have large side-chain diversity, and offer the unique advantage for controlling self-assembly due to the lack of backbone hydrogen bond donors, we recently demonstrated the introduction of a wide range of channel-forming molecules into these peptoid-based 2D nanomaterials, offering the potential of building artificial channels through peptoid self-assembly.

WEDNESDAY 11:00 am - Ballroom 4

26d - Scalable High-Performance Membranes with High Density Channel Protein-Polymer Nanosheets

Yu-Ming Tu (Penn State University)*, Woochul Song (Penn State University), Tingwei Ren (Penn State University), Yuexiao Shen (UC Berkeley), Ratul Chowdhury (Penn State University), Prasangi Rajapaksha (University of Kentucky), Drew Carson (Penn State University), Yuxuan Dai (Penn State University), Arwa Mukthar (Penn State University), Miaoci Zhang (Penn State University), Alina Thokkadam (Rutgers University), DB Bhattacharyya (University of Kentucky), Bill Phillip (University of Notre Dame), Robert Hickey (Penn State University), Yinai Wei (University of Kentucky), Manish Kumar (Penn State University)

Biological membranes containing channel proteins exhibit high permeability while maintaining high solute selectivity due to transport proteins with well-defined pore geometry and functionality to exclude or pass specific solutes. Nevertheless, membrane protein-based biomimetic membranes studied so far suffer from low protein insertion, the use of vesicular morphologies, and the use of poorly scalable dialysis-based self-assembly procedures in synthesizing such membranes. These challenges have resulted in much lower than anticipated improvements. In this presentation, we will present the development of scalable high-performance biomimetic membranes that are based on densely packed β-barrel channel proteins in block copolymers to create robust and flat nanosheets. These highly packed crystalline structures and well-ordered nanosheets were constructed by an innovative solvent evaporation method that can be completed within 2-hours and further supported on a porous substrate as high-performance selective layers by a modified layer-by-layer technique. Three passive β-barrel channel proteins with inherently narrow and uniform pore sizes of ~ 0.5 - 1.5 nm were tested in this work including a passive outer membrane protein F (OmpF), alpha-hemolysin (αHL) and a redesigned outer membrane protein mutFhuA (FhuA ΔC/Δ4L). OmpF, αHL and FhuA ΔC/Δ4L nanosheet-incorporated membranes maintained their unique molecular selectivity with molecular weight cutoffs of ~477, 927 and 1134 Da respectively, while demonstrating high water permeabilities of 322, 793 and 1092 L m^{-2} h^{-1} bar^{-1}, which are about 20-500 times greater than commercial nanofiltration membranes with comparable molecular exclusion limits. This approach of combining channel proteins and BCPs utilizing the scalable membrane fabrication technique could open new routes in developing high performance biomimetic membranes thus exploiting the exquisite specificity and functionality of various channel proteins.

WEDNESDAY 11:30 am - Ballroom 4

26e - Fabrication of aquaporin-based biomimetic membrane for seawater desalination

Ye Li (Nanyang Technological University)*, Rong Wang (Nanyang Technological University)

Aquaporins (AQPs), consisting of a family of water channel proteins, have been adopted to fabricate high performance membranes attributed to their superior water/solute selectivity. This study focuses on enhancing the mechanical strength of aquaporin-based biomimetic membranes for seawater desalination application. Aquaporin incorporated vesicles were embedded into the selective layer of the optimized thin film composite (TFC) membrane. The resultant membrane, denoted as ASW, exhibited a stable water flux around 20 L·m^{-2}·h^{-1} and 99% NaCl rejection at a constant pressure of 55 bar using 32,000 ppm NaCl solution as feed in reverse osmosis (RO) measurement. The robustness of the ASW membranes were evaluated by measuring the water flux change due to pressure compaction and characterized by Atomic Force Microscope (AFM). The water flux of ASW membrane was almost 100% enhanced compared with that of aquaporin-free control TFC membranes (i.e., a water flux of 11-12 LMH). The filtration performance of the ASW membrane was further evaluated by a seven-day desalination test using a real seawater secondary effluent collected from a desalination plant in Singapore as feed. To our best knowledge, this is the first report on the aquaporin-incorporated RO membrane that can be used for seawater desalination. A commercial SW30HR membrane was tested in parallel for comparison. The robust ASW membrane exhibited a nearly 80% higher water flux in comparison to the SW30HR membrane with a comparable overall solute rejection, suggesting the advantage and feasibility of Aquaporin based biomimetic membranes for seawater desalination.

WEDNESDAY 12:00 pm - Ballroom 4

26f - Biomimetic channel membranes: can they realize their full potential upon scale-up?

Viatcheslav Freger (Technion - IIT)*

The unique selectivity and permeability of biological water channels provide much inspiration for developing their synthetic mimics [1]. Recent progress in biomimetic channels has been remarkable and their performance approaches that of commercial reverse osmosis membranes [2]. However, most estimates of the potential performance are based on experiments with individual channels or vesicles or membrane patches of microscopic dimensions. Under such conditions, the concentration polarization (CP) is small, since the boundary layer has a spherical symmetry and its effective thickness is commensurate with the membrane size, which is at least an order of magnitude smaller than in standard membrane operations. As a result, the CP level is apparently negligible in micro-scale experiments and does not interfere with the channel performance. However, the geometry of the diffusion profile will change and CP level increase drastically in macroscopic membranes incorporating densely spaced channel arrays [3]. If not properly addressed in membrane design, the increased CP level in scaled-up channel-based membranes may greatly compromise the observed selectivity and reduce permeation rates due to increased osmotic pressure. Although re-setting the target of selectivity to a much higher value may partly offset this limitation, the presented analysis highlights a major engineering challenge towards scale-up and utilization of the biomimetic channels in next-generation water purification membranes. [1] JR Werber, M Elimelech, Science Advances 2018, 4 (6), eaar8266. [2] RH Tunuguntla et al., Science 2017, 357 (6353), 792-796; J Abraham et al, Nat. Nanotechnol. 2017, 12 (6), 546-550; Z Sun et al., Faraday Discuss., 2018, 209, 113-124; W Song et al., Faraday Discuss., 2018, 209, 193-204 [6] V Freger, Faraday Discuss., 2018, 209, 371

WEDNESDAY 9:30 am - Kings Garden 5

27a - Surface Nano-structuring with Hydrophilic Polymer Brush Layers for Membrane Performance Tailoring and Optimization

Yian Chen (UCLA)*, Soomin Kim (UCLA), Anditya Rahardianto (UCLA), Yoram Cohen (UCLA)

Membrane technologies are increasingly used for water treatment and purification to relieve global water scarcity. However, improving reverse osmosis (RO), nanofiltration (NF) and ultrafiltration (UF) membrane performances require one to effectively mitigate membrane fouling and mineral scaling, enable high membrane cleaning efficiency, and ensure high membrane permeability and solute rejection. Surface nano-structuring with tethered hydrophilic polymers is touted as an effective way of rendering membrane surfaces fouling/scale resistance while facilitating effective membrane cleaning with freshwater, thereby avoiding/reducing the need for chemical cleaning agents. In order to synthesize membranes with tethered hydrophilic polymers surface activation of the base membrane can be accomplished with Atmospheric Pressure Plasma-induced Graft Polymerization (APPIGP), followed by free radical graft polymerization. The above approach was investigated in the present work aiming to both optimizing membrane surface properties and correlating surface structure and interfacial interactions with (a) alteration of RO/NF and UF membrane characteristics and performance due to tethered polymer chains formed by APPIGP; (b) RO/NF and UF membrane fouling and cleaning mechanisms, and (c) scaleup of the process for production of high-performance spiral-wound RO/NF elements for a range of desalination applications.

WEDNESDAY 10:00 am - Kings Garden 5

27b - Roughness in Polyamide RO Membranes: Its Formation Mechanism, Control, and Implications

Chuyang Y Tang (University of Hong Kong / University of New South Wales)*

Water scarcity is one of the major global challenges of our time. Reverse osmosis technology plays a crucial role in providing alternative sources of water through seawater desalination and water reclamation. Thin film composite polyamide membranes are the gold standard for reverse osmosis technology, but tuning their permeability and selectivity remains a major challenge because of the inherent permeability-selectivity trade-off. The polyamide rejection layer of a typical reverse osmosis membrane contains a large number of discrete nano-sized voids occupying as much as 30% of its total volume. Despite the common belief that these nanovoids regulate the membrane water permeability and salt rejection, little is known about their formation mechanism. In this presentation, we show that these nanovoids are formed due to the degassing of nanobubbles from the monomer solution during the interfacial polymerization process to fabricate the polyamide selective film. Simple nano-foaming methods, such as addition of sodium bicarbonate and ultrasound application, can result in remarkable improvement in both membrane water permeability and salt rejection, thus overcoming the long-standing permeability-selectivity trade-off of desalination membranes. Our study further establishes a framework for the fundamental understanding of the structure-property-performance relationship of polyamide desalination membranes.

WEDNESDAY 10:30 am - Kings Garden 5

27c - Mechanism of chlorine attack on polyamide membranes studied by EIS

Mikhail Stolov (Technion - IIT), Viatcheslav Freger (Technion - IIT)*

The well-known and yet incompletely understood Achilles heel of polyamide membranes is low tolerance to chlorine. Much information on the chemical mechanism and transformations that take place upon chlorine attack has been obtained by post-mortem spectroscopic analytical methods, which limited the sensitivity and temporal resolution of the kinetic analysis. As an alternative method that may directly probe variations of transport characteristics and ion rejection properties of polyamide layer in real time and for low chlorine doses, we employed electrochemical impedance spectroscopy (EIS) to investigate the mechanism and kinetics of chlorine attack. In this way, we could focus on early stages of the chlorination and low chlorine concentrations (15 to 197 ppm). EIS results partly conform to established two-stage mechanism that proceeds as N-chlorination followed by either C-chlorination in acidic conditions or amide bond scission in alkaline conditions. However, early-time kinetics in acidic conditions shows inconsistencies with this model and is best explained by direct ring chlorination that proceeds in parallel to (reversible) N-chlorination, possibly obscured by a finite rate of polymer relaxation. The data also indicate that N-chlorination reduces membrane polarity and ion permeation, yet its effect is eventually reversed by C-chlorination that has an opposite effect. Thereby, the ultimate chemical transformations (C-chlorination at low pH or amide bond cleavage at high pH) result in an irreversible increase in membrane polarity and loss of ion rejection. The results highlight the potential of EIS as a powerful and sensitive tool that may assist in assessing membrane stability and developing new chlorine-resistant membranes [1]. [1] M. Stolov, V. Freger, Degradation of polyamide membranes exposed to chlorine: an impedance spectroscopy study, Environ. Sci. Technol., in press, DOI: 10.1021/acs.est.8b04790

WEDNESDAY 11:00 am - Kings Garden 5

27d - Physicochemical and electrochemical characterization of cation-exchange membranes modified with polyethyleneimine for elucidating enhanced monovalent permselectivity of electrodialysis

Wenbin Jiang (New Mexico State University)*, Lu Lin (New Mexico State University), Xuesong Xu (New Mexico State University), Pei Xu (New Mexico State University)

Monovalent permselective cation-exchange membrane (CEM) CR671 was developed by coating polyethyleneimine onto the normal grade CEM CR67. Pilot-scale electrodialysis of brackish groundwater demonstrated excellent monovalent permselectivity of the CR671 as compared to the CR67. Advanced analytical approaches were employed to characterize the physicochemical and electrochemical properties of the CEMs to elucidate the mechanisms of permselectivity. Ion-exchange capacities of the CR671 and CR67 were measured to be 2.0 and 2.1 meq/g dry membrane, respectively. Zeta-potential analysis revealed that the CR671 surface was positively charged as a result of polyethyleneimine coating. Electrochemical impedance spectroscopy (EIS) data indicated larger impedance for the CR671 and fitted well to the Maxwell- Wagner model, which provided an equivalent electric circuit for the CEMs, as well as indication of the existence of polyethyleneimine layer. The time constants of different ions transporting through polyethyleneimine modification layer were calculated based on the EIS responses. The results revealed the polyethyleneimine modification layer led to longer transport time for Ca2+ ions through the CR671 than for Na+ ions, supporting the pilot-scale testing results that the CR671 improved the Na+ removal by selectively rejecting Ca2+ ions. However, the difference in ion transport time became less significant with increasing ionic strength of the feed water. The EIS results suggest the monovalent permselectivity decreased during treatment of higher salinity water, supporting the finding that the CR671 exhibited higher monovalent permselectivity during electrodialysis of brackish groundwater than reverse osmosis concentrate in which the salt concentration was 5.7 times higher than the brackish groundwater. The study demonstrates that ion transport time constant is a better indicator for permselectivity of modified ion-exchange membranes than electrical resistance measured by EIS.

WEDNESDAY 11:30 am - Kings Garden 5

27e - Unraveling the Film-Formation Kinetics during Interfacial Polymerization via Low Coherence Interferometry

Xin Liu (Southern University of Science and Technology)*, Weiyi Li (Southern University of Science and Technology), Zhuo Li (Southern University of Science and Technology), Anthony Fane (Singapore Membrane Technology Centre), Baolin Deng (Southern University of Science and Technology)

Interfacial polymerization (IP) is one of the most important methods for fabricating thin film composite (TFC) membranes. Understanding the underlying mechanisms of the IP-film formation is of great value for the development of membranes with enhanced performance. The current work employed a novel method to in-situ characterize the film-formation kinetics. This method exploited low coherence interferometry (LCI) to in-situ measure the variation in the optical properties at the interface where polymerization was induced by bring together two monomers (1,3,5-benzenetricarbonyl trichloride, TMC 0.15 wt% in hexane, and m-phenylenediamine, MPD 2 wt% soaked in the polysulfone substrate membrane). A numerical algorithm was developed to mitigate the effects of the perturbed interface and therefore enabled the surface-averaged intensity (SAI) profiles to resolve different kinetic regimes. To the best of our knowledge, this is the first demonstration of applying LCI to the characterization of an IP process. All the characterization results not only confirm the theoretical model but also provide direct evidence for the different polymerization processes at varied depths (i.e., the asymmetric growth of the IP film).

WEDNESDAY 12:00 pm - Kings Garden 5

27f - Operando Membrane Spectroscopy for Elucidating Transport Mechanisms in Membranes

Casey O'Brien (University of Notre Dame)*

Operando spectroscopy is a technique that is commonly used in catalysis to elucidate catalytic reaction mechanisms. It involves the spectroscopic characterization of the catalytic material coupled with simultaneous measurement of catalytic reaction rates under realistic reaction conditions. By characterizing the structure of the catalyst and measuring reaction rates simultaneously, one can obtain detailed structure-function relationships and elucidate reaction mechanisms in more detail than is possible with ex-situ characterization techniques. While operando spectroscopy has been used almost exclusively to study catalytic materials, other fields such as gas separation membranes would benefit from the operando spectroscopy tools and techniques. Towards this end, our research group has dedicated the last three years to developing operando spectroscopy tools for elucidating transport mechanisms in gas separation membranes. In this talk, I will discuss our results with two types of gas separation membranes: (1) Pd-based H_2 separation membranes, and (2) PVAm facilitated transport membranes. Pd-based membranes are attractive for H_2 separation applications, however, other gases that are common to H_2-containing gas streams, such as CO, C3H6, and H_2S, can severely inhibit hydrogen transport. The fundamental mechanisms by which these gases inhibit hydrogen transport is not well understood. To address this problem we have developed an operando spectroscopy tool that can monitor the reactions occurring on the surface of Pd-based membranes using polarization-modulation infrared reflection absorption spectroscopy and simultaneously measure hydrogen permeation rates. I will demonstrate that this tool can elucidate unknown poisoning mechanisms in great detail. We have also developed an operando spectroscopy tool for elucidating CO_2 transport mechanisms in PVAm FTMs using transmission FTIR spectroscopy.

WEDNESDAY 12:00 pm - Ballroom 3

28a - Biofouling resistant membranes for wound healing

Antoine Venault (Chung Yuan Christian University)*, Hao-Tung Lin (Chung Yuan Christian University), Yung Chang (Chung Yuan Christian University)

Low-biofouling porous membranes are widely applied to wastewater treatment, but their inherent properties make them ideal potential candidates for a totally different use: wound healing. An ideal wound dressing should maintain a moist environment above the wound. Besides, it should be highly porous to promote gas exchanges and exudates drainage, resistant to bacterial attachment to prevent wound infection and more generally resist to the attachment of proteins and cells (so called non-sticky property) to avoid secondary injuries following dressing removal/change. Low-biofouling porous membranes meet all of these requirements. We have applied different surface modifications to polytetrafluoroethylene membranes using PEGylated, zwitterionic or pseudo-zwitterionic materials, and studied their ability to resist biofouling by proteins, blood cells, bacteria or fibroblast cells. Then, applied as bandages onto wounds on model rats, we observed significant improvement of wound healing kinetic/quality, as compared with wounds covered with commercially available bandages. Parallelly, we worked with chitosan soft-membranes and enhanced the biocompatibility of the natural polymer via a proper surface modification starring a copolymer of glycidyl methacrylate units and sulfobetaine methacrylate moieties. We applied the grafted chitosan in the healing of diabetic wounds. This work was supported by the Minsitry of Science and Technology of Taiwan (project grant MOST 108-2923-E-033 -001)

WEDNESDAY 2:30 pm - Ballroom 4

28b - Impacts of Bio-inspired Zwitterionic Membranes for Health Care Applications, Yung Chang (R&D Center for Membrane Technology, CYCU)*

Bio-inspired zwitterionic materials are the latest generation of biomaterials for nonfouling interfaces and membranes in health care applications. This talk summarizes our fundamental developments in R&D Center for Membrane Technology (Taiwan) related to the functionalization of interfaces and membranes using zwitterionic materials. Our molecular designs of zwitterionic polymers and copolymers, sulfobetaine-based, carboxybetaine-based, or phosphobetaine-based, will be reviewed. Then, the strategies used to functionalize surfaces/membranes by coating, grafting onto, grafting from, or in situ modification will be introduced, and the important part of this talk will be the focus to key medical applications of zwitterionic membranes. Finally, some potential future directions for molecular designs, functionalization processes, and applications will be summarized. This work was supported by Ministry of Science and Technology in Taiwan. (MOST 108-2923-E-033 -001) References [1] C. C. Lien, L. C.Yeh, A. Venault, S. C. Tsai, C. H. Hsu, G. V. Dizon,Y. T. Huang, A. Higuchi and Y. Chang, Journal of Membrane Science, 565, 119-130 (2018). [2] Y. W. Chen, A. Venault, J. F. Jhong, H. T. Ho, Y. Chang, Journal of Membrane Science, 537, 355-369 (2017). [3] A. Venault, K. M. Trinh, Y. Chang, Journal of Membrane Science, 501, 68-78 (2016). [4] A. Venault, M. R. B. Ballad, Y. H. Liu, P. Aimar, Y. Chang, Journal of Membrane Science, 477, 101-114 (2015). [5] S. H. Chen, Y. Chang, K. R. Lee, J. Y. Lai, Journal of Membrane Science, 450, 224-234 (2014). [6] A. Venault, Y. Chang, H. H. Hsu, J. F. Jhong,J. Huang, Journal of Membrane Science, 439, 48-57 (2013).

WEDNESDAY 3:00 pm - Ballroom 4

28c - Biomimetic membrane systems utilizing electro-dynamic interfaces

Bruce Hinds (University of Washington)*

An important challenge for the membrane community is to mimic natural protein channels that outperform, by orders of magnitude, man-made systems based on pore size and coarse chemical selectivity. To mimic protein channel pumping on a robust engineering membrane platforms, applied bias can be used to actuate charged gatekeepers and induce ionic pumping. CNT membranes, with tips functionalized with charged molecules, is a nearly ideal platform to induce electro-osmotic flow with high charge density at pore entrance and a nearly frictionless surface for the propagation of plug flow. Use of the electro-osmotic phenomenon for responsive/programmed transdermal drug delivery devices for nicotine addiction. Electrophoresis and electroosmosis phenomena are discussed in detail through CNT conduits. Another approach is to mimic natural protein channel transport cycles with binding/transport/release/reset events. Porous alumina (AAO) membranes have top and bottom electrodes coated with thin Au layers with pore dimension tuned to match protein dimensions. At this thin layer at pore entrances, Ni-NTA is able to bind to hys-tag residue on target protein, as is commonly employed in chromatography. A binding voltage pulse attracts anionic target protein to top electrode and blocking the pore, while repelling the cationic imidazole release agent. The second voltage cycle attracts cationic release agent to top of membrane while pumping anionic target protein to bottom permeate and resetting the pumping cycle. The separation efficiency of 1cm^2 of membrane was comparable to convention 1cm^3 volumes used in chromatography. Further modeling of system kinetics shows order of magnitude increase in throughput with sub-second cycle times enabled with high imidazole concentrations in a continuous cross flow operation.

WEDNESDAY 3:30 pm - Ballroom 4

28d - Responsiveness and function of DNA-gated membranes

Beñat Olave (Polymat, University of the Basque Country), Iliane Rafaniello (Polymat, University of the Basque Country), Thomas Schäfer (Polymat, University of the Basque Country)*

DNA-aptamers are oligonucleotides that can be designed to specifically interact with a molecular target coinciding with a conformational change. We have previously shown that such DNA-aptamers can function as "gate keepers" in nanoporous membranes where a molecular recognition event modulates membrane permeability. The gating function of the aptamer was proven to be reversible, specific, and target concentration dependent. However, a significant drawback of the system consisted in the necessity of fine-tuning between pore size of the membrane support and the DNA-aptamer dimensions in order to fulfil its gating function. A way to overcome this limitation is taking advantage of the versatility of DNA and create DNA-aptamer sandwich structures which are scalable in size. We monitored the build-up of such structures in real time using quartz crystal microbalance with dissipation measurement and multi-parameter surface plasmon resonance using a multiple wavelength mode. Both techniques allowed quantifying the length-scales of the DNA-sandwich created and, in this way, optimize systematically the design of the responsive membrane. The resulting responsive DNA-nanostructure was widely independent of the pore size of the membrane support and the flexibility of the DNA-based responsive membrane was significantly increased. We investigated the responsiveness and thermodynamic basics of the DNA-gating in more detail. Attention was focused on whether there exists a trade-off between membrane stability and break-up upon target recognition. We studied in how far a change in sensitivity of the DNA-aptamer occurred when incorporated in a sandwich structure. We found that creating responsive DNA-nanostructures based on aptamers is not as straightforward as can be found reported in literature. However, the versatility of the DNA-building blocks offer ample possibilities for optimisation, strategies for which will be critically discussed.

WEDNESDAY 4:00 pm - Ballroom 4

28e - Nature-inspired Coating-free Membranes for Desalination

Himanshu Mishra (King Abdullah University of Science and Technology), Ratul Das (King Abdullah University of Science and Technology), Sankara Arunachalam (King Abdullah University of Science and Technology), Zain Ahmad (King Abdullah University of Science and Technology), Edelberto Manalastas (King Abdullah University of Science and Technology), Jamilya Nauruzbayeva (King Abdullah University of Science and Technology)*

Membrane distillation (MD) processes typically utilize membranes with perfluorinated coatings to extract pure water vapor from hot saline feeds. This choice of chemicals is based on the conventional wisdom that water-repellent coatings prevent the intrusion of water due to capillary forces. However, the time-dependent loss of wetting of those coatings due to the deposition of amphiphilic foulants and/or coating-damage under high operational temperatures or abrasion is limiting their scope. Thus, coating-free membranes that could prevent liquid intrusion are desirable, but they have remained unachievable with intrinsically wetting materials so far. Here, we report on the development of biomimetic coating-free gas-entrapping membranes (GEMs) inspired by the cuticles of springtails and hairs of Halobates germanus, both of which exhibit mushroom-shaped features. We demonstrate that, against common expectation, intrinsically wetting materials can reliably prevent the intrusion of liquids after carving those biomimetic microtextures onto them; the solid-liquid-vapor system is trapped into metastable Cassie-states whose life-time could be tuned by the shape and size of the pores. We used silica and poly(methyl methacrylate) (PMMA) as model systems to realize GEMs by carving arrays of microscale cylindrical pores with mushroom-shaped inlets and outlets. Remarkably, both silica- and PMMA-GEMs robustly entrapped air in upon immersion in water at 293 K and 1 atm, which stayed intact in the pores for over six weeks. When our PMMA-GEMs (porosity, φ= 0.1) were used to separate a solution of ~0.6 M NaCl at 333 K from deionized water at 288 K, the vapor flux of the desalinated water was $J \sim 1$ L-m^{-2}-h^{-1}. To our knowledge, this is the first demonstration of coating-free desalination by MD. Our bio-inspired strategy might advance the rational design of coating-free membranes, from common materials, exhibiting superior resilience against higher temperatures and fouling.

WEDNESDAY 4:30 pm - Ballroom 4

28f - Carbon nanotube porins as model biomimetic membrane nanopores

Aleksandr Noy (Lawrence Livermore National Laboratory)*

Controlling ion and water transport on a molecular scale is fundamental for membrane applications ranging from industrial water treatment, to chemical separations, to advanced batteries and fuel cells. Living systems move ions and small molecules across biological membranes using protein pores that rely on nanoscale confinement effects to achieve efficient and exquisitely-selective transport. I will show that carbon nanotube porins-pore channels formed by ultra-short carbon nanotubes assembled in a lipid membrane-can exploit similar physical principles to transport water, protons, and small ions with efficiency that rivals and sometimes exceeds that of biological channels. I will discuss the role of molecular confinement and slip flow in these pores and show how it can enhance water and proton transport efficiency and influence the mechanisms of ion selectivity in these pores. Overall, carbon nanotube porins represent simple and versatile biomimetic membrane pores that are ideal for studying nanoscale transport phenomena and building the next generation of separation technologies.

WEDNESDAY 2:00 pm - Kings Garden 5

29a - A novel method to distinguish between crystal and defect transport in zeolite membranes

David Carter (University of Ottawa), Shaaima Al Akwaa (University of Ottawa), Boguslaw Kruczek (University of Ottawa)*, Handan Tezel (University of Ottawa)

Zeolite-based inorganic membranes have an extraordinary potential to separate gas mixtures at modest temperatures and pressures. However, this potential has not yet been realized, in part because of difficulty in forming a continuous, defect-free layer of zeolite crystals. The presence of defects in zeolite membranes arises from, for example, incomplete zeolite crystal growth and thermally induced linear expansion differences between zeolite crystals and the membrane support. The presence of defects is manifested by a decrease in membrane selectivity; however, the selectivity of the zeolite membrane alone does not allow quantitative determination of defect characteristics, without which any attempts to improve membrane synthesis protocols are highly empirical. In this paper we present a novel method that distinguishes between the crystal and defect transport of gases in zeolite membranes. The proposed method relies on single gas permeation tests with He and N_2 before and after calcination of the membrane, and knowledge of the adsorption isotherms of these gases on the zeolite crystals incorporated into the membrane. The validity of the method is verified by comparing the experimental and predicted permeances of He and N_2 before and after membrane calcination at different feed pressures. In addition, knowing the surface diffusion coefficient ratios from literature of N_2 with other adsorbing gases, for example CO_2 and CH_4, the proposed method allows prediction of membrane permeances of these gases. These permeances can then be compared with the experimental data, thus serving as another verification for the method. The application of the proposed method is demonstrated using silicalite zeolite membranes, prepared by the pore-plugging technique, inside porous TiO_2 supports.

WEDNESDAY 2:30 pm - Kings Garden 5

29b - Study on antifouling behaviors of GO modified nanocomposite membranes through QCMD and surface energetics analysis

Amin Karkooti (University of Alberta), Neda Nazemifard (University of Alberta), Mohtada Sadrzadeh (University of Alberta)*

Membrane fouling is the main challenge that limits the practical application of membranes in water reclamation. Development of nanocomposite membranes may provide an effective solution to improve membrane-fouling resistance. Membrane surface properties are significant factors affecting initial membrane fouling, where hydrophilic surfaces are generally less prone to fouling. The extent of adsorption of foulants to the membrane surface depends on membrane-foulant affinity, which relies on the nature of the foulant and membrane material. In the present work, an extended DLVO (aka XDLVO) interaction energy analysis was employed to explain the adsorption of organic foulant to the surface of the ultrafiltration membrane with different degree of hydrophilicity. The fundamental interactions (van der Waals, electrostatic, and acid-base interaction) that control organic fouling were evaluated. In addition, a quartz crystal microbalance with dissipation monitoring (QCM-D) was employed to investigate organic matter adsorption/desorption to/from a thin layer of membranes. The amount of adsorption/desorption and the mass of deposited structures on membranes were measured by real-time monitoring of resonance frequency and energy dissipation changes (Δf and ΔD). To obtain different hydrophilicity of membranes, graphene oxide (GO) nanosheets and polyvinylpyrrolidone (PVP) were added into the polyethersulfone (PES) casting solution. Finally, in order to validate the XDLVO and QCM-D results the surface properties of the membranes were characterized by contact angle analyzer, fourier transform infrared spectroscopy (FTIR), X-ray photoelectron spectroscopy (XPS), atomic force microscopy (AFM), zeta potential measurement, and flux recovery ratio (FRR) tests. The information on membrane-foulant interaction is critical to optimize use of additives and to reduce the membrane fouling.

WEDNESDAY 3:00 pm - Kings Garden 5

29c - 3D characterization of polyamide reverse osmosis membranes

Tyler E Culp (Penn State University)*, Manish Kumar (Penn State University), Enrique Gomez (Penn State University)

Polyamide-based thin-film composite membranes used for reverse osmosis (RO) are at the forefront of water desalination and purification technologies due to their high salt rejection, high energy efficiency, and ease of operation. Nevertheless, in spite of the benefits of RO membranes, many open questions about the internal nanostructure of the membrane active layer remain, such as the distribution of voids, surface area and mass. We demonstrate that with transmission electron microscopy, where electron tomography is used to obtain a three-dimensional (3D) reconstruction of the polyamide active layer, these parameters can be quantified to a series of RO membranes. The application of quantitative 3D characterization tools provides a strategy for linking the morphology to water transport properties of RO membranes.

WEDNESDAY 3:30 pm - Kings Garden 5

29d - Selective Ion Transport Properties of Membrane Materials

Anita Hill (CSIRO)*, Aaron Thornton (CSIRO), Cara Doherty (CSIRO)

Closed loop material reuse and recycling is enabled by separations technology. Mechanisms to separate ions and molecules rapidly, efficiently, and with high purity are achievable; however, currently these methods are found only in nature. Progress in fast selective molecular transport within membranes has been achieved via free volume elements with characteristics, such as those found in zeolites of well-defined cavity and neck topologies, that allow multiple wide pathways for flow connected by smaller sieving pathways. These topologies mimic those of biological membrane channels. The fundamental principles underlying fast selective transport in biological channels originate in the architecture of the pore. Advances in materials chemistry and characterization have made it possible to produce porosity in synthetic materials that can imitate these biological channels. The need for membrane materials with the ability to perform low energy molecular and ionic separations is increasing as the world transitions to a circular economy. For example, the separation and recovery of critical metals from process streams using low energy membrane technology not only contributes to recycling but also provides resiliency in the face of national shortages of these critical metals. We examine free volume elements, as measured by positron annihilation lifetime spectroscopy, in a range of materials that enable outstanding molecular and ionic transport and separation performance.

WEDNESDAY 4:00 pm - Kings Garden 5

29e - Membrane-based oil-gas separation for dissolved gas-in-oil extraction: gas transport properties of ceramic supported Teflon membranes

Liang-Chih Ma (Arizona State University)*, Chuan Chen (Global Energy Interconnection Research Institute Co., Ltd.), Jerry Lin (Arizona State University)

Dissolved gas extraction from transformer oil is an indispensable part in dissolved gas analysis (DGA) for transformer fault monitoring and diagnosis. Membrane-based oil-gas separation represents a promising approach for dissolved gas-in-oil extraction in terms of operation and maintenance, long-term stability, and cost competitiveness. However, no comprehensive study available in the literature on gas transport properties of membranes during the oil-gas separation process. Therefore, the objective of the present study is to clarify gas transport properties of our developed Teflon AF2400/ceramic composite membranes during the oil-gas separation process, focusing on both kinetics and thermodynamics aspects. In the study, effects of operating temperature, oil feed flow rate as well as stirring rate on the equilibrium time and the equilibrium level for 7 characteristic gases (i.e. H_2, CO, CO_2, CH_4, C_2H_6, C_2H_4, and $C2H_2$) have been investigated. The gas-dissolved oil used for oil-gas separation experiments was prepared by the standard mixed gas which comprises 7 characteristic gases with 5,000 ppmv balanced with Ar. The results show that the rate-limiting step in the oil-gas separation process is gas diffusion through the oil instead of the membranes. Reduced equilibrium time can be achieved by increasing oil feed flow rate and/or stirring rate due to enhanced gas diffusivity in the oil. In addition, it was found that the driving force for gas permeation is the gas concentration difference between the oil and the permeate side at 1 atm and thus the equilibrium time strongly depends on the Henry's law solubility constant. Except H_2, since the solubility constant decreases with increasing temperature, the equilibrium time would decrease with increasing temperature even though the gas diffusivity in the oil increases. Finally, the equilibrium level, which is determined by the solubility constant, cannot reach 100% due to the driving force limited in the dilute system.

WEDNESDAY 4:30 pm - Kings Garden 5

29f - Porous support of TFC membranes: Does it truly have negligible resistance?

Masoud Aghajani (University of Colorado Boulder)*, Alan R. Greenberg (University of Colorado Boulder), Yifu Ding (University of Colorado Boulder)

For thin film composite (TFC) membranes, the polyamide barrier layer is generally believed to be responsible for the entire resistance to water permeation, while the resistance of the porous support is considered as negligible. However, this understanding is based on the "separate" rating of the resistance of the porous support under high trans-membrane pressure (TMP), which is distinctively different from the situation during high-pressure reverse osmosis (RO). Specifically, during RO, the porous support is under much higher mechanical loading (the overall TMP of the TFC membrane) but a significantly lower permeation rate (low TMP across the support). Such high mechanical loading, e.g. 7 MPa (>1000 psi) during seawater desalination is above the yield strength of the porous support and can cause substantial mechanical deformation, especially over an extended period of time. To measure the true resistance of the porous support during high-pressure RO, we developed a new measurement method that uses a series arrangement of a TFC membrane on top of a typical porous support. Our results show that, under high overall TMP (> 5 MPa), the plastic deformation of the additional porous support is as much as 3X larger than that of the same support subjected to identical TMP but without the TFC membrane on top. A direct relationship is observed between the applied TMP and the plastic deformation of the porous supports. Moreover, DI water permeation experiments revealed an order of magnitude increase in resistance for the deformed porous supports in comparison to the pristine (undeformed) ones. Most importantly, our analysis indicates that the resistance of a representative commercial porous support used to fabricate TFC membranes can account for up to 33% of the overall TFC membrane resistance during high pressure permeation (5.5-8.2 MPa).

WEDNESDAY 2:00 pm - Kings Garden 4

30a - Numerical and experimental studies on the deposition of sticky particles near a membrane surface

Seon Yeop Jung (Seoul National University)*, Kyung Hyun Ahn (Seoul National University)

Deposition of sticky particles onto a membrane surface is studied by numerical simulation and experiment. During a membrane separation process, which is a core unit operation in water purification, bioprocessing, or pharmaceutical manufacturing, particle deposition and cake layer formation are induced by the continuous accumulation of rejected matters onto the membrane surface. Although deposition of particles is affected by complex interactions such as hydrodynamic, particle-particle, and particle-membrane interactions, comprehensive modeling and simulation of particle deposition dealing with those interactions have rarely been conducted. In addition, a prescribed velocity field was often assumed in the majority of previous studies which neglected the influence of particle cake layer onto the surrounding flow behavior. In this study, we propose a simple algorithm to roughly capture the interplay between cake layer formation and flow structure in the implementation of a modified Brownian Dynamics (BD) simulation, which tracks floating particles exerted by various forces. A surface spring was introduced in our simulation to model complex attractive interactions between particles. We defined a dimensionless number named S, defined by the ratio of the attractive force by the surface spring to the Stokes' drag force near a membrane surface, to manipulate the extent of attractive forces. A distorted flow field was generated near a particle cluster and it affected the progress of particle deposition. We compared the configuration of deposited particles with and without this new algorithm for the credible reproduction of particle deposition phenomenon. A microfluidic experiment with microalgae was conducted to capture the particle deposition and cake layer formation onto a surface of a constructed porous region. A qualitative agreement with the numerical simulation and experiment was identified in the growth of the cake layer and local movement of floating particles.

WEDNESDAY 2:30 pm - Kings Garden 4

30b - Characterization of the Striping Phenomenon during Membrane Fouling via Optical Coherence Tomography

Weiyi Li (Southern University of Science and Technology)*, Xin Liu (Southern University of Science and Technology), Anthony Fane (Singapore Membrane Technology Centre)

Mitigating membrane fouling entails powerful tools to study the complex transport phenomena occurring in the polarized boundary layer and thereby provide deeper insights into the development of a fouling layer. The current study employed a novel method to in-situ characterize a fouling process during ultrafiltration. This characterization method was based on an advanced optical technique, that is, optical coherence tomography (OCT). The focus of the current study was to dynamically capture the formation of 'stripes', which are dissipative structures in the boundary layer discovered about 30 years ago but attracted little attention in the field of membrane separations. Various algorithms were developed to analyze the OCT datasets and generate a series of digitalized fouling layers for resolving the evolution of the fouling layer on a micron scale. It was revealed that the striping phenomenon should play an important role during both the initial and long term fouling; the vortices induced by the wall permeation could be a critical mechanism accounting for the deposition and removal of the foulants at the liquid-membrane interface. Useful implications were obtained for the system design to intensify the secondary flows in the boundary layer and thereby alleviate the membrane fouling.

WEDNESDAY 3:00 pm - Kings Garden 4

30c - A closer look at biofouling: time-lapse optical microscopy of biofilm formation in wastewater reuse

Emily W Tow (Olin College)*, Behzad Rad (Lawrence Berkeley National Laboratory), Robert Kostecki (Lawrence Berkeley National Laboratory)

Biofouling is commonplace in wastewater reuse, but we rarely get the chance to see inside the slime. Biofilms tend to form on the microfiltration and reverse osmosis membranes utilized in advanced treatment of wastewater due to its high organic and microbial content. Biofouling raises the cost of water reuse by necessitating frequent membrane cleaning and, in some cases, irreversibly damaging membranes. Although cost-effective fouling mitigation requires understanding of biofilm behavior in wastewater reuse, existing biofilm visualization research focuses on other applications. Therefore, in this study, we use confocal microscopy to capture time-lapse images of biofouling on wastewater reuse membranes. Secondary effluent from municipal wastewater treatment (prior to disinfection) obtained from California's East Bay Municipal Utility District was used as the feedwater. It was stained using fluorescent wheat germ agglutinin to label extracellular substances as well as Hoechst 33342 and propidium iodide probes to differentiate live from dead cells. A miniature cross-flow membrane cell was created to operate microfiltration and ultrafiltration and at typical flow rates and transmembrane fluxes, with and without periodic backwashing, for periods of 1-5 days. Meanwhile, the biofilms that formed on the membranes were imaged in situ with a confocal laser scanning microscope. Imaging was also conducted during several clean-in-place procedures. Although single-cell resolution was attained in some images, larger maps were also obtained to observe heterogeneity on the millimeter scale. Images demonstrate biofilm accumulation and removal mechanisms including cell proliferation and EPS production as well as film detachment, wrinkling, sloughing and re-deposition. These images provide insight into biofilm behaviors specific to wastewater reuse and inform the development of targeted biofouling mitigation strategies.

WEDNESDAY 3:30 pm - Kings Garden 4

30d - Improved graphene oxide membrane increases membrane distillation desalination of RO concentrate

Samuel Olatunji (Texas A&M University-Kingsville), Lucy M Camacho (Texas A&M University-Kingsville)*

Commercialization of membrane distillation (MD) as a feasible alternative for desalination is limited due to the inexistence of membranes specifically designed to meet the needs of the MD desalination process. The ideal properties of a MD membrane are fundamentally different from those of reverse osmosis (RO). The main difference is based on the hydrophobicity of the MD membrane. In this research, a dry-wet phase inversion method was used to synthesize an optimal graphene oxide (GO) membrane that is suitable for MD desalination. The effect of pH of the GO sheets used as source material for the GO membrane was also evaluated. SEM studies showed that the hydrophobic properties of the synthesized GO membrane varies with the dry-wet phase inversion method as compared to the method of dry- or wet-phase inversion. The pH of the GO sheets also plays a critical role on the properties of the membrane. The new GO membrane was tested with simulated and real brackish RO concentrate in a portable pilot MD system. It was found that the new membrane can produce a permeate flux of 85 $L/m^2 \cdot h$, which is higher than the flux produced by a commercial polypropylene (PP) membrane tested with MD under the same operational conditions. The results indicated that the GO membrane prepared with the wet-dry phase inversion method increases the water recovery of the RO system by more than 98%. The MD process using the proposed synthesized GO membrane shows promising water recovery, salt rejection, and brine reduction from brackish RO concentrate.

WEDNESDAY 4:00 pm - Kings Garden 4

30e - Automated real-time membrane biofouling assessment using microbial enzyme activity

Babar K Khan (King Abdullah University of Science and Technology)*, Luca Fortunato (King Abdullah University of Science and Technology), TorOve Leiknes (King Abdullah University of Science and Technology)

The advent of advanced membranes for water desalination over the past several decades have greatly improved water purification processes. Biofouling is a major recurring cost driver for membranes used in desalination and play a major role in membrane life-cycle assessment. A number of methods have been developed to detect bacterial growth and provide cleaning guidance, but most suffer from either being unviable in-situ or are not capable of early biological detection. An automated at-line sensor was devised and operated with real seawater to measure biofouling over time in NF and RO membrane modules. The system samples the membrane module in-situ by injecting artificial fluorogen bound to substrate (fluorogen-substrates) at discrete time points and measures fluorescence from the liberated fluorogen in real-time. The signal has been correlated to microbial extracellular enzyme activity. Our results demonstrate the capability to track microbial biomass from initiation of colonization to biofouling on the membrane surface. This approach has the advantage of indirectly measuring bacterial abundance on membranes not readily adaptable for direct visualization (RO membrane modules), without the need for sophisticated techniques found in other molecular approaches (ATP). Our results indicate that fluorogen-substrate cleavage can be used in early biofouling detection as a prognostic tool to guide cleaning protocols for membranes in aqueous systems. In this study, fluorogen-substrate cleavage by bacterial extracellular enzymes was used to monitor microbial colonization proximal to the surface of membrane filtration systems non-invasively using an automated system. Biofouling is a major impediment to the lifecycle of membranes, therefore measuring enzymatic activity by fluorogen-substrates cleavage can be utilized to monitor membrane health for guiding anti-biofouling practices, enabling both assessment, targeted cleaning, and therefore extension of the membrane lifespan.

WEDNESDAY 4:30 pm - Kings Garden 4

30f - Interactions between extracellular polymeric substances and zwitterionic hydrogels as a designing tool for antifouling surfaces

Angelin Ida Vedhamanickam (Ben-Gurion University of the Negev)*, Nir Goldberg (Ben-Gurion University of the Negev), Ronit Bitton (Ben-Gurion University of the Negev), Moshe Herzberg (Ben-Gurion University of the Negev)

Membrane fouling by microbial biofilm components, mainly extracellular polymeric substances (EPS) is a major impediment of the membrane filtration process, which can significantly reduce membrane performance. Common promising approach for biofouling mitigation focuses on reducing adsorption of EPS to membrane surface by surface modification using zwitterionic polymers. However, favorable interactions between the EPS and the hydrogel coating, in our case poly-sulfobetaine (poly-SPE), will reduce the coating anti-fouling efficiency. Currently, the common design of hydrogel-coated membranes with poly-zwitterions is based on "trial and error" approach in a sense of analyzing membrane performance during filtration tests of different fouling and biofouling agents, for a certain coating condition. The overall aim of the proposed research is to utilize a detailed characterization of the poly-zwitterions (poly-SPE in our case), EPS components and interactions between them, on the surface and in solution, as a tool to tailor better performance of the anti-fouling, hydrogel-coated membranes. The effects of EPS model compounds on the polySPE in solution (bulk-physical properties) using small angle x-ray scattering (SAXS) were compared to adsorption experiments of the EPS model compounds to pristine and modified polysulfone (PSF)-coated sensors of quartz crystal microbalance with dissipation monitoring (QCM-D). The positive correlation between the SAXS and QCM-D results, in which fruitful interactions were observed, indicating interactions in solution can be used as a tool to predict interaction of foulants and the protective polySPE layer, under variety of aquatic conditions. At last, the predicted successes and limitations of the well characterized polySPE protective layer were confirmed by fouling experiments of pristine and modified ultrafiltration (UF) membranes.

WEDNESDAY 2:00 pm - Kings Garden 3

31a - Optimizing Microfluidic Nanoparticle Capture from Biofluids on Ultrathin Silicon Nanomembranes: A Computational and Experimental Analysis of Tangential Flow Analyte Capture (TFAC)

Kilean Lucas (University of Rochester)*, Danial Ahmad (University of Rochester), Mehdi (Aslan) Dehghani (Rochester Institute of Technology), Thomas Gaborski (Rochester Institute of Technology), Richard Waugh (University of Rochester), James McGrath (University of Rochester)

Extracellular vesicles (EVs) are lipid vesicles ranging from 30 - 1000 nm in diameter secreted by cells to serve a variety of functions. Certain members of the EV family, such as exosomes, are thought to be important for intercellular communication. Exosomes contain cell specific biomarkers (i.e. miRNA, protein) that are useful for point-of-care (POC) diagnostics. A variety of techniques are used to isolate EVs, including ultracentrifugation, affinity-based capture, precipitation and filtration, but these often require large sample volumes (50 - 300 mL) which are not feasible for a POC system. Additionally, these isolations are often contaminated with protein, leading to difficulty in the identification of biomarkers. Therefore, an alternative method of isolating EVs from small volumes is required for POC. Silicon nanomembranes are ultrathin (50-100 nm) nanoporous membranes discovered during the rapid crystallization of pure silicon films. These membranes are manufactured with a variety of pore diameters (~ 5 nm - 100 nm) and have high porosities (20% - 40%) and exceptional permeabilities. These membranes are manufactured in a 'chip' format that is 5.4 mm x 5.4 mm, making them perfect for small volume applications such as microfluidic filtration. In this work, we present a novel exosome capture method for small volume (~ 1 mL) samples. We call this method tangential flow analyte capture (TFAC). After demonstrating the capture of exosomes from undiluted serum, we present a computational model for TFAC that is validated with 60 nm gold nanoparticles with and without the addition of serum albumin (2 mg/ml). We use the model to define optimal capture conditions, and then demonstrate the capture of > 75% of exosomes from defined protein/exosome sample and compare these results to capture from raw urine.

WEDNESDAY 2:30 pm - Kings Garden 3

31b - Development and scale-up of laterally-fed membrane chromatography for the purification of therapeutic viruses

Karina Kawka (McMaster University)*, Pedram Madadkar (McMaster University), Shabnam Shoaebargh (McMaster University), Natasha Kazhdan (McMaster University), Maria Fe C. Medina (McMaster University), Brian Lichty (McMaster University), Raja Ghosh (McMaster University), David Latulippe (McMaster University)

Viruses are a new class of biotherapeutics with potential to treat a number of diseases including cancer and there exists a number of different types depending on the application. While adenoviruses are the most used type in gene therapy, rhabdoviruses have potent oncolytic (cancer-killing) activity and are currently under clinical trials in Canada. Limitations with respect to large-scale virus purification (i.e. removal of host-cell proteins and DNA) are one of the major obstacles between these therapeutics and patients. Although membrane chromatography has great potential to address this issue, the radial-flow configuration most commercial MC devices are currently designed causes non-uniform flow distribution, affecting the performance of the separation. Our group has demonstrated the advantages of the laterally-fed membrane chromatography (LFMC) over radial flow MC, which include lower sample dilution and higher virus purities when 5 mL of adenovirus culture were purified using 1 mL devices (i.e. membrane volume). Recent scale-up experiments with 50 to 100 mL of adenovirus culture using the Sartobind Q anion-exchange membrane and 1 mL LFMC devices resulted in virus recoveries close to 100% and removal of contaminant proteins and DNA reaching 90%. However, only 40% recovery was obtained when 200 mL of culture was processed because the binding capacity of the membrane was exceeded. For this reason, a 10 mL LFMC device was developed to target the purification of up to 1 L of adenovirus culture. As a second objective, the LFMC at 1 mL scale was investigated for the purification of rhabdoviruses. Early results indicate less than 10% virus recovery using the same conditions previously applied for adenoviruses. Even after adjustment of pH and elution conditions, no significant improvement was observed. Therefore, other types of membrane adsorbers based on hydrophobic and pseudo-affinity interactions were tested to maximize rhabdovirus recovery and purity.

WEDNESDAY 3:00 pm - Kings Garden 3

31c - Characterization of EV Secretion at Single Cell Resolution

Gerardo Mauleon Ramos (University of Chicago)*, Kilean Lucas (University of Rochester), Vladimir Riazanski (University of Chicago), James McGrath (University of Rochester), Deborah Nelson (University of Chicago)

Extracellular vesicles (EVs) are lipid bilayer enclosed nanocarriers secreted by cells. They are a heterogeneous group of particles generally divided into exosomes, microvesicles or apoptotic bodies depending on their cellular origin, dimension, and tetraspanin surface expression. Circulating EVs carry transmembrane proteins and luminal cargo including soluble cytoplasmic proteins and nucleic acids to other cells. The actual mechanism by which EVs deliver their cargo is still unknown, but, EV cargo is believed to mediate transfer of biologically active molecules between cells. As such, EVs have gained attention as a player in therapeutic delivery. Before EVs can take this step, information on size, concentration of secreted EVs, heterogeneity, and the overall factors that control EV release from secretory cells must be resolved. To investigate these unknown factors, we developed an isolation protocol that allows us to study EV secretion from single cells as a response to a chemical stimulus. Our initial assay uses a novel microfluidic platform that was engineered specifically to allow the measurement of released EVs by cultured cells. Additionally, we are using a filtration based approach to selectively capture secreted EVs based on their size using a novel ultrathin (50nm thick) nanoporous silicon nitride membrane. In order to differentiate between EVs with an intracellular origin from shedding vesicles, we express a fluorescently labelled PI(3,5)P2-specific probe which localizes to the multivesicular bodies and lysosomes. We have deployed this system to establish optimal conditions for EV secretion. Using this combined assay, we have established that the EV population shows at least a 3:1 ratio of membrane shedding vesicles versus exosomes. Additionally, we should be able to use this platform to design screening strategies exploring some of the unknown factors in stimulus-secretion coupling in EV release; size, concentration, secretion dynamics and heterogeneity.

WEDNESDAY 3:30 pm - Kings Garden 3

31d - Charge, size distribution and hydrophobicity of viruses: Effect of the virus purification method

Hang Shi (Michigan State University), Vlad Tarabara (Michigan State University)*

Viruses propagated and purified by different methods may exhibit different adsorption, aggregation and electrokinetic behaviors. The differences are caused by residual impurities, either in the dissolved phase or associated with virions. The effects of growth and purification protocols on the isoelectric point, surface tension, and hydrophobicity have not been explored and are the focus of the present work. The study comparatively evaluated two virus propagation methods (in broth and on double agar overlay) and three purification procedures (CsCl density gradient centrifugation, PEG precipitation, and centrifugal diafiltration) in terms of their impact on the properties of resulting virus stocks. Agar-propagated MS2 purified by PEG precipitation or centrifugal diafiltration (100 kDa) included large (~ 0.5 μm) aggregates that persisted through multiple cycles of post-filtration with a 0.1 μm pore size membrane. Regardless of the purification method, such interferences were not observed for MS2 propagated in E. coli broth. Accordingly, the degree of bacteriophage purity achievable with each of the three purification methods was compared based on the data for MS2 and P22 propagated in broth. CsCl density gradient centrifugation produced virus suspensions of highest quality. The impurities remaining in the virus suspension after PEG precipitation and centrifugal diafiltration broadened the size distribution and interfered with either electrophoretic mobility measurements or hydrophobicity characterization or both. The impact on the free energy of virus-virus interaction was especially dramatic with P22 bacteriophage appearing hydrophilic in the presence of impurities and hydrophobic in better purified stocks. The trends in measured physicochemical properties were rationalized by considering impurity-coated virions as permeable soft particles. The obtained data were used to rank the purification methods according to six criteria of likely practical relevance.

WEDNESDAY 4:00 pm - Kings Garden 3

31e - Risk Mitigation Strategies for the use of polymeric consumables for manufacturing of Anti-Sense Oligonucleotides

Anuradha Vaidya (Biogen), Bill Scott (Biogen)*, Dave Kolwyck (Biogen)

Due to increased demands on throughput efficiencies, single use systems (SUS) have become widely popular in the Biologics manufacturing industry. However, new modalities such as Anti-Sense Oligonucleotides (ASOs) entail the use of non-conventional upstream and downstream manufacturing conditions involving organic solvents and harsh chemical reagents. As a result, these new modalities present a whole new set of challenges with respect to the use of SUS. In some cases, the worst-case testing conditions historically used to assess the chemical compatibility and performance of SUS are milder than the actual manufacturing conditions. In this presentation, we will present data on the specific risks associated with the application of SUS in Anti-Sense Oligonucleotide manufacturing and suitable mitigation strategies based on comprehensive risk assessments, chemical compatibility studies and leachable studies.

WEDNESDAY 4:30 pm - Kings Garden 3

31f - Use of Nanosphere Self-Assembly to Pattern Nanoporous Membranes for the Study of Extracellular Vesicles

Marcela Mireles (University of Rochester)*, Cody Soule (Rochester Institute of Technology), Mehdi Dehghani (Rochester Institute of Technology), Thomas Gaborski (Rochester Institute of Technology)

Nanoscale biocomponents naturally released by cells, such as extracellular vesicles (EVs), have recently gained interest due to their therapeutic and diagnostic potential. Purification and labeling techniques as well as commercially available cell culture membranes have been utilized in an effort to directly or indirectly study exosomes and their effects. Nevertheless, improved platforms for the study of exosomes (30 - 150 nm EVs) are still needed. Adequate membranes, for separation and cell culture, require pore sizes to reach into the nanoscale. These pore sizes cannot be achieved through traditional lithographic techniques and commercially available thick nanoporous membranes commonly exhibit low permeability. Here we utilized nanospheres, similar in size and shape to the targeted EVs, as patterning features for the fabrication of freestanding SiO2 membranes released through a sacrificial ZnO layer and integrated into silicone-based devices. Through filtration of fluorescent polystyrene beads and exosomes we evaluated the size cut-off characteristics of the membranes as well as their potential for selective permeability of exosomes based on size. The membranes presented here achieved pore sizes of 100 or 250 nm and a thickness of 160 or 300 nm, and demonstrated their potential to allow permeability of exosomes while preventing traffic of microvesicles. This selectivity is crucial for filter-based purification as well as for the modulation of EVs exchange in a co-culture set-up. Overall, this work provides a platform for the direct and indirect study of EVs for which interest will continue to grow.

WEDNESDAY 2:00 pm - Ballroom 3

32a - A Holistic Approach to Explore Interfacially Confined Ionomers Designed for Energy Conversion Device

Shudipto K Dishari (University of Nebraska-Lincoln)*

Understanding and improvement of ion transport limitations at nanothin polymer-catalyst interface is critical to design next generation energy conversion and storage devices. In nanoscale confined ionomeric systems, ion conduction is controlled by ionic domain characteristics, mobility and viscoelastic properties of ionomer layer in a complex manner. In this work, we studied four major fluorocarbon and hydrocarbon based fuel cell ionomers with varied backbone and side chain structures under thin film nanoconfinement. We measured proton conductivity values of sub-micron thick, spin-coated, ionomer films on substrates using electrochemical impedance spectroscopy as a function of film thickness and relative humidity. In addition, the viscoelastic properties of the same films were obtained from peak frequencies of the 1st mode of vibration of AFM tips (in contact with the films) using contact resonance atomic force microscopy (CRFM) and subsequent modeling efforts. The ion conductivity and viscoelastic properties were correlated with thin film morphology (transmission electron microscopy (TEM)), and, ionic domain size and spacing (Small angle X-ray scattering (SAXS)). Finally an innovative approach was adopted to probe zone specific ionomer properties in thin films using fluorescence confocal microscopy. Combining all the information helped us to visualize what is happening with fuel cell ionomers (with varied chemical structure) at nanoscale, and, identify the most important parameters controlling interfacial ion conductivity.

WEDNESDAY 2:30 pm - Ballroom 3

32b - Anion Conducting Membranes Based on Poly(norbornene): High Conductivity, Chemical Stability and Fuel Cell Performance >2.5 W/cm^2

Mrinmay Mandal (Georgia Institute of Technology)*, Garrett Huang (Georgia Institute of Technology), Wanting Chen (Georgia Institute of Technology), Paul A Kohl (Georgia Institute of Technology)

Anion-exchange membranes (AEMs) have widespread applications in electrochemical devices. The development of AEMs with long-term alkaline stability and high hydroxide ion conductivity is of current interest. The use of non-precious catalysts, facile reaction kinetics and reduced fuel crossover are some of the AEM benefits compared to proton-exchange membranes. Previous AEM materials have had low ionic conductivity, poor stability at high pH, and high water uptake. A series of tetrablock copolymers containing non-hydrolysable backbone and long-chain head-group tethers based on vinyl addition and ring-opening metathesis polymerization (ROMP) of norbornenes were synthesized for use in AEM fuel cells and electrolyzers. Control over the free and bound water uptake, and high hydroxide mobility were achieved by creating phase segregation between the polymer blocks. Polymers with very high IEC values (ca. 3.46 meq/g) were cross-linked with N,N,N',N'-Tetramethyl-1,6-hexanediamine (TMHDA) in order to mitigate unwanted water uptake. The membranes with light cross-link density had record high conductivity, 198 mS/cm at 80°C, for a chemically stable AEM. The AEMs had long-term alkaline stability, as demonstrated by <1% conductivity loss in over 1200 h at 80°C. Ultrathin (ca. 10 μm thick) composite membranes were cast using polytetrafluoroethylene reinforcement. The peak power density of a H_2/O2, hydroxide conducting fuel cell was >2.5 W/cm^2 at 70°C.

WEDNESDAY 3:00 pm - Ballroom 3

32c - High Temperature Fuel Cells with Ion-Pair Membranes and Phosphonated Ionomers

Michael R Hibbs (Sandia National Laboratories)*, Cy Fujimoto (Sandia National Laboratories), Ehren Baca (Sandia National Laboratories), Albert Lee (Los Alamos National Laboratory), Yu Seung Kim (Los Alamos National Laboratory)

Recently, Los Alamos National Laboratory in collaboration with Sandia National Laboratories developed phosphate-quaternary ammonium ion-pair coordinated proton exchange polymers. These ion-paired polymers can be used in fuel cell membranes and they have very strong interactions between phosphate ions and quaternary ammonium groups. This enhances water tolerance under high temperature and low relative humidity fuel cell operating conditions. The ion-pair coordinated membrane provides good mechanical properties with a substantially low phosphoric acid doping level and allows for the possibility of polymer membrane fuel cells that operate at temperatures above 200°C. Ion-pair coordinated polymers still have a risk that the acid will be partly evaporated at high temperature operating conditions (> 200°C) or be partly leached out, especially within the electrodes where water is produced. Phosphonic acids can be covalently attached to polymers in order to prevent evaporating or leaching. Because of the impossibility of acid evaporation with phosphonated polymers, the proton conductivity of phosphonated polymers increases with temperature until thermal degradation of the polymers occurs. In this project, we are aiming to develop advanced phosphonated ionomers for ion-pair coordinated polymer electrolyte membrane fuel cells to further improve fuel cell performance. Ionomer synthesis and ion-pair membrane fuel cell results will be included in the presentation. Sandia National Laboratories is a multi-mission laboratory managed and operated by National Technology and Engineering Solutions of Sandia, LLC., a wholly owned subsidiary of Honeywell International, Inc., for the U.S. Department of Energy's National Nuclear Security Administration under contract DE-NA-0003525.

WEDNESDAY 3:30 pm - Ballroom 3

32d - High temperature, anhydrous proton conducting membranes and micropatterned bipolar membranes for electrochemical energy conversion and storage technologies

Christopher G Arges (Louisiana State University)*

This talk primarily focuses on mild temperature (100 to 250°C) proton conductors composed of polycation polymers, typically used as anion exchange membranes, blended with polybenzimidazole (PBI). These moderate temperature membranes are envisioned for fuel cell systems that operates with hydrogen containing carbon monoxide - produced from steam-reformed methane. Evaluating polycation membrane variants alone demonstrated lower proton conductivity when compared to PBI doped with phosphoric acid (H3PO4). PBI doped with phosphoric acid, on the other hand, was observed to be unstable above 160°C because of H3PO4 boiling and evaporation. Plus, the PBI-H3PO4 membrane loses its acid due to leaching if condensed water is present (i.e., lack of robust operation if the cell drops below 100°C). To overcome these problems, polycation polymers, were blended with polybenzimidazole (PBI) and these membranes routinely showed conductivity values as high as 0.18 S cm^{-1} and were stable under dry and wet conditions and with thermal stability up to 250°C. The talk will close with our recent efforts to pattern polymer electrolyte membrane surfaces to make high surface area bipolar junction interfaces. Larger interfacial surface areas in the bipolar junction region are more effective for water splitting. It is envisaged that an acidic anode and alkaline cathode with a bipolar membrane offers a pathway forward for very low platinum loading fuel cells.

WEDNESDAY 4:00 pm - Ballroom 3

32e - Membrane through-plane alignment of ion-conducting channels

Xin Liu (Tianjin University), Yi Li (Tianjin University), Jiandang Xue (Tianjin University), Weikang Zhu (Tianjin University), Junfeng Zhang (Tianjin University), Yan Yin (Tianjin University), Michael D Guiver (Tianjin University)*

A strategy to make membranes with short-pathway through-plane orientated ion-conductive channels is described [1]. Previously, this has only been accomplished in a magnetic field by proton nonconductive metal oxide fillers. Here, under a strong magnetic field, a proton-conducting paramagnetic complex based on ferrocyanide-coordinated polymer and phosphotungstic acid is used to prepare composite membranes with highly conductive through-plane-aligned proton channels. This approach simultaneously overcomes the high water-solubility of phosphotungstic acid, preventing its leaching out of the membrane. The ferrocyanide groups in the coordination polymer continuously consume free radicals by a redox cycle, thus improving the long-term in situ membrane durability. The composite membranes exhibit outstanding proton conductivity, fuel cell performance and durability, compared with other types of hydrocarbon membranes and industry standard Nafion 212. [1] Nature Communications, in press. Publication date February 19th, 2019: DOI: 10.1038/s41467-019-08622-2

WEDNESDAY 4:30 pm - Ballroom 3

32f - Novel sulfonated aromatic polymer membranes for breaking the proton selectivity-conductivity trade-off limitation in vanadium redox flow battery

Tongshuai Wang (University of Illinois, Chicago), JunYoung Han (Rensselaer Polytechnic Institute), Kihyun Kim (Rensselaer Polytechnic Institute), Andreas Muenchinger (Max-Planck-Institute for Solid State Research), Klaus-Dieter Kreuer (Max-Planck-Institute for Solid State Research), Chulsung Bae (Rensselaer Polytechnic Institute), Sangil Kim (University of Illinois, Chicago)*

Vanadium redox flow batteries (VRFBs) have received extensive attention due to its attractive features for the large-scale energy storage because of their design flexibility, absence of intercalation/deintercalation and stress build-up in electrodes, and safe operation. One of the most critical components in VRFB is an ion exchange membrane which can prevent cross mixing of the multivalence vanadium ions, while it still allows the transport of protons (or anions) to complete the circuit. However, the trade-off relationship of proton exchange membrane limits the energy efficiency (EE) of the VRFB due to low coulombic efficiency (CE) caused by vanadium ion cross over through the membrane or high membrane resistances. Anion exchange membrane suffers from poor chemical stability and low voltage efficiency due to low ion conductivity. In this study, we synthesized a series of novel sulfonated aromatic polymer membranes having high proton conductivity as well as selectivity which surpass the conventional proton conductivity-selectivity trade-off relationship. The effect of the pendent group structure on the membrane ion selectivity and battery performance was investigated. It was found that the incorporation of aromatic structures on the pendent groups effectively improved the proton conductivity and selectivity. The membrane with aromatic pendent groups (BPAr-F4) shows 3 times higher proton/vanadium ion selectivity than Nafio$N_2$12. As a result, the VRFB cell equipped with the 80um-thick BPAr-F4 membrane shows significantly better performance in comparison to the cell with Nafion membranes, as high as 97.18% CE, 95.91% VE and 93.20% EE (vs. 75.14% CE, 96.12% VE and 72.22% EE of Nafio$N_2$12; 91.30% CE, 95.63% VE and 87.62% EE of Nafion117) at 20 mA/cm^2. Small-angle X-ray scattering studies indicates that their unique morphologies with narrow aqueous ionic domains which are well organized with acidic -SO3H groups results in the high proton selectivity and VRFB performance.

POSTER SESSION

Monday, May 13, 7:00 pm - 10:00 pm | Ballrooms 1 & 2

Poster Session Chairs

Prity Bengani-Lutz (Repligen Corp) plutz@repligen.com
Ramesh R Bhave (Oak Ridge National Lab) bhaverr@ornl.gov
Yuexiao Shen (Uni of California Berkeley) syxbach@gmail.com

Presenting authors are indicated by an asterisk *

SESSION A - Innovation in Bio-Inspired and Bio-Applied Membranes

A1 **High-Capacity Multimodal Anion-Exchange Membranes for Purification of Biologics,** Joshua Osuofa (Clemson University)*; Scott Husson (Clemson University)

Multimodal membranes are a novel class of chromatography media that bind proteins and bioprocess impurities through combinations of Coulombic interactions, hydrophobic interactions and hydrogen bonding. Because of the multiple modes of binding, these membranes can operate over a wider range of conditions than conventional membranes, enabling their use in chromatographic purification steps with minimal or no pretreatment of the feed. In this study, we are evaluating the performance of Purilogics' multimodal anion exchange membrane in bind-and-elute mode under a variety of industrial relevant conditions. In this presentation, we will present performance data from dynamic binding capacity (DBC) measurements for bovine serum albumin (BSA), human immunoglobulin G (hIgG), and salmon sperm DNA (ssDNA) for a range of conductivities (0 -30 mS/cm) and pH values (5-9). We will show that these novel membranes have high capacities at short residence times even under conditions of high ionic conductivity. Dynamic binding capacities for BSA, IgG and ssDNA varied with pH and ionic conductivity. The isoelectric point of the protein determined the trend of DBC with pH. DBC also varied with ionic conductivity, often showing a maximum at 5.7 mS/cm (50 mM NaCl). Based on comparisons to commercial membrane products, these high-capacity membranes can be an alternative for purifying large bioprocess volumes using short residence times without dilution or desalting of the feed.

A2 **High throughput screening of ultra-filtration membranes with mixed dextrans in 96 well plate,** Masha Khazan (Pall Corporation)*; James Hathcock (Pall Corporation); Jian Qiu (Pall Corporation)

High Throughput Screening (HTS) of ultra-filtration (UF) membranes was developed to eliminate membrane testing bottleneck and to allow assessment of membrane uniformity on a small scale. Using custom-made infrastructure to quickly assemble up to 24 various UF membranes into 96 well plates, we utilized commercially available automated liquid handling system to verify 96 well plate's integrity and to perform dextran challenge. Membrane dextran selectivity was assessed based on HPLC analysis of permeates performed directly from the 96 well collection plate. Dextran challenge conditions for 96 well plate were optimized to achieve Molecular Weight Cut Off (MWCO) correlation with stirred cells. Dyes were used to quantify volumetric water and dextran flux in 96 well plate test. During assay validation phase, we demonstrated uniformity of dextran selectivity within and between plates as well as ability to differentiate "good" vs. "bad" media on the same plate. Implementation of this HTS test for preliminary screening of UF membranes allows to evaluate fine scale membrane uniformity, speed up membrane performance testing, and free up operator time.

A3 **Impacts of Bio-inspired Zwitterionic Membranes for Health Care Applications,** Yung Chang (R&D Center for Membrane Technology, CYCU)*

Bio-inspired zwitterionic materials are the latest generation of biomaterials for nonfouling interfaces and membranes in health care applications. This talk summarizes our fundamental developments in R&D Center for Membrane Technology (Taiwan) related to the functionalization of interfaces and membranes using zwitterionic materials. Our molecular designs of zwitterionic polymers and copolymers, sulfobetaine-based, carboxybetaine-based, or phosphobetaine-based, will be reviewed. Then, the strategies used to functionalize surfaces/membranes by coating, grafting onto, grafting from, or in situ modification will be introduced, and the important part of this talk will be the focus to key medical applications of zwitterionic membranes. Finally, some potential future directions for molecular designs, functionalization processes, and applications will be summarized. This work was supported by Ministry of Science and Technology in Taiwan. (MOST 108-2923-E-033 -001) References [1] C. C. Lien, L. C.Yeh, A. Venault, S. C. Tsai, C. H. Hsu, G. V.

Dizon,Y. T. Huang, A. Higuchi and Y. Chang, Journal of Membrane Science, 565, 119-130 (2018). [2] Y. W. Chen, A. Venault, J. F. Jhong, H. T. Ho, Y. Chang, Journal of Membrane Science, 537, 355-369 (2017). [3] A. Venault, K. M. Trinh, Y. Chang, Journal of Membrane Science, 501, 68-78 (2016). [4] A. Venault, M. R. B. Ballad, Y. H. Liu, P. Aimar, Y. Chang, Journal of Membrane Science, 477, 101-114 (2015). [5] S. H. Chen, Y. Chang, K. R. Lee, J. Y. Lai, Journal of Membrane Science, 450, 224-234 (2014). [6] A. Venault, Y. Chang, H. H. Hsu, J. F. Jhong, J. Huang, Journal of Membrane Science, 439, 48-57 (2013).

A4 Enhanced Membranes for Biogas Upgrading, Ning Shangguan (Compact Membrane Systems)*; Sudip Majumdar (Compact Membrane Systems); Kenneth Pennisi (Compact Membrane Systems)

Every year over 1 billion tons of fresh animal manure is produced in US and it can be a pollution source if not handled properly. "Anaerobic digestion" is an environmental friendly and energy efficient way to dispose animal manure. Biogas is a mixture of 60% methane, 40% carbon dioxide and trace amounts of other contaminant gasses such as hydrogen sulfide. Considered to be highly corrosive and toxic, biogas is often burned immediately as a low grade fuel and there is no storage. If the non-methane components can be removed from biogas, biogas will be chemically identical to natural gas and can be put into many advanced uses, which will increase the economic value of the anaerobic digestion. A low cost solution to upgrade raw biogas (remove CO_2, H_2S and H_2O) is much needed to improve the economics of this conversion from animal manure to bioenergy. Compact Membrane Systems is proposing a membrane separation technology which can efficiently upgrade biogas (remove CO_2, H_2S and H_2O). We have developed membranes with very high CO_2 flux and high CO_2/CH_4, H_2S/CH_4 and H_2O/CH_4 selectivity. These highly fluorinated membranes have shown excellent long term resistance to high concentration H_2S biogas. Economic analysis shows a significant cost saving compared with conventional CO_2 gas membrane separation used in biogas upgrading. We scaled up the biogas membranes, made membrane modules and are designing /building pilot membrane systems. Field test will be conducted on real biogas streams in farms. We expect this technology will promote the use of anaerobic digestion to treat animal manure for biogas production. This will result in less water pollution from animal manure, less greenhouse gas emission from animal manure degradation, more renewable energy from biogas production and more jobs in rural areas.

A5 Monitoring of Membrane Fouling by the Zeta Potential, Thomas Luxbacher (Anton Paar GmbH)*; Hermina Buksek (University of Maribor); Irena Petrinic (University of Maribor); Elisa Innocenti (GSK Vaccines S.r.l.); Krzysztof Trzaskus (Aquaporin A/S); Maria Salud Camilleri-Rumbai (Aquaporin A/S)

The tendency of a membrane towards fouling is influenced by the properties of the membrane, the composition of the feed solution, and the filtration conditions. Among the properties of a membrane, parameters which describe the interaction of membranes with their environment are of major importance for studies of membrane fouling. The membrane zeta potential gives direct information on electrostatic interactions between the membrane surface and compounds in the feed solution, which contribute to membrane fouling. The surface charge of a membrane can also be tuned to favor or suppress certain interactions with components in the feed solution. Surface charge analysis can also be used to optimize membrane cleaning by evaluating the effects of different cleaning agents on the membrane surface. We report on the zeta potential analysis of the inner surface of Aquaporin Inside® hollow fiber forward osmosis (HFFO) membranes, which were employed for the up-concentration of fermentation broths from vaccine production processes. Aquaporin Inside® HFFO fibers assembled in a lab-scale filtration module offers ideal conditions for the surface zeta potential analysis by the streaming potential method. The isoelectric points (IEPs) of pristine and fouled HFFO membranes were assessed by a pH scan of the zeta potential and compared with the IEP of fouled HFFO membranes after rinsing with deionized (DI) water. The preliminary zeta potential results suggest a temporary deposition of a fouling layer by organics in the fermentation broth that is successfully removed after rinsing. The efficient removal of HFFO membrane fouling is confirmed by the same flux of DI water determined before and after filtration and rinsing. The streaming potential technique is a unique method for the non-destructive analysis of the inner surface of hollow fiber membranes. It allows monitoring of fouling layer deposition on the membrane surface and supports quantification of the cleaning efficiency.

A6 Tangential flow microfluidics for the capture and release of nanoparticles and extracellular vesicles on conventional and ultrathin membranes, Mehdi Dehghani (Rochester Institute of Technology)*; Kilean Lucas (University of Rochester); James McGrath (University of Rochester); Thomas Gaborski (Rochester Institute of Technology)

Extracellular vesicles (EVs) contain nucleic acids and protein cargo of their cells of origin and accumulate in body fluids. The role of EVs in intercellular communication has received increased attention for diagnostic and therapeutic applications and the development of many alternative purification techniques. While membrane based separations have been widely used in the biomedical area due to their robustness and simplicity, these are not commonly used for EV purification. Here, we develop a method for the capture and release of nanoparticles and EVs in a novel microfluidic device and flow scheme. We term our novel purification strategy Tangential Flow for Analyte Capture (TFAC). In this approach, the sample is pushed tangentially through the channel of the microfluid device, particles are captured by the pores of the membrane surface due to a controlled rate of ultrafiltration. We show that that these particles can be visualized, counted and later released by simply reversing the transmembrane flow. We compare the capture and release of fluorescent nanoparticles using conventional thick (10 μm) polymeric track-etched (TE) membranes to ultrathin (100nm) silicon nitride nanomembranes. We show that silicon nitride nanomembranes exhibit better capture and release and less irreversible pore clogging compared to TE membranes under the same flow conditions. This result suggests that TFAC on silicon nitride nanomembranes is a viable new method for the isolation and purification of EVs from small fluid volumes.

A7 Low-cost ultrathin porous polymeric membranes with precise control over pore size and thickness gradient for investigation of leukocyte transmigration through barrier models, Shayan Gholizadeh (Rochester Institute of Technology)*; Alec Salminen (University of Rochester); Zahra Allahyari (Rochester Institute of Technology); Robert Carter (Rochester Institute of Technology); Henry Chung (Rochester Institute of Technology); Marcela Mireles (University of Rochester); Thomas Gaborski (Rochester Institute of Technology)

Ultrathin porous membranes are currently used for a variety of biomedical applications, ranging from co-culture systems to tissue barrier models. In human barrier models, the intimate proximity between cell-types plays a significant role in cell-cell communication and transmigration across the barrier which cannot be physiologically modeled with conventional co-culture systems typically separating cell types by more than 10 μm. Importantly, regulated leukocyte transmigration across endothelial barriers can play a significant regulatory role in health and disease, but the physical constraints regulating this immune function are not fully understood. Here, we developed physiologically-inspired Parylene co-culture membranes with precise thickness gradients (from 200 nm to 1.5 μm) and highly controllable pore sizes on a single silicon wafer using a single patterned mask (obtaining pores ranging from 0.8 um to 1.7 um), which can result in porous free-standing membranes with a relatively wide range of pore sizes and thicknesses on a single silicon wafer. The process starts with coating water-soluble sacrificial layer on a 6" silicon wafer, followed by Parylene chemical vapor deposition on the physically confined silicon substrate. Using optical photolithography, distinct regions of the wafer is exposed with various user-defined exposure regimes, which would result in various pores with highly controllable pore sizes after RIE dry etching with a combination of Ar, SF6, and O2, and vigorous residual resist removal. These membranes enable leukocyte transmigration studies through defined pore sizes and thicknesses to identify critical dimensions. Moreover, the membranes can be used to understand the underlying mechanisms for leukocyte transmigration across barrier models including the blood brain barrier in neurodegenerative disorders.

~~**A8 Application of Novel Multimodal Anion-exchange Membrane Chromatography Columns in a Two-step mAb Purification Scheme - Aggregates and HCP Removal,**~~ ~~Daniel Henn (Purilogics, LLC)*; Anna Forsyth (Purilogics, LLC); Graham Temples (Purilogics, LLC); Jinxiang Zhou (Purilogics, LLC); Scott Husson (Clemson University)~~

~~Protein aggregation is a growing problem in the manufacturing of biologics due to increasing solution titers. Antibody aggregates and other impurities often are removed in a three or four-step process involving a bind-and-elute cation exchange chromatography step and additional hydrophobic interaction chromatography (HIC) columns after an initial Protein A capture step. We present case studies in which aggregates and host cell proteins (HCP) are removed in a simplified two-step monoclonal antibody (mAb) purification scheme. In this scheme, the mAb is isolated from supernatant using a Protein A chromatography step followed by a polishing step purification using Purilogics' multimodal anion-exchange membrane chromatography columns (MMM-AEX) operating in flow-through mode. In this study, a variety of mAbs were purified using the two-step purification strategy. Different buffer types (e.g. tris base, phosphate buffers) and buffer conductivities (1-15 mS/cm) were explored. HCP level and fractions of monomer, dimer and aggregates were compared before and after the polishing step. The results were compared with studies using two other leading commercial membrane products. Additionally, we will demonstrate the scalability and reusability of the novel MMM-AEX column.~~

A9 Effect of Nanomembrane Orientation and Flow Modality on Benchtop Urea Clearance, Sam Walker (The State University of New York at Buffalo)*; Kayli Hill (University of Rochester); Alec Salminen (University of Rochester); Dean G Johnson (University of Rochester)

Our lab has developed highly efficient nanomembranes (75 nm thick) that aim to replace center-based dialysis filters through the creation of a wearable hemodialysis device. Nanomembranes, formed from a thin-film stack on 5.4 mm by 5.4 mm chips, are exposed from the backside via a trench (2 mm by 0.7 mm) created in an etch process. Chips can be formed with either one trench or two giving two different membrane areas for dialysis. The different etching processes used on each side of the nanomembranes result in distinct surface characteristics and potential for anisotropic filtering. Benchtop single-pass hemodialysis experiments were designed to investigate whether membrane orientation affected urea clearance. Additionally, studies aimed to determine the effect of analyte to dialysate flow rate ratios on urea clearance were performed. Findings for the membrane orientation show that it is clear that neither chip configuration offers significantly better urea clearance than its counterpart under these experimental conditions. However, the analyte-trench combination is effective when clearing urea and should be implemented in future devices and experiments concerning blood analytes. Future studies will utilize blood samples to confirm this result. Results indicate a significant improvement in urea clearance when the flow rate ratio of analyte to dialysate is 1:2. The 2 μL/min : 4 μL/min flow rate combination offered significantly better urea clearance compared to that of 3 μL/min : 4 μL/min and 4 μL/min : 4 μL/min combinations. This is important to note as doubling membrane area offers considerably more than twice the urea clearance, under these experimental conditions. When compared to the previous results, the 2 μL/min (analyte) and 4 μL/min (dialysate) flow rate combination provided significantly better urea clearance. It is believed that faster analyte flow rates effectively decrease the volume of fluid near the membrane where diffusion has the potential to occur.

A10 Feed Spacer Modification for Listeria Control in Dairy Processing, Mainara Costa Teixeira (The University of Alabama)*; Preston Richier (The University of Alabama); Jake Colburn (The University of Alabama); Shelby Brooks (The University of Alabama); Will Baker (The University of Alabama); Thomas Hendrich (The University of Alabama); Ryan Summers (The University of Alabama); Stephen Ritchie (The University of Alabama)

Listeria monocytogenes (LM) is a serious foodborne pathogen that may be present in industrial dairy processes. Pasteurization is effective against the bacteria, but does not address downstream recontamination. Quaternary ammonium (QA) compounds are common antimicrobial agents. They can be immobilized on the feed spacer material that is used in spiral wound elements. Bacteria are adsorbed on the mesh surface due to electrostatic interaction between the positively charged QA head and negatively charged bacterial cell membrane. Permeation of the QA side chains then causes leakage of cytoplasmic material. Bacteria adsorption also reduces downstream concentrations. A modified feed spacer mesh material was produced to control bacteria growth. Vinylbenzyl chloride was reacted with octadecylamine to produce an aminated monomer. The resultant product was polymerized by photo-polymerization. We determined the anti-bacterial effectiveness of the polymer through biotesting with a surrogate for LM, Lactobacillus plantarum. Further testing showed that presoaking the material in water had a significant impact in delaying and reducing the growth of the gram-positive bacteria. Scanning electron microscope images before and after the biotest showed a high adsorption of bacteria to the surface. Preliminary results will also be presented on the functionalized polypropylene feed spacer and tests with LM.

A11 Critical Flux of Ultrathin Silicon Membranes in Tangential Flow Filtration of Protein Solutions, Danial Ahmad (University of Rochester)*; Kilean Lucas (University of Rochester); Mehdi (Aslan) Dehghani (Rochester Institute of Technology); Thomas Gaborski (Rochester Institute of Technology); James McGrath (University of Rochester)

Dead-end filtration of protein-rich fluids through ultrathin nanomembranes (<100nm thick freestanding) leads to rapid surface fouling due to their high permeability and lack of internal structure. By contrast, tangential flow operations with ultrathin membranes and complex biofluids (such as hemodialysis) have been found to proceed for days without a loss of membrane permeability, even in the presence of some ultrafiltration. Inspired by these observations, here we systematically test the hypothesis that ultrathin nanoporous membranes exhibit a higher 'critical flux' in tangential flow filtration (TFF) than conventionally thick membranes. After removing protein aggregates by centrifugation, we show that solutions of bovine serum albumin (BSA) as concentrated as 10 mg/ml achieve a critical flux of 1157 l/m^2 h in a microfluidic device featuring 18.8% porosity nanoporous silicon nitride (NPN) membranes. This critical flux is approximately five times higher than literature values achieved for TFF microfiltration, where a flux of 220 l/m^2 h was acquired with 0.2 μm track-etched (TE) membranes using 1 mg/mL BSA. Repeating our studies with 10 μm thick TE membranes with the same size (~50 nm diameter) and number (~5 x 10^7) of pores as NPN confirms our hypothesis that membrane thinness leads to a higher critical flux. Based on the fact that transmembrane pressures (TMP) in these studies are higher for TE vs. NPN with the same ultrafiltration rate we posit a mechanism for critical flux enhancement: the inherent thinness of NPN mitigates the formation of a stagnant protein 'cake layer' on the membrane surface and instead promotes a 'flowing protein cake'. More specifically, a lower TMP limits the fouling experienced by NPN versus TE membranes and this confers long term viability and performance.

A12 Grafting Zwitterionic Polymer and Poly (amino acid) on Polyamide Membranes: "Defending and Attacking" Strategies for Biofouling Control, Wen Ma (Concordia University)*; Liuqing Yang (Concordia University); Saifur Rahaman (Concordia University)

In this study, dual functions block copolymers with fouling resistant and antimicrobial properties were grafted on polyamide membranes by utilizing graphene oxide (GO) as modification media to control biofouling. The block copolymer consisted of bacteria-"defending" and -"attacking" moieties formed by zwitterionic polymer (poly(sulfobetaine methacrylate), PSBMA) and poly (amino acid) (poly methacryloyl-L-Lysine, PLysMA), respectively. To controllably grow the copolymer on membranes with desired architecture, the activators regenerated by electron transfer-atom transfer radical polymerization (ARGET-ATRP) technique was utilized. The successful polymerization of copolymers on membrane was examined by FTIR and SEM. The grafting density and thickness of polymer brush was optimized by controling the polymerization condition. Both PSBMA and PLysMA grafting significantly increased the hydrophilicity of the membrane and reduced its surface roughness and charge density. Combined PSBMA and PLysMA reducing the biofouling caused by gram positive (G+) bacteria, B. Subtilis, via bacterial-"defending" and -"attacking" synergistic effect. However, in case of biofouling caused by gram negative bacteria (G-), E.coli, "defending" functions played the predominant role. When the "defending"-PSBMA was grafted on the top and directly exposed to bacteria-containing suspensions, the membrane exhibited the lowest bacterial attachment and maintained a high water-permeation-flux along with biofilm growth. With a strong water affinity and fouling-resistance, the M-PLysMA-b-PSBMA membrane exhibited less flux reduction (27%) associated with biofilm growth in comparison to that of PA (80%) under the same condition. 100% flux recovery was observed on the M-PLysMA-b-PSBMA membrane after two cycles of a "fouling-cleaningâ€ procedure; while only 94% recovery of flux was observed on the control membrane.

A13 Monitoring of biological fluids with extraction membranes, Pinar Cay-Durgun (Arizona State University); Tianmiao Lai (Arizona State University); Nai-Yuan Liu (Arizona State University); Mark Sprowls (Arizona State University); Stewart Mann (Arizona State University); Leslie Thomas (Mayo Clinic in Arizona); Erica Forzani (Arizona State University); Mary Laura Lind (Arizona State University)*

Diagnosis and management of kidney and liver disorders could benefit from a point of care device for measurement of small molecules in biological fluids. Example conditions include ones which may alter systemic acid/base homeostasis, metabolic conditions, and other diseases of the kidney and liver. Practical and technical limitations of small molecule measurement have impeded the use of these molecules as a routinely employed biomarkers in medicine. We are developing membranes for selective extraction of small molecules from biological fluids for colorimetric detection. We will present our results on novel membranes for these separations.

~~**A14 Fine-Tuned Biohybrid Polymeric Particles Mimicking Biological Compartmentalization Prepared by Membrane Emulsification,** Emma Piacentini (National Research Council of Italy, Institute on Membrane Technology, CNR-ITM); Lidietta Giorno (National Research Council of Italy - Institute on Membrane Technology)*~~

~~Biomacromolecules such as enzymes have a considerable role at the heart of biotechnological processes. However, significant improvements in their formulation are required to broaden their application, in particular for the synthesis of pharmaceutical, food and cosmetic products. Enzyme immobilization might improve biocatalyst properties because substrate specificity might be enhanced and the effect of inhibitors might be reduced. However, many methods of immobilization and entrapment cause important structural deformation of the enzyme, leading to reduction in activity. The optimization of the immobilization method is therefore required to reach biocatalyst resistance to leaking, retention of enzyme activity over long-term storage, operational stability under adverse environmental conditions, accessibility to substrates, fast catalysis, and, in general, proper enzyme immobilization density with adequate orientation. Membrane emulsification (ME) is emerging as a very promising formulation method to produce monodisperse particles by using microporous membranes in mild operative conditions and low energy consumption. The use of ME for simultaneous preparation of polymeric microspheres and phase transfer biocatalyst immobilization will be described. Lipase is used as a model interfacial biomacromolecule to demonstrate the powerful of the use of ME mechanism in the production of carriers with tailored properties for amphiphilic molecule immobilization. Enzyme entrapment during particle formation permitted to orient the enzyme without affecting its intrinsic biocatalytic activity. Furthermore, it permitted an enzyme confinement with a good balance between molecular flexibility and rigidity. ME can potentially extend the development of bio-functionalized micro-nanostractured devices with improved functions, sensitivity, and stability for applications in biocatalysis, biosensors and biomolecule-responsive drug delivery systems.~~

A15 Bio-inspired immobilization of casein-coated silver nanoparticles on cellulose acetate ultrafiltration membranes: from laboratory to scale-up, Xiaobo Dong (University of Kentucky); Halle Shannon (University of Kentucky)*; Tequila Harris (Georgia Institute of Technology); Isabel Escobar (University of Kentucky)

Casein-coated silver nanoparticles (casein-AgNP) had been previously loaded on cellulose acetate (CA) ultrafiltration membranes for biofouling control[1]. However, when physically added to the membrane casting solution, AgNPs tend to leach off the membranes. Therefore, there is strong interest in developing methods to chemically bind the AgNPs to the membranes without losing their activity, and in making these methods scalable. Inspired by the affinity of silver to the thiol groups of cysteine proteins in bacteria, CA membranes were cast and imbedded with AgNPs via attachment with functionalized thiol groups using polymerized glycidyl methacrylate (GMA) and cysteamine. However, the chemical reactions required three steps after casting the membranes, and each step needs a large amount of solution bath. To minimize the materials and energy needed in the chemical attachment process and maximize the reaction efficiency, this study integrated all the chemical reactions in the stage of preparing dope solutions. The recipe and procedures changed due to the difference of dope solution and membranes. Membranes immobilized with AgNPs were successfully cast for laboratory-scale studies as well as for scale up studies. This method successfully prevented leaching of AgNPs during cross-flow and dead-end studies. Pseudomonas fluorescens Migula in brackish water was used for biofouling test and the membranes displayed a significant reduction in the accumulation of bacterial cells. This method provided a promising approach of casting CA membranes immobilized with AgNPs. 1. Sprick, C.; Chede, S.; Oyanedel-Craver, V.; Escobar, I.C. Bio-inspired immobilization of casein-coated silver nanoparticles on cellulose acetate membranes for biofouling control. Journal of Environmental Chemical Engineering 2018, 6, 2480-2491.

A16 Surface-Specific Thermoresponsive Coating of Membranes, Marcela Mireles (University of Rochester)*; Cody Soule (Rochester Institute of Technology); Luis Delgadillo (University of Rochester); Thomas Gaborski (Rochester Institute of Technology)

Label-free biological separations such as filtration hold great potential for the isolation and recovery of viable biological material such as circulating tumor cells (CTCs) or extracellular vesicles. However, recovery can be limited by non-specific adsorption of the targeted species onto the membrane surface and over the pores. Thermoresponsive coatings, such as poly(N-isopropyl acrylamide) (PNIPAM) which undergoes a conformational transition from extended to collapsed, could be used to increase collection efficiency by releasing the trapped targeted species. However, conformal PNIPAM coatings of a membrane lead to potentially undesirable changes in pore size and a reduction in the initial flux values due to coating of the pore walls. In this work, nanospheres were placed on a SiO_2 film and coated with Au:Cu followed by Al. A porous pattern on the metallic layer is revealed upon nanosphere removal and transferred through dry etching into the SiO_2. The remaining Al is etched to expose the Au:Cu and the samples are placed in an ethanolic solution containing acetic acid for PNIPAM grafting. This coating is limited to take place at the membrane surface only due to metal-sulfur interaction of the thiol-terminated PNIPAM, hence not coating the pore walls. Chemical characterization validates the presence of PNIPAM and environmental atomic force microscopy showed its response to temperature which was evidenced by a change in polymer coating stiffness. Our approach preserves the specific pore size and characteristics of the membrane, while promoting release of adsorbed and captured species over the pores.

A17 Pervaporation Separation of n-Butanol from Aqueous Solutions Using PDMS/Lignin Mixed Matrix Membranes, Ali Zamani (University of Ottawa); Handan Tezel (University of Ottawa)*; Jules Thibault (University of Ottawa)

The scarcity and rising prices of fossil fuels, coupled with environmental challenges, lead researchers to find out a microbial alternative. Biofuel has emerged as an attractive solution which can be effectively used in the current scenario. Among them, biobutanol is an important renewable biofuel. Butanol is produced via acetone-butanol-ethanol (ABE) fermentation from renewable resources such as agricultural by-products. The most important challenge to produce this alcohol as an economically viable biofuel, is improving fermentation performance. To improve the productivity, the in-situ removal of the fermentation products, especially butanol as the most toxic one, from fermentation broth is essential. In this study, pervaporation, a membrane-based process was studied for in-situ separation of butanol. The primary objective of this study was to improve the performance of the Polydimethylsiloxane (PDMS) membrane for the pervaporation separation and the recovery of butanol by adding fillers into its matrix. Lignin, a natural polymer resin, was used as the filler. Different lignin percentages and membrane thicknesses were used and compared to the neat PDMS. Different characteristics of lignin and its impact on the performance of the host membrane were discussed. Results showed that increasing filler particles up to 10 wt% will improve MMM performance by 280% and 60% in flux and selectivity, respectively. This enhancement was reasonable due to the presence of lignin, which enlarged the free volume within the PDMS matrix and provided more permeation pathways for penetrants. On the other hand, by increasing the filler particles to 15 wt%, flux was improved by 40% but selectivity was reduced dramatically by 70%. The lignin agglomeration and low filler-polymer bonds at excessive loadings should be responsible for the water molecules (with smaller kinetic diameter than butanol) to more easily bypass the hydrophobic lignin particles and make the membrane less selective.

A18 Rapid and Selective Water Permeation Across Biomimetic Membranes Through Artificial Water Channel Aggregates, Woochul Song (Penn State University)*; Ratul Chowdhury (Penn State University); Himanshu Joshi (University of Illinois at Urbana-Champaign); Joseph S. Najem (The University of Tennessee); Yue-xiao Shen (UC Berkeley); Chao Lang (Penn State University); Stephen Sarles (University of Tennesseee); Jun-li Hou (Fudan University); Aleksei Aksimentiev (University of Illinois at Urbana-Champaign); Manish Kumar (Penn State University)

Achieving high permselectivity of desalination membranes is of great interest in membrane research, due to ever increasing global demands on safe water supply and reliance on membrane-based desalination technologies. However, overcoming desalination membranes' water/salt permeability and selectivity tradeoffs has not been quite successful compared to overall advances in water purification technology during last few decades. As next generation materials, biomimetic membranes have rapidly emerged as a promising desalination membrane development platform, because they can accommodate specialized pore structures, water channels. Here, we present new synthetic molecular architectures as a novel class of artificial water channels. These channels self-aggregate inside biomimetic membranes and provide cooperatively enhanced water permeation paths. Molecular transport studies reveal that these channels have intrinsic water/salt permselectivities that exceed current polymeric desalination membranes by orders of magnitudes, highlighting their potential in future desalination membrane development.

A19 Can porous membranes be engineered to induce physiologically relevant cell-substrate interactions?, Zahra Allahyari (Rochester Institute of Technology)*; Stephanie Casillo (Rochester Institute of Technology); Shayan Gholizadeh (Rochester Institute of Technology); Henry Chung (Rochester Institute of Technology); Thomas Gaborski (Rochester Institute of Technology)

Despite the increasing use of porous membranes in co-culture systems and barrier models, and their well-known contribution in cell-cell communication and multi-tissue interfaces, their role in controlling cell behavior including cell adhesion, cell migration and extracellular matrix (ECM) formation, and their enabling mechanisms are not fully understood. In the present work, we have developed ultrathin (300nm) SiO_2 porous membranes with micron and submicron features and we investigated how these membranes regulate cell-substrate interaction and affect different cellular behaviors including cell spreading, ECM formation, and cell migration. We demonstrated that porous membranes weakened cell-substrate interaction similarly to soft substrates and biomimetic barrier models, and subsequently reduced focal adhesion formation and endothelial fibronectin fibrillogenesis, and increased cell motility and cell-cell interaction. We further investigated the underlying mechanism of cellular response to these substrates by developing non-fouling micropatterned poly polyethylene glycol (PEG) islands to resemble pores in our porous membrane but without pore edges in order to decouple the role of surface disruption and cell gripping to the pore edges. Similarly to the porous membranes, cells showed reduced fibronectin fibrillogenesis on PEG islands and they exhibit a similar trend of changes in migration speed and directionality on both substrates, which confirms the role of surface disruption in cell-substrate interaction. On the other hand, we found reduced cell spreading and lower magnitude of migration speed on these patterned substrates as compared to porous membranes, suggesting "cell gripping to the pore edges" as a deterministic mechanism for cell-substrates interaction on a porous membrane. These finding can be further employed to modulate cellular responses and mimic tissue and barrier formation by optimizing the design of porous membranes.

A20 High Productivity Harvest - Intensify Harvest and Displace Depth Filtration in Fed-Batch Cell Culture, Shashi Kudugunti (Repligen Corporation)*, Jyoti Amatya (Repligen Corporation), Jamie Peyser (Repligen Corporation)

A conventional production fed-batch process involves a multi-stage seed expansion and elaborate clarification steps to isolate the protein of interest from other process impurities. The current clarification technology utilizes centrifugation and depth filtration trains that are complex and limited in their ability to operate in a sterile or low bioburden manner. With an increase in demand for production of protein therapeutics, it is essential to simplify the traditional fed-batch processes to result in better yield, shorter time and smaller footprint. Repligen developed a simplified & scalable single-step clarification method: High Productivity Harvest (HPH) using the XCellTM ATF system. HPH helps maximize the yield of a fed-batch bioreactor in a sterile manner while eliminating centrifugation, depth, and sterile filtration steps. The current project explores the benefits of combining the HPH method with a perfusion enabled (n-1) seed train bioreactor. The case study focuses on performing (n-1) bioreactor in a perfusion mode that enabled to inoculate production fed-batch bioreactor at 10e6 cells/mL, 20x higher compared to traditional seeding density. Increased seed density facilitates reducing the fed-batch culture duration by half. Implementing the HPH clarification process using XCell ATF system resulted in an enriched environment with a 150% boost in the overall yield in half the fed-batch culture time. The HPH process is performed as a single step in a sterile manner with a 0.2μm clarified product as the output. The product is readily available for either batch or continuous chromatography without additional processing.

A21 Membrane Filtration of Flexible Particles: Rejection of Single Stranded DNA, Hossein Nouri Alavijeh (Clarkson University)*; Ruth E. Baltus (Clarkson University)

In recent decades, DNA purification has generated interest in studying the feasibility of implementing membrane filtration for these flexible particles in pharmaceutical processes. The ability of flexible particles to adopt conformations and enter into membrane pores influences their filterability. In this study, the effects of flexibility and transmembrane pressure (TMP) on the rejection coefficient of single-stranded DNA (ssDNA) were investigated. To characterize flexibility of DNA, Fluorescence Recovery After Photobleaching was integrated with Worm Like Chain polymer model to determine the persistence length. Persistence length is an important parameter that shows the degree of flexibility; it is ionic strength dependent. The results demonstrate that increasing salt concentration results in increasing DNA flexibility. The rejection coefficients of ssDNA with 30 bases in different solution ionic strength were measured using the track-etched polycarbonate ultrafiltration membrane. In addition, The effects of TMP on the rejection coefficients were studied experimentally. In fact, DNA deformability necessitates studying how transmembrane pressure affects DNA transportations through the membrane pores. Results generated from this study will be important for the optimal design of micro- and ultrafiltration systems for water and wastewater treatment and in biotechnological and pharmaceutical sectors.

SESSION B - Innovation in Membrane Materials, Synthesis, and Characterization: Gases

B1 Covalently modified graphene oxide incorporated in PIM-1 mixed matrix thin film composite membranes for gas separation, Elvin Aliyev (Helmholtz-Zentrum Geesthacht)*; Volkan Filiz (Helmholtz-Zentrum Geesthacht); Sergey Shishatskiy (Helmholtz-Zentrum Geesthacht)

PIM-1 has received great attention since its discovery and the gas transport performances of the membranes based on PIM-1 have been studied for extensively, exhibiting the moderate ideal gas selectivities (α) for gas pairs, e.g. α (CO_2/N_2) ≈ 20 and α (CO_2/CH_4) ≈ 13. Graphene oxide (GO), being a two-diemensional material, has also attracted enormous attention in membrane science due to oxygen functional groups distributed on the graphene layers and their possibility for chemical modification. In this study, mixed matrix membranes (MMM) consisting of graphene oxide (GO) and functionalized graphene oxide (FGO) incorporated in a polymer of intrinsic microporosity (PIM-1) serving as a polymer matrix have been fabricated by dip-coating method. Their single gas transport properties were investigated. The pristine GO acts as an effective barrier material for single gas transport through the PIM-1/GO selective layer with both permeance and ideal selectivity decreasing with increase of the GO loading. The gas transport properties of GO embedded into the PIM-1 matrix are much different from properties of functionalized GO-AEDPPF and GO-DCIBAO. These show very good compatibility to the PIM-1 at 9 and even at 33 wt.% loading the MMMs show ideal selectivities overpassing those of the pure PIM-1 TFC membrane. Acknowledgment The authors acknowledge financial support from the Helmholtz Association of German Research Centres through the Helmholtz Portfolio MEM-BRAIN and MOL-FIL Carbon nano-membrane project and German Academic Exchange Service (DAAD).

B2 Tunable Interlayer Channels in Graphene Oxide Membranes for Molecular Separations, Shaofei Wang (King Abdullah University of Science and Technology)*; Zhongyi Jiang (Tianjin University); Michael D Guiver (Tianjin University); Suzana Nunes (King Abdullah University of Science and Technology)

Two-dimensional (2D) lamellar membranes are highly advantageous in molecular separations. However, tuning the interlayer channel to realize high separation is challenging. In this presentation, we will show how the ions, molecules and nanoparticles work on channel-tuning to obtain high performance graphene oxide (GO) membranes for CO_2 separation and organic solvent nanofiltration. 1) By intercalating ion cross-linker, borate ions, GO membranes with favorable physical, chemical and water microenvironment for CO_2 molecules are created, rendering ordered size-selective nanochannels, plenty of fixed facilitated transport carriers and moderate water content to implement efficient CO_2 transport. 2) By intercalating molecular cross-linker, poly(ethylene glycol) diamines (PEGDA), we report GO membranes containing CO_2-philic and non-CO_2-philic nanodomains in the interlayer channels. Owing to the orderly stacking of nanochannels as well as the distributed nanodomains with moderate CO_2 affinity, high performance of CO_2 separation membrane in the dry state was reported. 3) By the in-situ intercalating and cross-linking of SiO_2 nanoparticles in the GO interlayers, we design dual-spacing channel GO membranes, that give high permeance, high rejection and high stability in organic solvent nanofiltration. The hydrophilic nanoparticles locally widen the interlayer channels to enhance the solvent permeance. In the alternating nanoparticle-free areas, the GO layers simultaneously bend and the π-π interactions keep the narrow and hydrophobic channel, promoting a high solute rejection. References: 1) S. Wang, M. D. Guiver*, Z. Jiang*, et al. Energ. Environ. Sci. 2016, 9, 3107. 2) S. Wang, M. D. Guiver*, Z. Jiang*, et al. Angew. Chem. Int. Ed. 2017, 56, 14246. 3) S. Wang, S. Nunes*, et al. J. Mater. Chem. A, (2019). DOI: 10.1039/C8TA10872B.

B3 Nanoporous two-dimensional nanosheets and their self-assembled gas-sieving membranes, Kumar Varoon Agrawal (École polytechnique fédérale de Lausanne)*; Mostapha Dakhchoune (École polytechnique fédérale de Lausanne); Luis Francisco Villalobos (École polytechnique fédérale de Lausanne)

An innovative route for the fabrication of the gas separation membrane is to synthesize nanoporous two-dimensional (N_2d) nanosheets hosting size-sieving nanopores [1]. Subsequently, N_2d can be self-assembled leading to thin yet permselective membranes, improving the scalability of nanoporous membranes. However, barring metal-organic frameworks nanosheets [2], this has not been demonstrated. In this presentation, I will report two examples of the top-down synthesis of N_2d materials (sodalite precursor RUB-15 [3] and polytriazine imide or PTI [4]) hosting crystalline nanoporous structures for sieving of H_2 from CO_2, N_2 and CH_4, as confirmed by our ab-initio molecular simulations. Highly crystalline layers were synthesized using optimized hydrothermal and solid-state synthesis routes for RUB-15 and PTI, respectively, leading to high-aspect-ratio layered plates. Exfoliation of these nanosheets into highly crystalline (as confirmed by high-resolution transmission electron microscopy) and predominantly single-layer nanosheets (as confirmed by atomic force microscopy) was achieved using swelling of the layers, melt compounding, and by the intercalation of solvents. Finally, packing of these nanosheets into compact films by self-assembly produced molecular-sieving membranes with attractive gas-sieving performances. For example, zeolite nanosheet membranes yielded H_2/CO_2, H_2/N_2 and H_2/CH_4 selectivities of 13, 93.7 and 116.6, respectively, while the PTI nanosheets membranes yielded H_2/CH_4 and H_2/N_2 selectivities of 41 and 93, respectively. Overall, the synthesis of molecular-sieving membranes by the colloidal self-assembly can address the scale-up issues related to the nanoporous membranes. References [1] Tsapatsis and coworkers, Science 2011, 334, 72. [2] Yang and coworkers, Science 2014, 346, 1356. [3] Gies and coworkers, Angew. Chemie Int. Ed. 1996, 35, 2869. [4] Schnick and coworkers, Chem. - A Eur. J. 2011, 17, 3213.

B4 Thermally Rearranged Semi-interpenetrating Polymer Networks (TR-SIPNs) for Olefin/Paraffin Separation Applications,

Won Hee Lee (Hanyang University)*; Young Moo Lee (Hanyang University); Jong Geun Seong (Hanyang University); Ho Hyun Wang (Hanyang University); Sun Ju Moon (Hanyang University)

In olefin/paraffin separation, there has been many reports on excellent separation using inorganic and organic membranes. However, there are some issues on large scale manufacturing and low separation performance far beyond the commercial level. Polymer membranes have exhibited potentials due to their good processabilities and economic benefits, and some of them have shown improved gas transport properties such as TR polymers1 and their derivatives2. However, further enhancements of gas separation performances have been still needed for polymer membranes to apply on practical hydrocarbon gas separation applications. Herein, we demonstrated TR-SIPNs incorporating network structures for olefin/paraffin separations. Free-standing films of TR-SIPNs were facilely fabricated via consecutive thermal treatments on polymer precursor solution mixture casted onto the glass plate. Hydroxyl polyimides and bismaleimide-based monomer building blocks in homogeneous casting solutions were used for linear and network polymers in TR-SIPNs, respectively. Chemical structures and chain configurations of TR-SIPNs were demonstrated by FT-IR, NMR, WAXD and SAXS. TR-SIPNs exhibited enhanced BET surface area and higher glassy transition temperature compared to reported TR polymers indicating microporous and rigid polymer structures. Pure gas transport behaviors of the resulting TR-SIPNs were explored for ethylene, ethane, propylene and propane gases. They showed far increased olefin permeability than linear TR polymers, resulting in pass the upper bound plots of ethylene/ethane and propylene/propane gas pairs. Moreover, TR-SIPNs showed reinforced plasticization resistances for all controlled hydrocarbon gases due to their rigid polymer frameworks. References 1. H.B. Park et al., Science 2007 318 (5848), 254-258 2. Y.S. Do and W.H. Lee et al., Chem. Commun., 2016 52 (93), 13556-13559

B5 Design of new polymeric materials for membrane gas-separation based on norbornenes bearing hydrocarbonic groups,

Alyona Wozniak (A.V. Topchiev Institute of Petrochemical Synthesis, Russian Academy of Sciences); Evgeniya Bermesheva (A.V. Topchiev Institute of Petrochemical Synthesis, Russian Academy of Sciences); Danila Bakhtin (A.V. Topchiev Institute of Petrochemical Synthesis, Russian Academy of Sciences); Ilya Borisov (A.V. Topchiev Institute of Petrochemical Synthesis, Russian Academy of Sciences); Alexey Volkov (A.V. Topchiev Institute of Petrochemical Synthesis, Russian Academy of Sciences); Eugene Finkelshtein (A.V. Topchiev Institute of Petrochemical Synthesis, Russian Academy of Sciences); Maxim Bermeshev (A.V. Topchiev Institute of Petrochemical Synthesis, Russian Academy of Sciences)*

Recently it has been shown that Si-containing polynorbornenes are promising materials for membrane separation of hydrocarbons or CO_2/N_2. Depending on the nature of Si-containing side substituent and the rigidity of polymer chain, it is possible to obtain high free volume polymers displaying extremely high gas permeability (like polyacetylenes or PIM's) or polymers exhibiting high C4H10/CH_4 and/or CO_2/N_2 selectivities. It is less known about gas-transport properties of polynorbornenes with hydrocarbon substituents. We have synthesized a group of such polymers and studied their gas-transport properties. The advantage of the considered polymers is their easy synthesis from dicyclopentadiene and α-olefins as well as the availability of the feed for their preparation. Desired homopolymers can be obtained with demanded molecular weights in high yields in the presence of common Pd-catalysts and appropriate cocatalysts. Addition polynorbornenes with side hydrocarbonic substituents are glassy and fully saturated, therefore, they exhibited good thermal stability. Being low free volume polymers, they are not prone to ageing. The obtained polynorbornenes displayed enhanced C4/C1 selectivity relative to Me3Si-substituted counterparts and polydimethylsiloxane. The some reduced permeability is attributed to the formation of a more dense packing of polymer chains in comparison with Me3Si-substituted addition polynorbornenes. This was supported by WAXD, fractional free volume and diffusion coefficients measurements. The combination of ease of preparation, good selectivity and moderate hydrocarbons permeability makes these polymers attractive materials for the following and detail studies. The work was supported by Russian Science Foundation, grant #17-19-01595.

B6 Cake layer characterization in Activated Sludge Membrane Bioreactor: Real-time analysis, Luca Fortunato (King Abdullah University of Science and Technology)*; TorOve Leiknes (King Abdullah University of Science and Technology)

Over the past two decades, the MBR technology that combines biological activated sludge process with membrane filtration has emerged as an innovative and alternative technology for tertiary wastewater treatment and reuse. Fouling represents one of the major drawback and disadvantages of the process. In this study, the suitability of real-time monitoring performed with Optical Coherence Tomography (OCT) to monitor fouling due to cake layer development in AS-MBR was evaluated. The in-situ observation was performed directly in the MBR reactor treating primary effluent to assess the online fouling development over a UF flat sheet membrane. Cake layer properties were evaluated using both 2D and 3D image analysis. The non-invasive nature of OCT imaging enabled monitoring fouling development over time, where an increase of thickness and a decrease of roughness was observed. The 3D OCT image analyses were also analysis was compared with the 3D CLSM image analyses performed at the end of the study. Results demonstrate that OCT imaging can be applied for on-line, real-time monitoring and analysis of fouling behavior in AS-MBR systems.

B7 Facilitated Transport Membrane for CO_2 Capture from Flue Gas: Module Fabrication and Scale-Up, Kai Chen (Ohio State University)*; Witopo Salim (Ohio State University); Yang Han (Ohio State University); Dongzhu Wu (Ohio State University); Winston Ho (Ohio State University)

In order to modularize a composite membrane with an ultrathin selective layer, it is a challenge to scale up the membrane fabrication procedure from lab to pilot scale. Herein, a knife-coating method was successfully integrated into a roll-to-roll fabrication machine, making a 165-nm thick facilitated transport membrane for CO_2 capture from flue gas with 14 inches in width and > 1200 feet in total length. The scale-up membranes were subsequently used to make spiral-wound (SW) elements and modules. To increase the membrane packing density in a single element, a detailed multi-leaf rolling procedure was developed to scale up the elements efficiently. Although the fabrication of a multi-leaf SW membrane module has been well recorded in the literature, the majority of the methods had all the membrane leaves directly attached to the central permeate tube, thus, the number of leaves was limited by the central tube surface area. In order to improve the fabrication, the glue lines were modified to allow an arrangement of a stack of staggered membrane leaves not attached to the central tube before rolling. Moreover, the spacer thickness was reduced to as low as 6 mil, which further increased the membrane area to 2.94 m^2 in the most recent modules. The modules also reproduced the transport performance of the flat-sheet membrane under various testing conditions. For instance, a CO_2 permeance of 1450 GPU and a CO_2/N_2 selectivity of 185 were showed with a simulated flue gas at 67°C. In addition, satisfactory pressure drops were also achieved. From the first successful SW element made in house to the most recent ones, a scale-up factor of >70 was achieved. A predictive model was developed to relate the operating conditions to the module separation performance, accounting for the limitations set by concentration polarization and pressure drop. The experience gained is helpful to design a full-sized element (40 inches in length and 8 inches in diameter) for industrial application.

B8 Interfacial phenomenon-driven hierarchical structuring of hard templated asymmetric ultrathin carbon molecular sieves films, Megha Sharma (Lehigh University)*; Mark Snyder (Lehigh University)

Inorganic membrane materials such as carbon molecular sieves (CMS) show promise in transforming gas and liquid separations, however, they struggle from the well-known trade-off between molecular selectivity and permeability. This motivates efforts to reduce defect-free membrane thickness while simultaneously tailoring membrane texture and function. In this talk, we present a versatile silica colloidal crystal based nanocasting approach that yields ultra-thin microporous polyimide-derived CMS self-supported on three-dimensionally ordered mesoporous (3DOm) carbon layers. We show how the large area associated with the template-replica interface in our 3DOm carbon films offers a previously unexploited handle for simultaneously tailoring microstructure of the replica phase. Specifically, we elucidate how temperature-tunable silanol surface chemistry of the template can be used to tailor the graphitic character (e.g., graphitic content relative to turbostratic carbon) of CMS replicas under mild carbonization conditions (600-900°C) and in the absence of specific metal catalysts. The interfacial origin of this effect is confirmed through studies showing a sensitivity of graphitic character to film thickness. Studies also show enhancement of graphitic character in thin carbon films as compared with untemplated carbon molecular sieves, with maximum sensitivity near the beginning of the carbonization process (around 600°C). We hypothesize that the interface-mediated tunability of the replica graphitic character derives from adsorption-induced orientation of pyromellitic dianhydride-co-4,4'-oxydianiline (PAA) precursors once the surface silanol density is reduced to the point of approximate registry (e.g., hydrogen-bonding) with PAA molecular dimensions. Taken together, this templating approach offers a strategy for multi-scale control over film structure, which can help in achieving high-flux, high selectivity separations.

B9 Layer-by-Layer Assembled Metal Organic Framework Nanosheets with Polymer, Christy Koerner (U.S. Department of Energy National Energy Technology Laboratory)*; Fangming Xiang (U.S. Department of Energy National Energy Technology Laboratory); Eric Popczun (U.S. Department of Energy National Energy Technology Laboratory); David Hopkinson (U.S. Department of Energy National Energy Technology Laboratory)

Metal-organic framework (MOF) nanosheets have been widely used in gas separation, sensing, and catalysis. They have been added as functional fillers into polymers due to their unique two-dimensional shape and tunable porous structure. However, the resultant composites usually showed limited improvement over the neat polymers that they were based on, due to misalignment and aggregation of MOF nanosheets. In this study, layer-by-layer (LbL) assembly was used to produce a polymer composite with exfoliated and highly aligned leaf-like zeolitic imidazolate nanosheets (ZIF67-L). Polyethylenimine (PEI) was used to replace the surface ligands on ZIF67-L through coordination bonding, due to its higher affinity with the cobalt ions in the MOF. This functionalization process was designed to add positively charged amine groups onto MOF nanosheets and enabled them to ionically bond with negatively charged carboxylate groups on poly(acrylic acid) (PAA). The successes of this functionalization strategy and the LbL assembly process were confirmed using Fourier-transform infrared spectroscopy (FTIR). The resultant polymer/MOF multilayer membrane featured exfoliated MOF nanosheets with parallel in-plane orientation, as evidenced by scanning electron microscope images as well as the preferential expression of {h00} peaks in the X-ray diffraction patterns. The method described here demonstrates for the first time how to produce polymer composite membranes containing exfoliated and aligned MOF nanosheets, which could be of great interest to the polymer composites, gas separation, and MOF communities.

B10 Highly oxygen-rich rubbery polymers for membrane CO_2/CH_4 separation, Krysta Clark (The State University of New York at Buffalo)*; Junyi Liu (Air Liquide. Inc); Gengyi Zhang (The State University of New York at Buffalo); Haiqing Lin (The State University of New York at Buffalo)

Membrane materials with high CO_2/CH_4 selectivity and CO_2 permeability are in pursuit for the separation of CO_2 from natural gas. Glassy polymer-based membranes are one of the leading technologies for CO_2 capture from natural gas. These glassy polymers display high CO_2/CH_4 diffusivity selectivity, but the heavy hydrocarbons in natural gas decrease CO_2 permeability. To counter these issues of glassy polymers, rubbery solubility-selective polymers were synthesized. To maximize the ether/ester ratio cationic ring opening polymerization 1,3-dioxolane and 1,3,5-trioxane was used. The ether/ester groups allow the polymer to be CO_2-philic, leading to high CO_2/gas solubility selectivity. The macromonomer and polymers were characterized for physical properties and chemical structures using NMR, FTIR, MS, DSC, etc. The synthesized polymer with an O:C ratio of 0.71 (P71) exhibits a mixed-gas CO_2 permeability of 320 Barrers and CO_2/CH_4 selectivity of 21 at 50°C. These oxygen-rich polymers display robust CO_2/CH_4 separation properties that are better than conventional polyimides and cellulose acetate and are above the Robeson's upper bound.

B11 Adsorptive and Destructive Mixed Matrix Membranes for Chemical Protection, Yufeng Song (New Jersey Institute of Technology)*; John Chau (New Jersey Institute of Technology); Kamalesh K Sirkar (New Jersey Institute of Technology); Gregory Peterson (U.S. Army Edgewood Chemical Biological Center); Uwe Beuscher (W.L. Gore & Associates Inc.)

Flexible semi-permeable membrane filters which selectively allow moisture transport and block or destroy toxic gases/vapors are of significant interest for protective garments. Adsorptive and reactive nanostructured materials such as metal organic frameworks (MOFs) and metal oxides/hydroxides when incorporated in membranes can create a mixed matrix membrane (MMM) having those types of capabilities. We will report the results of our efforts to incorporate MOFs such as UiO-66-NH_2 in a polymeric membrane. We will also report results of barrier performance of such MMMs vis-a-vis toxic gases e.g., ammonia, chlorine from an environment along with our results of moisture vapor transmission rate through such a protection medium.

B12 Maximizing Ether Oxygen Content in Polymers for Membrane CO_2 Removal from Natural Gas, Junyi Liu (American Air Liquide. Inc)*; Gengyi Zhang (The State University of New York at Buffalo); Krysta Clark (The State University of New York at Buffalo); Haiqing Lin (The State University of New York at Buffalo)

Membrane materials for CO_2 removal from natural gas are based on glassy polymers with high CO_2/CH_4 diffusivity selectivity. However, these polymers suffer from a permeability/selectivity trade-off, competitive sorption by heavy hydrocarbons that decreases CO_2 permeability, and physical aging that reduces gas permeability with time. We circumvent these issues by designing rubbery, solubility-selective polymers with a ratio of ether/ester oxygen to carbon as high as 0.8 through the use of 1,3-dioxolane and 1,3,5-trioxane. The ether/ester oxygen groups interact favorably with CO_2 but do not interact with CH_4, leading to high CO_2/gas solubility selectivity that is unaffected by heavy hydrocarbons in the raw natural gas. These polar groups are incorporated in short branches to yield amorphous and rubbery nature, leading to high gas permeability that is stable over time. An exemplary polymer of P71 with an O:C ratio of 0.71 shows a mixed-gas CO_2 permeability of 320 Barrers, CO_2/CH_4 selectivity of 21, which is above upper bound and CO_2/C2H6 selectivity of 12 at 50°C. The long term stability is also studied in this presentation. The success of these highly oxygen-rich polymers may stimulate a pursue of functional polymers with superior solubility selectivity for membrane gas separation.

B13 Novel CO_2 gas separation membranes prepared by chemical modification of highly permeable polymers, Eugenia Mariana Sandru (SINTEF AS); Arpenik Kroyan (Norwegian University of Science and Technology); Nebosja Simic (Norwegian University of Science and Technology); Jing Deng (Norwegian University of Science and Technology); Liyuan Deng (Norwegian University of Science and Technology); Marius Sandru (SINTEF AS)*

Most of the approaches of current gas separation membranes are based on increasing either the CO_2 solubility or CO_2 diffusivity over N_2 by using different methods to increase the free volume of polymers and their sieving capacity. An approach employed in the POLYMEM project is based on bulk modification with CO_2 reactive groups in a controlled manner and high CO_2 permeable but low selective polymers such as poly(trimethylslilylproyne) (PTMSP). This represents a new approach in gas separation membranes that mitigates the risks associated with use of polymer blends and addition of various CO_2 enhancement materials (nanoparticles, etc): incompatibility between materials, poor and limited dispersion of particles, reproducibility and difficulty in preparing membranes. In this paper, we will report membrane preparation by using bulk modification of poly(trimethylslilylproyne) (PTMSP) polymer using CO_2 -philic groups (amines and carboxyl) and will discuss the influence of various parameters on membranes structure and their separation performances. The modified polymers structure and their degree of modification is characterised by 1HNMR, FTIR, DSC and TGA. The structural changes of polymers (PTMSP) were confirmed by 1HNMR and FTIR and the amine content was estimated as well. The amine content of modified PTMSP was low, indicating that a small percentage of the monomer units in the polymer chains were modified. The membranes structure was characterized by SEM pictures and FT-IR and was correlated with the results from single and mixed gas permeation testing using a synthetic flue gas: 10% CO_2 in N_2, fully humidified. Depending on the polymer and modification method, significant CO_2/N_2 selectivity enhancement was documented. Acknowledgement This work is supported by the Research Council of Norway through CLIMIT program ("POLYMEM" project, No. 254791)

B14 Intrinsically Microporous Pentiptycene-based Polymers for Enhanced Gas Separation Performance and Physical Aging Resistance, Tanner Corrado (University of Notre Dame)*; Ruilan Guo (University of Notre Dame)

High-free-volume glassy polymers such as PIM-1 have shown exceptional promise for advancing current gas separation membrane technology. However, high-free-volume polymers are generally limited by drastic declines in permeability over time due to physical aging reducing their largely conformation-based free volume. Here, we present an innovative macromolecular design of ladder-like copolymers that exploit the unique molecular architecture of pentiptycene as building blocks for non-collapsible, configuration-based free volume with tunable microporosity. The structural hierarchy of the pentiptycene unit imbues intrinsic free volume into the polymer for increased size sieving while providing a versatile framework that allows tunable ultra-microporosity. Monomers utilizing different shapes of the pentiptycene framework (i.e., S- or C-shaped) were synthesized with tailorable substituent groups and incorporated into copolymers with PIM-1 to study the effect of configurational free volume on gas transport. Achieved copolymers have shown remarkable potential as gas separation membranes, surpassing the 2008 upper bound for several gas pairs. For example, incorporation of 33 mol % of the S-shaped pentiptycene unit provided exceptional combinations of permeability and selectivity for O_2/N_2, with an O_2 permeability of 2370 Barrer and an O_2/N_2 selectivity of 4.6. Additionally, a thin film integrated with only 17 mol % of the C-shaped pentiptycene unit had a hydrogen permeability of 5900 Barrer with upper bound selectivities for typical hydrogen-based separations (i.e., $\alpha(H_2/CH_4) = 4.9$; $\alpha(H_2/N_2) = 7.7$). A preliminary aging study showed that the 17 mol % C-shape polymer had 30% increases in permeability across all gases after aging 120 days, while maintaining selectivity. In this talk, synthesis and characterization of these new copolymers will be presented. Discussions will focus on elucidating the fundamental structure-property relationships within the series.

B15 Structural Designs of Cross-linked Polymer Network and Ionic Liquids for Ion Gel Gas Separation Membranes, Victor A Kusuma (U.S. Department of Energy National Energy Technology Laboratory)*; James S Baker (U.S. Department of Energy National Energy Technology Laboratory); Megan Macala (U.S. Department of Energy National Energy Technology Laboratory); Samir Budhathoki (U.S. Department of Energy National Energy Technology Laboratory); David Hopkinson (U.S. Department of Energy National Energy Technology Laboratory)

Certain ionic liquids were found to be excellent plasticizers for cross-linked poly(ethylene oxide) (PEO) polymers. 1,3-substituted imidazolium bis(trifluoromethylsulfonyl)imide ionic liquids can be compatibilized at high loadings to increase gas permeability of these polymers, and their low vapor pressure allows long term stability of the resulting ion gels. In this talk, we will discuss various strategies we have taken to improve the performance of these ion gels. We evaluated the effect of incorporating different functional groups on the imidazolium cation on gas permeability and found that the simplest alkyl substituents were more effective than having long alkyl, long alkoxy or a polar terminal group. While incorporating polar substituents resulted in better CO_2/N_2 selectivity, this was offset by the selectivity loss due to incorporation of the anion. One of the best ion gels we developed exhibited pure gas CO_2 permeability of 530 barrer coupled with CO_2/N_2 selectivity of 31 at 40°C. The second strategy we evaluated was modifying the cross-link density of the polymer network, either by changing the cross-linker or incorporating additional branching by introducing co-monomers. We found that increasing cross-link density reduces the amount of ionic liquids that can be incorporated and also reduces membrane performance. Molecular dynamics simulation of CO_2 in ionic liquid mixtures with low molecular weight PEO was employed to confirm that PEO-CO_2 interaction remained the dominant driving force for separation in these ion gels. We will also present preliminary results from our performance testing with high humidity CO_2/N_2 mixed gas.

B16 Scalable hydrogen-bonded polyimide/metal-organic framework hybrid membranes for ultra-fast separations of multiple gas pairs, Canghai Ma (Lawrence Berkeley National Laboratory)*; Jeffrey Urban (Lawrence Berkeley National Laboratory)

Membranes can effectively reduce the enormous amount of energy required for gas separations, compared to conventional separation processes, such as cryogenic distillation. Currently, polymers dominate gas separation membrane markets due to their appealing manufacturability and performance. However, polymeric membranes suffer from the productivity-selectivity tradeoff, i. e., the Robeson upper bound. To overcome this challenge, growing attention has been shifting towards hybrid materials synergistically utilizing advantages of each component in the hybrid membranes, i.e., easy processing of polymers and molecular sieving of fillers, such as metal-organic framework (MOFs). However, only a few MOF-based hybrid membranes perform above current Robeson upper bounds, predominantly relying upon rigid Polymers of Intrinsic Microporosity or aggressive thermal annealing as high as 240 C, adding complications for scaling-up. Moreover, most hybrid membranes are specialized for a particular gas pair (e.g. CO_2/N_2), lacking a general approach to separating multi-component gas mixtures. This work demonstrates readily scalable polyimide/MOFs hybrid membranes residing above current Robeson upper bounds via a generalizable approach. The hydrophilic amino ($-NH_2$) groups in MOFs intensively interact with carboxylic moieties in a 6FDA-based polyimide via hydrogen bonds to foster a firm interphase adhesion. The polyimide/MOFs hybrid membranes demonstrate a CO_2 and H_2 permeability of over 2400 and 2900 Barrers, respectively, with a CO_2/CH_4, H_2/CH_4 and H_2/N_2 selectivity of over 29, 34, and 24 respectively, surpassing current Robeson upper bounds. The permeability was enhanced by record-high 14 times comparing with the neat polymer. With a processing temperature as low as 80 C and performance continuously exceeding the upper bounds for over 5300 h, the membrane is readily scalable and compatible with the state-of-the-art membrane manufacture processes.

B17 Performance Testing of Polyphosphazene-Based Membranes for Post-combustion Carbon Capture with Humidified Gas, Zi Tong (U.S. Department of Energy National Energy Technology Laboratory)*; David Hopkinson (U.S. Department of Energy National Energy Technology Laboratory); Victor A Kusuma (U.S. Department of Energy National Energy Technology Laboratory)

Polyphosphazene is a range of polymers with different side groups and a phosphorus-nitrogen backbone composition. This structure allows heteropolymers to be easily obtained at desired compositions, which makes polyphosphazene an excellent candidate membrane material to achieve high gas separation performance. However, in post-combustion carbon capture, the presence of moisture can significantly influence the membrane performance. Although the dry gas performance of polyphosphazene has been reported, there are few reports available using wet gas. In this work, the effect of humidity on polyphosphazene-based membrane CO_2/N_2 separation performance was investigated by testing under different humidity levels in our newly developed automated mixed gas permeation testing system. The designed humidification method allowed the system to be operated under a broad range of relative humidity, from dry to 95% relative humidity. Details of the humidified testing system are also shared in this report.

B18 Systematic Optimization of MOF-based Mixed-Matrix Membranes: Surface Functionalization, Particle Size, and Loading, Patrick Muldoon (Oak Ridge Institute for Science and Education)*

A number of promising examples of mixed-matrix membranes (MMMs) for gas separations have been published in which the incorporation of metal-organic frameworks (MOFs) enhances selectivity and/or permeability relative to pure polymer-based membranes. However, it is unclear the extent to which the interior pore chemistry and pore metrics of the MOFs is responsible for these changes when we consider the concurrent morphological changes to the polymer matrix that inevitably result from the incorporation of crystalline filler particles. Without understanding the nature of these competing effects, we cannot form a rational basis for the design of high-performance MMMs. In this study, three factors thought to affect integration of MOF and polymer are systematically tuned for MMMs comprised of UiO-66-NH_2 (University of Oslo) and PIM-1 (Polymers of intrinsic microporosity) -- namely, surface functionalization, particle size, and particle loading. Resulting MMMs are then evaluated for carbon capture applications.

~~**B19 Membrane heat exchanger for novel heat recovery in post-combustion carbon capture,**~~ ~~Shuaifei Zhao (Dalian Maritime University)*; Shuiping Yan (Huazhong Agricultural University); Qiufang Cui (Huazhong Agricultural University)~~

~~Post-combustion carbon capture using aqueous alkanolamine solvents is considered as one of the most mature methods to reduce CO_2 emissions. However, Post-combustion carbon capture still requires huge energy inputs for absorbent regeneration and CO_2 desorption. In this work, we demonstrate a commercial ceramic membrane heat exchanger (CMHE) with an average pore size of 4 nm for novel heat recovery in post-combustion carbon capture. The CMHE shows superior performance over a conventional stainless steel heat exchanger (SSHE) with the same dimensions in recovering heat from the stripped gas mixture ($H_2O(g)/CO_2$) on top of the stripper in a monoethanolamine-based rich-split carbon capture process. Due to the coupled mass and heat transfers of water vapor through the membrane, the CMHE has higher heat flux, heat recovery and overall heat transfer coefficient than the SSHE. Liquid water transfer dominates the mass transfer mechanism in the CMHE. Thermal conduction contributes to more than 80% of the total heat transfer, dominating the heat transfer through the membrane heat exchanger. Our study demonstrates that membrane heat exchangers can be excellent candidates for heat recovery in post-combustion carbon capture. In further research, more types of membranes with higher thermal conductivities (e.g., porous metal membranes with lower porosity and smaller thickness) should be fabricated and tested for further performance enhancement.~~

B19 High temperature gas separation properties of sub-micron polybenzimidazole membranes, Melanie M Merrick (The University of Texas at Austin)*, Benny Freeman (The University of Texas at Austin)

Polybenzimidazoles (PBIs) are a class of polymer materials that have the potential to improve and expand the applications of commercial gas separation membranes. While many commercial membranes can only perform gas separations up to 100°C, PBIs are both chemically and thermally stable at temperatures up to 400°C. A potential application of PBI membranes is the high temperature hydrogen-hydrocarbon separations proposed by the NSF Engineering Research Center for Innovative and Strategic Transformation of Alkane Resources (CISTAR). Two novel aspects of this work are the previously unstudied ethane and ethylene gas separation properties at elevated temperatures and the temperature dependence of gas transport in sub-micron films over an expansive temperature range. Hydrocarbon gas separation properties of Celazole® PBI membranes were characterized at temperatures up to 190°C. Gas permeabilities, gas selectivities, activation energies of permeation, and physical aging rates differed significantly between thin sub-micron and bulk 20-micron Celazole® films. In particular, the activation energies of permeation were 1.5 to 4 times higher in an aged 450-nm thin film than in a 27-micron bulk film, suggesting a denser membrane structure and greater size-sieving ability. The first known evidence of physical aging in PBIs showed an 80 percent decrease in methane permeability over a 1000-hour period at 190°C for a 450-nm Celazole® film. This study elucidates the effects of elevated operating temperature on thin films by encompassing a broader and higher temperature range than previous studies of flat, sub-micron polymer membranes.

B20 CANAL Ladder Polymers for Membrane Gas Separation, Holden W. H. Lai (Stanford University)*; Francesco Benedetti (University of Bologna); Zach Smith (MIT); Yan Xia (Stanford University)

Ladder polymers represents an interesting polymer architecture consisting of continuously fused rigid and contorted rings in the backbone, leading to inefficient chain packing and abundant microporosity. These unique features make ladder polymers promising materials for membrane gas separation, combining high permeability with moderate permselectivity. Despite their great potential, only two types of ladder bakbones, benzodioxane and Troger's base, have been extensively investigated for their gas separation performance. Our lab recently developed the catalytic arene-norbornene annulation (CANAL) for the synthesis of a diverse array of nobornyl benzocyclobutene ladder polymers, which has enabled us to elucidate structure-property relationship of ladder polymers for important gas separation applications. The effect of chemical functionalities and chain conformation on solid-state packing and the transport of small molecules was investigated. The diffusion mechanism of gas molecules in these ladder polymers will also be discussed.

B21 Fabrication of CO_2 selective miscible polyimide blend membranes for gas separation applications, Chamaal Karunaweera (The University of Texas at Dallas)*; John Ferraris (The University of Texas at Dallas); Kenneth Balkus, Jr. (The University of Texas at Dallas); Inga Musselman (The University of Texas at Dallas); Samitha Panangala (The University of Texas at Dallas); Shahed Haghiri (The University of Texas at Dallas)

Synthesis of block-co-polyimides is widely used to obtain polymers with enhanced properties. However, complex syntheses procedures limit the applications of these polymers. Polyimide blends can be useful, low-cost alternatives for block-co-polyimides. Although, most polyimides are immiscible with each other, compatibilization can be utilized to enhance the properties of these immiscible blends. Herein we report how to convert an immiscible polyimide blend to a miscible blend using a small molecule additive. Scanning electron microscopy and differential scanning calorimetry were utilized to detect the miscibility of these blends. Enhanced CO_2 permeabilities and CO_2/CH_4 ideal selectivities were obtained for polymer blend membranes in comparison to membranes of the individual polymers.

B22 On development of constant-volume permeation system for monitoring upstream pressure decay, Haoyu Wu (University of Ottawa)*; Jules Thibault (University Ottawa); Boguslaw Kruczek (University of Ottawa)

Dynamic gas permeation experiments, which provide basis for the time-lag measurements, are a well-known method for the determination of permeability, diffusion and solubility coefficients of gases in polymer films. These dynamic permeation experiments are normally performed in a constant volume (CV) system, in which time-dependent gas permeation across the membrane is obtained from the rate of pressure increase in a receiving volume downstream from the membrane. In principle, one could monitor time-dependent gas permeation into the membrane based on the rate of pressure decay in a fixed upstream volume. However, since the resolution of an absolute pressure transducer is inversely related to the upper limit of the pressure transducer, the latter is not used in practical dynamic permeation experiments. On the other hand, there are potential advantages of dynamic permeation experiments relying on the upstream pressure decay. For example, when testing barrier films, it may take hours or even days before any gas molecules appear at the permeate side of the membrane. At the same time, the maximum rate of the upstream pressure decay exists right after initiation of a dynamic permeation experiment. Moreover, combining the capability of monitoring of the upstream pressure decay with monitoring of the downstream pressure-rise could provide invaluable information for the mechanism of gas transport in membranes. In this paper, we present the progress in the development of a constant-volume system capable of monitoring the rate of pressure decay. The latter is measured based on a pressure difference between a working part of the upstream volume, in which the pressure decreases because of gas permeation into the membrane, and the reference volume in which the pressure is constant. This allows overcoming the resolution limit of high-pressure absolute pressure transducers. Examples of experimental results will be presented and the remaining challenges will be discussed.

B23 Two-dimensional-material mixed-matrix membranes for gas separation, Gongping Liu (Nanjing Technical University)*

Two-dimensional (2D) materials, with atomic thickness and micrometer lateral size, are emerging building blocks for separation membranes. Extraordinary molecular separation properties for purifying water and gases have been demonstrated by developing 2D-material membranes, which attract a huge surge of interest during last a few years. As a prominent characteristic of 2D-material membranes, inter-layer space has been proved to play a significantly important role in molecular transport. We introduced mixed-matrix approach to develop 2D-material membranes for gas separation, with particular focus on tailoring sub-nanosized interlayer spaces between nanosheets for fast and selective transport of gases. Two typical 2D-materials will be demonstrated in this presentation: graphene oxide (GO) and MXene. First, membranes with fast and selective CO_2 transport channels of GO laminates were proposed based on the construction of hydrogen bonding between GO nanosheets and PEBAX chain. Notably, incorporating only 0.1 wt% GO could double the CO_2 permeability and CO_2/N_2 selectivity of pristine PEBAX membrane, transcending the 2008 Robeson upper-bound. Our explorations (e.g., TEM, positron annihilation and sorption) found that molecular-sieving of the interlayer spaces and CO_2-philic pathways of the GO laminates play critical roles in the enhancement of gas transport properties. Second, borate and amine intercalated MXene membranes with ultra-thin skin layer (~20 nm) were fabricated by a facile spin-casting assisted assembly technique. With finely tuned interlayer spacing and introduced CO_2-falicitate transport groups, the intercalated MXene membranes show excellent CO_2 separation performance that overcomes the CO_2/CH_4 upper-bound. The transport mechanism was also understood by analyzing the coefficients of sorption and diffusion. Our results demonstrate that 2D-material mixed-matrix membranes offer exciting opportunities for efficient gas separation.

B24 Synthesis of crosslinked polyether-based membranes for gas separations, Malgorzata Chwatko (The University of Texas at Austin)*; Christina Rodriguez (The University of Texas at Austin); Caitlin Bentley (The University of Texas at Austin); Nathaniel Lynd (The University of Texas at Austin); Benny Freeman (The University of Texas at Austin)

Membrane technology is increasingly used to separate gases in industrial settings due to low footprint, efficient energy use and low capital expenditure.(1) In the area of CO_2 separations, state-of-the-art membranes are based on poly(ethylene oxide) (PEO). The performance of PEO is proposed to be derived from a specific interaction between the ether oxygens within the PEO repeat units and carbon dioxide.(2,3) However, while PEO is widely available, there are two potential drawbacks to using PEO as a membrane material: crystallinity and hydrophilicity. Crystallinity limits transport as crystalline domains are inaccessible to gases.(4) While, the hydrophilic nature of PEO could cause a reduction in selectivity due to plasticization by water vapor. Nonetheless, PEO is still a commonly used material due to its excellent performance relative to incumbents, but there are open possibilities for improvement. Polyethers based on terminal epoxides possess the same advantageous polyether backbone as PEO, but offer tunable properties via a pendant side group. The polyethers have not been studied in detail for gas separations due to limitations in synthetic methodoloty. In this study, we developed a new synthetic platform to overcome past synthetic challenges and create crosslinked polyether networks using monomers such as n-butyl glycidyl ether and 1,2-epoxydodecane. These crosslinked polyether-based membranes contain controllable degrees of hydrophobicity and crosslinking. The gas transport characteristics of the membranes was tested with the following gases nitrogen, oxygen, methane, hydrogen, and carbon dioxide. References: 1) Reijerkerk SR, et al. J Membrane Sci 2011, 367 (1-2), 33 2) Kalakkunnath S, et al. Macromolecules 2005, 38 (23), 9679 3)Lin H, Freeman BD. J Membrane Sci 2004, 239 (1), 105 4) Lin H et al. Macromolecules 2005, 38 (20), 8381

B25 Novel polyimides with bulky tert-butyl and fluorine-containing side groups: gas permeation parameters and free volume, Susanta Banerjee (Indian Institutes of Information Technology); Nikolay Belov (A.V. Topchiev Institute of Petrochemical Synthesis, Russian Academy of Sciences); Roman Nikiforov (A.V. Topchiev Institute of Petrochemical Synthesis, Russian Academy of Sciences); Mikhail Mazo (Institute of Chemical Physics, Russian Academy of Sciences); Ivan Strelnikov (Institute of Chemical Physics, Russian Academy of Sciences); Yury P Yampolskii (A.V. Topchiev Institute of Petrochemical Synthesis, Russian Academy of Sciences)*

Novel polyimides (PI) containing a diamine with bulky tert-butyl and CF3 side groups and several common dianhydrides were prepared and studied. Gas permeation parameters (permeability (P), diffusion (D) and solubility coefficients) were estimated for a wide range of gases (He, H_2, O_2, N_2, CO_2, CH_4). Free volume in these polymers was investigated by means of Molecular Dynamics simulation. It was shown that the level of the P and D values of novel PIs is higher than those of majority of polymers prepared using the same dianhydrides apparently due to the presence of bulky tert-butyl and CF3 groups in the structure of novel PIs. Size distribution of free volume elements in novel PIs was obtained using gas probes of different size from H_2 to Xe. Size distribution curves have a Gauss-type shape with extended tails and maxima of accessible cavity in the vicinity of 1.5-2 Å. It was shown that the average accessible volume fraction decreases with probe (penetrant) radius r and its limiting value for zero r is about 40%. Comparison with the literature data indicates that similar values are observed in many glassy polymers and this apparently characterizes a general chain packing in glassy state. This works has been supported by Russian Science Foundation grant N18-19-00258 and Russain Foundation for Basic research grant N 17-58-45045 and Russian Foundation for Basic Research grant No 17-58-45045 .

B26 New mixed matrix composite membrane PEBAX based membrane incorporated with metal organic framework for CO_2 separation, Tarik Eljaddi (Laboratoire Réactions et Génie des Procédés)*; Julien Bouillon (Normandie University); Laurent Lebrun (Normandie University)

The membrane technology presents only 10 % of the techniques used for carbon dioxide separation. For this reason, more investigations are needed to develop new membranes with high permeability and selectivity. Recently, several studies show promising results, when organic and inorganic materials were combined. In our work, new mixed matrix membranes (MMMs) were prepared by incorporating ZIF and HKUST metal organic frameworks into poly (ether-block-amide) as an active layer on the top surface of polymer porous support (PVDF). The membranes were prepared by solvent evaporation technique. The structure of the membranes was studied according to the thickness of the active layer, the quantity of fillers and the presence of additives. The membranes were characterized by TGA, FTIR, SEM and gas permeation measurements. The preliminary permeance tests revealed that incorporation of ZIF in PEBAX matrix with dispersing additives increase the CO_2/N_2 selectivity.

B27 Dual-Layer MOF Composite Membranes with Tunable Gas Transport Properties for Post-combustion CO_2 Separation, Sameh K Elsaidi (U.S. Department of Energy National Energy Technology Laboratory)*; David Hopkinson (U.S. Department of Energy National Energy Technology Laboratory); Surendar Venna (U.S. Department of Energy National Energy Technology Laboratory); Mona Mohamed (University of Pittsburgh)

We present a novel concept for the fabrication of custom-designed dual-layer metal organic framework (MOF) membranes that show great promise in improving permeability and selectivity performance for CO_2/N_2 separation as compared with the individual constituent MOF materials. A highly selective thin MOF layer with narrow pore size (shell layer) was grown on the surface of a highly permeable MOF layer with large pore size (core layer). Most importantly, the interaction between the two layers was optimized through the chemical binding between the two MOF materials affording continuous, defect-free membranes. The two MOF@MOF membranes: SIFISIX-3-Ni@SIFISIX-1-Cu and SIFISIX-3-Ni@HKUST-1, represent the first examples of inorganic membranes constructed by dual-layer MOF composites. The CO_2/N_2 selectivities of the HKUST-1 and SIFSIX-1-Cu membranes were improved from 5.2 to 13 and 3.4 to 7.8, respectively, after the growth of a selective layer of SIFSIX-3-Ni. This work emphasizes the importance of the custom-design of the membrane material rather than the screening approach of dozens of materials.

B28 Performance of Pd-Based Membranes and Effects of Various Gas Mixtures on H_2 Permeation, Kourosh Kian (Worcester Polytechnic Institute)*

H_2 permeation and separation properties of two Pd-based composite membranes were evaluated and compared at 400°C and at a pressure range of 150 to 600 kPa. One membrane was characterized by approximately 8 μm-thick Pd-Au layer deposited on an asymmetric microporous SiO_2 substrate; the other membrane consisted of approximately 11 μm-thick pure Pd layer deposited on a YSZ support. At 400°C and with a trans-membrane pressure of 50 kPa, the membranes showed a H_2 permeance of 8.42×10^{-4} mol/m^2·s·Pa$^{0.5}$ and 2.54×10^{-5} mol/m^2·s·Pa$^{0.5}$ for Pd-Au and Pd membranes, respectively. Pd-Au membrane showed infinite ideal selectivity to H_2 with respect to He and Ar at 400°C and a trans-membrane pressure of 50 kPa, while the ideal selectivities for the Pd membrane under the same operating conditions were much lower. Furthermore, the permeation tests for ternary and quaternary mixtures of H_2, CO, CO_2, CH_4, and H_2O were conducted on the Pd/YSZ membrane. The H_2 permeating flux decreased at the conclusion of the permeation tests for all mixtures. This decline however, was not permanent, i.e., H_2 permeation was restored to its initial value after treating the membrane with H_2 for 7 hours. The effects of gas hourly space velocity (GHSV) and the steam-to-carbon (S/C) ratio on H_2 permeation were also investigated using simulated steam methane reforming mixtures. It was found that H_2 permeation is highest at the greatest GHSV, due to a decline in the concentration polarization effect. Variations in S/C ratio however, showed no significant effect on the H_2 permeation. The permeation characteristics for the Pd/YSZ membrane were also investigated at temperatures ranging from 350 to 400°C. The pre-exponential factor and apparent activation energy were found to be 5.66×10^{-4} mol/m^2·s·Pa$^{0.7}$ and 12.8 kJ/mol, respectively. SEM and XRD analyses were performed on both pristine and used membranes, and no strong evidence of the formation of Pd-O or any other undesirable phases was observed.

SESSION C - Innovation in Membrane Materials, Synthesis, and Characterization: Liquids

C1 Layer-by-layer modification of aliphatic polyamide anion-exchange membranes to increase chloride/sulfate selectivity, Muhammad Ahmad (University of Notre Dame)*; Chao Tang (University of Notre Dame); Liu Yang (University of Notre Dame); Andriy Yaroshchuk (Polytechnic University of Catalonia); Merlin Bruening (University of Notre Dame)

Anion-exchange membranes (AEMs) with high selectivities among anions may expand the applications of these materials in salt separations. We coated commercially available Fujifilm type-1 AEMs with 5.5 bilayers of poly(4-styrenesulfonate) (PSS) and protonated poly(allylamine) (PAH) to increase their Cl-/SO42- selectivities and found that the increase in selectivity varies greatly with the source-phase salt concentrations. The selectivity of coated membranes increases dramatically from 5.3 to 140 in diffusion dialysis (DD) and from 7.4 to 70 in Electrodialysis (ED) when the source-phase NaCl and Na2SO4 concentrations increase from 0.01 M to 0.1 M. On the other hand, the selectivity of bare AEMs increases from 1.7 to 13 in DD and from 1.1 to 11 in ED for the same range of source-phase salt concentrations. In addition, much higher selectivities occur with AEMs coated only on the receiving side than those coated only on the source side. Partitioning experiments and modeling give evidence for potential mechanisms behind such remarkable selectivities. At higher source-phase salt concentrations, the partitioning experiments show an increased Cl-/SO42- partitioning ratio. Additionally, modeling suggests that electromigration greatly decreases SO42- flux in the AEM.

C2 Development of electrically conductive hollow fiber membranes, Melissa J Larocque (McMaster University)*, David R Latulippe (McMaster University), Charles-François de Lannoy (McMaster University)

Membrane fouling in municipal and industrial wastewater treatment applications incurs higher operational costs associated with lost productivity, membrane cleaning, and antifoulant usage. Recently, there has been considerable interest in the development of electrically conductive membranes due to their ability to mitigate fouling and biofilm development [1]. To date, all of the current research and development has been on the development of electrically conductive membranes in a flat-sheet format. However, hollow fiber (HF) membranes are the preferred geometry for large-scale treatment applications. In this work, we have developed the first-ever electrically conductive HF membranes by adhering a thin layer of carbon nanotubes (CNTs) onto the surface of commercially available HF membranes according to a protocol adapted from a seminal work on flat sheet membranes [2]. CNT suspensions were formed from functionalized single-wall CNTs, surfactants (e.g. sodium dodecyl sulfate) and chemical cross-linkers (e.g. polyvinyl alcohol, suberic acid) then pressure deposited onto the active side of different HF membranes. We then cured the cross-linked CNTs to adhere them to the membrane. This process formed porous electrically conductive coatings. We studied this technique on outside-in PVDF (SUEZ, 0.04 μm) and inside-out PES (Spectrum, 0.2 μm) HF membranes. We observed significant variations in 1) membrane permeance, 2) coating conductivity, and 3) coating stability on curved HF surfaces as a function of material and synthesis parameters. We quantified the impact of curing temperature, CNT concentration, chemical cross-linking density, and polymeric support on these performance metrics. [1] F. Ahmed, B. S. Lalia, V. Kochkodan, N. Hilal, and R. Hashaikeh, Desalination, vol. 391, pp. 1-15, 2016. [2] C. F. de Lannoy, D. Jassby, D. D. Davis, and M. R. Wiesner, J. Memb. Sci., vol. 415-416, pp. 718-724, 2012.

C3 Structure Formation in Isoporous Hollow Fiber Membranes: An In Situ SAXS study, Kirti Sankhala (Helmholtz-Zentrum Geesthacht)*; D. C. Florian Wieland (Helmholtz-Zentrum Geesthacht); Joachim Koll (Helmholtz-Zentrum Geesthacht); Maryam Radjabian (Helmholtz-Zentrum Geesthacht); Clarissa Abetz (Helmholtz-Zentrum Geesthacht); Volker Abetz (Helmholtz-Zentrum Geesthacht)

Isoporous block copolymer membranes hold the potential of providing energy-efficient separations.1 Self-assembly of block copolymers in combination with non-solvent induced phase separation (SNIPS) is required for the formation of mesoscopically porous flat sheet and hollow fiber membranes.2-5 The objective of this work is to understand the kinetics of evaporation-induced self-assembly of block copolymers and non-solvent induced phase separation (SNIPS) for fabrication of isoporous membranes, mainly isoporous hollow fiber membranes (HFM). Weakly and highly segregated polystyrene-block-poly(4-vinylpyridine) (PS-b-P4VP) diblock copolymer solutions as pristine and with additive magnesium acetate (MgAc2), respectively are used to study the structure formation in HFM. The self-assembly of block copolymers during HF spinning is investigated by conducting in situ small-angle X-ray scattering (SAXS) experiments for different block copolymer solutions and the spinning parameters.6 The structural features obtained from in situ SAXS are correlated with the structure in the block copolymer solutions in the absence of shear and in the flat sheet and the HFM morphologies obtained by ex situ SEM.6 [1] V. Abetz. Macromolecular Rapid Communication 2015, 36:10. [2] K.-V. Peinemann, V. Abetz, P.F.W. Simon. Nature Materials 2007, 6:992. [3] M. Radjabian, J. Koll, K. Buhr, U. Vainio, C. Abetz, U.A. Handge, V. Abetz. Polymer 2014, 55:2986. [4] K. Sankhala, J. Koll, M. Radjabian, U. A. Handge, V. Abetz. Advanced Materials Interfaces 2017, 4:1600991. [5] K. Sankhala, J. Koll, V. Abetz. ACS Macro Letters 2018, 7:840. [6] K. Sankhala, D. C. F. Wieland, J. Koll, M. Radjabian, C. Abetz, V. Abetz. Nanoscale 2019, DOI: 10.1039/C8NR06892E

C4 Microscopic membrane fouling characterization, Nandini Debnath (University of Alberta); Aloke Kumar (Indian Institute of Science); Thoams Thundat (The State University of New York at Buffalo); Mohtada Sadrzadeh (University of Alberta)*

This work studies the microscopic fouling in a membrane mimic microfluidic device due to the complex interplay between hydrodynamic interactions and surface adsorption. Our microfluidic membrane mimic device consists of a staggered arrangement of pillars which enables real-time visualization and analysis at the pore scale. Different fouling scenarios are investigated by conducting constant-pressure and constant-flow rate experiments. Fouling experiments are performed with three different types of foulants: polystyrene particle solution (colloidal fouling), polyacrylamide polymer solution (organic fouling) and a mixture of these two solutions (combined fouling). Four major categories of microscopic fouling are observed: cake filtration (at upstream), pore blocking (inside the pores), colloidal aggregation (at downstream) and streamer fouling (at downstream). Experiments show that downstream colloidal aggregation and streamer formation have a significant effect on overall membrane fouling which was not studied before.

C5 New Insights into the Impact of Nano-scale Surface Heterogeneity on the Wettability of Polymeric Membranes, Md Farhad Ismail (University of Alberta); Behnam Khorshidi (University of Alberta); Mohtada Sadrzadeh (University of Alberta)*

Polymeric membrane surfaces often contain nanoscale heterogeneity. The role of this nanoscale heterogeneity on the membrane wettability, as well as surface tension, is still elusive. Conventional surface tension measurement includes quantifying the dispersive and polar components. Here, we showed that dispersive surface tension components may also alter due to the existence of nanoscale heterogeneity which in turn affects the overall solid-liquid interaction process. By analyzing five different commercial membranes with various non-polar probe liquids, we observed a linear dependency of the wettability on surface tension and consequently, and thus we could quantify the degree of heterogeneity. Further, the AFM and SEM results were used to validate this observation. This study provides valuable insight into tuning the membrane surface properties to obtain desired wettability and anti-fouling property.

~~**C6 A comparison of water and solute transport in ion exchange and desalination polymers,** Ryan Kingsbury (Membrion, Inc.); Jingbo Wang (UCLA); Mikayla D Armstrong (University of North Carolina at Chapel Hill)*; Orlando Coronell (University of North Carolina at Chapel Hill)~~

~~In electro-membrane processes, non-ideal water and salt transport across ion exchange membranes (IX) can severely compromise energy efficiency (Tedesco et al, 2017, J. Membr. Sci., 531, 172-182). However, a clear understanding of the factors underlying non-ideal transport is lacking. Accordingly, we measured the water and salt transport properties of 20 commercial IX membranes and evaluated their correlation to membrane physical and chemical properties. Results showed that water and salt transport in the membranes was governed primarily by the microstructure of the membranes, not the polymer type. This finding raised questions about how the transport characteristics of IX membranes compare to those of membrane polymers used in other applications such as desalination, and whether other polymers would be suitable for use as IX membranes or vice versa. Thus, we compared the water and salt transport properties of IX membranes with those of commercial desalination membranes and other desalination polymers. We used data collected in our laboratory and from literature. Results showed that the water and salt transport characteristics of IX membranes fall within similar ranges as those of desalination polymers. Further, IX membranes were found to have excellent water-salt partitioning selectivity, while polyamide active layers were found to have exceptional diffusion selectivity. On the basis of these findings, we propose a membrane architecture (i.e., polyamide coated with IX polymers) that combines the relative strengths of IX and desalination polymers to obtain superior water permeability selectivity compared to either polymer alone.~~

C7 Water permeation through nanoporous particles in thin film nanocomposite membranes, Pinar Cay-Durgun (Arizona State University)*; Mary Laura Lind (Arizona State University)

Most advanced reverse osmosis (RO) desalination membranes are composed of thin-film composite (TFC) membranes and nanoporous materials as fillers to form thin-film nanocomposite (TFN) membranes. The addition of the nanoporous fillers into the membranes can address the well-known permeability/selectivity trade-off by introducing molecular sieving transport mechanism in addition to the solution-diffusion transport governed by polymeric RO membranes. In literature, there are several TFN membrane studies with different particle types, sizes, and contents. Although these TFN membranes have enhanced water permeability compared to the corresponding TFC membranes, they exhibit inconsistent performance improvement trends with particle addition. Therefore, it would be useful to investigate the individual water transport contributions of the polymer, particle, and polymer/particle interface for the pressure-driven desalination to understand the behavior of TFN membranes. In this study, we investigated water permeation through single-crystal molecular sieve zeolite A particles in TFN membranes. First, we embedded different sizes of non-porous (pore-closed) and nanoporous (pore-opened) particles into conventionally polymerized TFN membranes. These membranes have some synthesis challenges such as particle aggregation and non-uniform particle dispersion the membranes. To overcome the challenges, we developed a TFN-model membrane synthesis method that immobilizes the particles onto the support membranes surface before the polymerization reaction. Both conventional TFN and TFN-model membranes have improved water permeability without significant change in salt rejection. Later, we analyzed the results using two different resistance models to quantify the possible water transport pathways in these membranes. Our analysis indicated that the nanoporous particles increase water permeability in TFN membranes primarily through increased water transport at the interface.

C8 Printing Membranes: Enabling the Use of Both Conventional and Novel Polymers in High Performance Membranes, Jeffrey R McCutcheon (University of Connecticut); Ayse Asatekin (Tufts University); Maqsud Chowdhury (University of Connecticut); Xin Qian (University of Connecticut); Tulasi Ravindran (University of Connecticut)*; Samuel Loudner (Tufts University)

Our recent work on printing polyamide membranes demonstrated the ability to make desalination membranes of exceptional smoothness and controllable thickness. The approach required the electrospray deposition of an amine monomer and an acid chloride crosslinker onto a substrates where they react to form an aromatic polyamide. The technique allows for exceptional control of the reaction and makes it advantageous over conventional interfacial polymerization. Conventional interfacial polymerization is a relatively uncontrollable process and has few options for independently tailoring selectivity and permeance. We can adapt this approach to making membranes of novel polymers as well, many of which cannot be formed easily into thin film or composite structures with high permeance and high selectivity. We demonstrate the use of the printing approach with several emergent polymer systems, including self-assembled materials, to demonstrate the value of this approach.

C9 Functionalization of PVDF Membranes with Thiol Groups for Heavy-Metal Capture, Ronald Vogler (University of Kentucky)*; Md. Saiful Islam (University of Kentucky); Evan Hatakeyama (Chevron Energy Technology Company); DB Bhattacharyya (University of Kentucky)

Heavy-metal ions, such as Hg^{2+}, pose a severe hazard to the environment and human health when present in water. Membranes are an emerging technology in the water-treatment field and can be functionalized with thiol groups for the capture of these ions from water. In this study, thiol-functionalized membranes were fabricated by first functionalizing the pores and surface of commercial polyvinylidene fluoride (PVDF) membranes with carboxyl groups via the in-situ polymerization of polyacrylic acid (PAA); the addition of PAA was confirmed by measuring the zeta potential of the surface of a PAA-functionalized membrane. Following functionalization with PAA, the membranes were functionalized with thiol groups using EDC/NHS coupling to bind cysteamine to their carboxyl groups. The convective-flow capture of Ag^+ was quantified for the resulting membrane and the focused-ion-beam technique along with energy-dispersive X-ray spectroscopy (EDS) were utilized to determine the location of this capture in the membrane. A shift in surface charge of a PVDF-PAA membrane near the pKa of PAA (~ 4.5) confirmed the presence of PAA on the surfaces of these membranes. After the successful incorporation of thiol groups into a membrane, its capture of Ag^+ was 71% of its estimated capacity during the convective flow of Ag+ through the membrane. EDS performed on this thiol-functionalized membrane indicated that Ag^+ capture occurred all throughout the membrane pores (~170 μm deep). This capture of Ag^+ at all depths within the thiol-functionalized PVDF membrane indicated the potential of this membrane for the capture of Hg^{2+} at a high capacity. Future work with this type of membrane will involve an investigation of the impact that hardness and residence time have on its capture of Hg^{2+} from refinery water. This research was supported by NIEHS-SRP, NSF-EPSCoR, and Chevron Corporation.

C10 Pure and mixed fluid sorption and transport in Celazole® polybenzimidazole: the effect of plasticization, Kelly P. Bye (University of Oklahoma); Michele Galizia (University of Oklahoma)*

The vast majority of industrial chemical synthesis occurs in organic solution. Solute concentration and solvent recovery consume approximately 50% of the energy required to produce chemicals and pose problems that are as relevant as the synthesis process itself. Novel, energy-efficient technologies based on polymer membranes are emerging as a viable alternative to distillation. Despite organic solvent nanofiltration (OSN) could revolutionize the chemical industry, its development is still in its infancy for two reasons: i) the instability of traditional polymer materials in chemically challenging environments, and ii) the lack of fundamental knowledge of elemental transport phenomena in OSN membranes. Most of available transport data refer to composite membranes, where the presence of a fabric backing makes it difficult to provide a fundamental description of solvent and solute sorption and transport in the active layer. The lack of fundamental information has hampered the development of rational methods to design better materials for OSN application. Tough polymers are often reported as ideal membrane materials for separations in harsh environments. This study focuses on the solubility of pure and mixed organic liquids in Celazole®, a commercial polybenzimidazole. Methanol was selected as a model penetrant to run sorption and diffusion experiments in Celazole® in the activity range 0-1. The role of polymer-penetrant interactions, membrane degree of swelling and penetrant clustering on small molecule sorption and transport in Celazole® was discussed and several structure-property correlations were identified. Remarkably, methanol produces a severe matrix plasticization, which helps explain the mixed methanol/PEG400 sorption behavior. Finally, the effect of solvent sorption on Celazole® mechanical properties was investigated.

C11 Enhanced Water Interaction and Antifouling Compatible Zwitterion-PVDF Membrane Designs, Mahboobeh Maghami (University of Saskatchewan); Amira Abdelrasoul (University of Saskatchewan)*

The most common form of water toxicity occurring in Western Canada is sulfate toxicity. Elevated sulfate levels in water have become a major challenge because of the excessively high amounts of various sulphate salts present in water. The major goal of this study is to perform a systematic study on novel zwitterions into Poly(vinylidene fluoride) (PVDF) membrane to enhance hydrophilicity and salt rejection. The Zwitterionic-PVDF (ZW-PVDF) membranes have drawn much attention because of the promising performance of Zwitterionic polymer with respect to pH responsive, reducing metal ion adsorption, and significantly enhancing membrane hydrophilicity and membrane fouling resistance. In this study, density-functional theory (DFT) was used to simulate several modification designs of PVDF with compatible Zwitterionic polymer for enhanced water interactions and minimized sulfate interactions. Furthermore, fragment molecular orbital (FMO) and the Pair Interaction Energy Decomposition Analysis (PIEDA) were carefully interrogated. The mechanism of water transport inside these ZW-PVDF membranes were extensively investigated. These types of analyses included the inter fragment interaction energy (IFIE), like the electrostatic (ES), exchange repulsion (EX), and charge-transfer and mixing term (CT+mix). Furthermore, FMO and PIEDA results showed enhanced water transport though hydrogen bonding and interactions between water molecules and ZW function groups at different alignments. The influence of Zwitterionic on reducing fouling in nanofiltration of sulfate wastewater will be discussed.

C12 Novel strategy to develop a composite membrane for desalination by membrane distillation, Tarik Eljaddi (Laboratoire Réactions et Génie des Procédés - UMR 7274)*; Deisy Mejia Mendez (Laboratoire Réactions et Génie des Procédés - UMR 7274); Eric Favre (Laboratoire Réactions et Génie des Procédés - UMR 7274); Denis Roizard (Laboratoire Réactions et Génie des Procédés - UMR 7274)

Membrane distillation (MD) is a promising technology to get freshwater. However, MD is still confined at labscale up to now. Indeed, MD can be easily coupled with RO to treat high concentrated brines with relatively low energy consumption. However, as a main drawback, permeability of conventional porous membranes is not stable with time because the pores will face the wetting issue. Thus, the separation efficiency by vaporization will rapidly decrease. Whereas most of current strategies to develop MD membranes are based on the chemical modification of membrane surface to generate hydrophobic properties, our group has studied a distinct approach. This strategy intends to use a thin, dense coating to fully prevent wetting hazard of the membrane pores. Hence, conversely to the usual approach focusing on membrane hydrophobicity, this approach is very different from the theoretical point because the intention is to protect the mouth of the pore from wetting by dense layer coating.In literature, pioneering trials in that direction concluded that this approach was not valid because the added transport resistance would preclude the application. Indeed, the permeance of a given membrane varies with two parameters: intrinsic permeability and layer thickness. Considering the wide scope of existing materials, in particular super glassy polymers, it can be simulated that with some materials, dense thin layers would have the same transfer resistance than porous membranes. With these membranes, the mechanism changes from vaporization into the pores to thermopervaporation with vaporization at the downstream side of the dense layer. But this modification doesnot affect energy needed for the water transport. But interestingly, the salts cannot anymore enter into the pore and only pure water is recovered.In the second part of our work we successfully prepared and tested such composite membranes as proof of concept. Even under harsh conditions of wetting, 100% salt rejection was reached.

C13 One-Step Tailoring Surface Roughness and Surface Energy to Prepared Superhydrophobic Polyvinylidene Fluoride (PVDF) Membranes for Enhanced Membrane Distillation Performances, Weihua Qing (The University of Hong Kong)*; Xiaonan Shi (New Jersey Institute of Technology); Chuyang Tang (University of New South Wales)

Superhydrophobic polyvinylidene fluoride (PVDF) membrane is a promising material for membrane distillation. Existing approaches for preparing superhydrophobic PVDF membrane often involve separate manipulation of surface roughness and surface energy. Here we report a one-step approach to simultaneously manipulate both the surface roughness and surface energy of PVDF nanofibrous membranes for enhanced direct-contact membrane distillation (DCMD) performances. The manipulation was realized in a unique solvent-thermal treatment process, during which a treatment solution containing alcohols was involved. We demonstrate that by using different chain-length alcohols in the treatment solvent, surface roughness can be promoted by creating nanofin structures on the PVDF nanofibers using an alcohol which has moderate affinity with the PVDF material. Meanwhile, surface energy can be tuned by adjusting the fraction distribution of crystal phases (mainly nonpolar α phase and polar β phase) in the membrane using different alcohols. PVDF membranes with different surface wettabilities were used to evaluate the effects of surface roughness and surface energy on the DCMD performances. Combining both low surface energy and multi-scale surface roughness, pentanol-treated PVDF membrane achieved best anti-water property (water contact angle of 164.1° and sliding angle of 8.1°), and exhibited superior water flux and enhanced anti-wetting ability to low-surface-tension feed in the DCMD application.

C14 High-Performance Thin Film Composite (TFC) Membranes: Design Consideration Underneath the Polyamide Thin Film,
Wangxi Fang (Suzhou Institute of Nano-Tech and Nano-Bionics, Chinese Academy of Sciences)*; Jian Jin (Suzhou Institute of Nano-Tech and Nano-Bionics, Chinese Academy of Sciences); Rong Wang (Singapore Membrane Technology Centre, Nanyang Technological University)

Despite the commercial success of thin-film composite (TFC) technique for reverse osmosis (RO) and nanofiltration (NF) membrane production, overcoming the longstanding "trade-off" relationship between permeability and selectivity of TFC membranes remains a global challenge. Extensive research efforts have been put on altering the layer formation chemistry or intrinsic nanostructure to simultaneously improve its permeability and selectivity, but the critical role of the underlying porous substrate on the membrane separation performance was often overlooked. In this study, we firstly demonstrated the relationship between the substrate surface pore profile and the TFC membrane separation behavior. Manipulating the substrate surface pore profile i.e. surface porosity, pore number density, etc. for water flow path improvement was proved efficient in enhancing the TFC membrane water permeability. Afterwards, we extended the idea of flow path improvement from conventional polymeric substrates to nanomaterial-based membrane structure. Although the involvement of nanomaterials has been proved to significantly improve the TFC membrane permeability and selectivity, the role of the nanomaterial in such improvement is still under debate. Via controlling the lateral position of the incorporated amphiphilic ABA copolymer-based hollow nanospheres within the TFC membrane, we found that the water permeability enhancement only occurred when the nanoparticles were positioned underneath the polyamide layer. Therefore, other than reducing the water permeation resistance inside the polyamide layer, the role of the nanoparticle is more towards improving the water flow path from the polyamide layer to the underlaying substrate. This study clarifies longstanding misunderstandings concerning the effects of the underlying porous substrate as well as the nanomaterial incorporation on TFC membrane performance, and provides insight into designing highly permeable and selective TFC membranes.

C15 Wafer-Scale, Vertically-Aligned SWCNT Composite Membranes for Nanofiltration, Melinda L Jue (Lawrence Livermore National Laboratory)*; Chiatai Chen (Cornell University); Steven Buchsbaum (Lawrence Livermore National Laboratory); Eric Meshot (Lawrence Livermore National Laboratory); Sei Jin Park (Lawrence Livermore National Laboratory); Kuang Jen Wu (Lawrence Livermore National Laboratory); Francesco Fornasiero (Lawrence Livermore National Laboratory)

Carbon nanotubes (CNT) have attracted significant interest as nanopores for separation applications due to their ability to sustain ultra-fast fluid transport rates that exceed those of other similarly-sized porous materials. When formed into oriented composites with high nanotube density, the well-defined pore size allows for both superior flux and selectivity for nanofiltration applications. Unfortunately, achieving large-area CNT membranes with a high density of conductive channels remains challenging. Here we demonstrate the synthesis of ultra-high density, vertically-aligned single-walled carbon nanotubes (SWCNT) to create high flux nanofiltration membranes with large areas up to 4-in in diameter. Chemically resistant parylene-N is used to form the SWCN composite, which remains stable even in the presence of aggressive cleaning agents. These membranes exhibit exceptionally high water flux while completely rejecting nanometer-sized particles and dyes. In addition, the membranes preferentially sieve out multivalent ions, thus showing potential for water softening. To our knowledge, this work represents the best co-optimization of SWCNT diameter, density, and membrane area demonstrated so far in the literature.

C16 Vapor Phase Infiltration of Metal Oxides into Microporous Polymers for Solvent Stable Nanofiltration Membranes, Fengyi Zhang (Georgia Institute of Technology)*; Emily McGuinness (Georgia Institute of Technology); Yao Ma (Georgia Institute of Technology); Mark Losego (Georgia Institute of Technology); Ryan Lively (Georgia Institute of Technology)

Linear microporous polymers (e.g., PIM-1) possess microporosity and solution processability, which positions them as promising candidate materials for challenging molecular separations. However, these membranes are susceptible to swelling and plasticization in the presence of various organic liquids. This solvent-induced mobility of the polymer structure impairs the organic solvent separation performance of PIM materials. Pyrolytic carbonization and crosslinking are two prominent post-fabrication methods for polymer stabilization; however, it requires a near-complete transformation of the precursor in the former case and chemical/thermal treatments in the latter case. Here, we develop a straightforward vapor-phase post-fabrication modification technique to stabilize microporous polymers. Two cycles of vapor phase infiltration of metal-organic precursors and water into the PIM-1 polymer network create interpenetrating atomic-scale networks of metal oxide and PIM-1, which retain the microstructure and macroscale form factors and exhibit excellent solvent resistance to strong solvents for PIM-1 (chloroform, dichloromethane, and tetrahydrofuran). Pristine PIM-1 membranes ineffectively reject polystyrene oligomers (i.e. molecular weight cutoffs in excess of 1200 g/mol are observed) from solvents such as ethanol. However, the AlOx/PIM-1 membranes exhibit consistent molecular weight cutoff below 204 g/mol in various solvents (e.g., ethanol, n-heptane, toluene, tetrahydrofuran, etc.). Moreover, the hybrid membranes are capable of solvent-solvent separations (e.g., alcohol mixtures, aromatic mixtures), which position these materials as good candidates for organic solvent reverse osmosis separations. Importantly, the vapor phase infiltration process can be directly applied to state-of-the-art hollow fiber membrane modules or spiral-wound membrane modules, which suggests that this approach has the potential to be useful for large-scale manufacturing of advanced membranes.

C17 Understanding the Interlayer-spacing and Mass Transport Nexus of Graphene Oxide Membrane for Organic Solvent Nanofiltration, Sunxiang Zheng (UC Berkeley)*; Baoxia Mi (UC Berkeley)

Ultrathin membranes formed by stacking two-dimensional (2D) graphene oxide (GO) hold great promises for improved permeability and separation capability in organic solvent nanofiltration (OSN). However, analogous to the scenarios in aqueous phase, the separation capability of a layer-stacked GO membrane for OSN can be significantly limited by its natural tendency to swell, that is, absorb organic solvents into the GO channel and form an enlarged interlayer spacing (d-spacing). In this study, the d-spacing of a layer-stacked GO membrane was experimentally characterized by liquid-phase ellipsometery, which allows precise quantification of the d-spacing when the membrane is soaking in its working environment. We found that solvents with high dipole moments (e.g., dimethylformamide, n-methyl-2-pyrrolidone) tends to cause significant swelling of GO with d-spacing increasing up to 2.74 nm, while non-polar solvents such as hexane and toluene only cause subtle swelling. To understand such unique swelling behavior, the solubility parameter of GO was experimentally determined to mathematically describe the affinity between GO and the testing solvents, which is believed to be strongly related to the swelling degree. Separation performance of the GO membrane in different organic solvents were tested to understand the mass transport in confined GO nanochannel. The layer-stacked GO membrane could achieve > 90% rejection of small organic dye molecules (e.g., rhodamine B and methylene blue) in ethanol and acetone without crosslinking due to insignificant swelling. Additionally, it was found that solvents with different polarity shows distinct transport resistance, which was further investigated by molecular simulation.

C18 Organic-inorganic hybrid separation membrane having high chemical tolerance, Koichi Takada (TORAY Industries, Inc.)*; Takahiro Tokuyama (TORAY Industries, Inc.); Hiroki Minehara (TORAY Industries, Inc.); Takafumi Ogawa (TORAY Industries, Inc.); Masahiro Kimura (TORAY Industries, Inc.)

In late years, water problems, such as occurrence of the extraordinary drought by the global climate change, are actualized in addition to shortage of water and water pollution due to a rapid increase of the world population and industrial development. The expectation to a membrane technology increases as a water treatment technique to secure safe and stable enough water resources. Conventional reverse osmosis (RO) membrane consists of cross-linked aromatic polyamide or cellulose acetate (CA). Cross-linked aromatic polyamide membrane has especially high property (flux and rejection), on the other hand, CA membrane has high chlorine tolerance. The purpose of this research is creating a membrane which has both features of cross-linked aromatic polyamide membrane and CA membrane. In this research we report membrane property and the result of chlorine tolerance test of organic-inorganic hybrid membrane. On conventional support layer, a thin separation functional layer (≒200 nm) based on organic-inorganic hybrid material is prepared by in situ polymerization and cross-linking. The hybrid membrane has high flux (1.0 m^3/m^2/day @0.75 MPa, NaCl 500ppm) and high NaCl rejection (more than 90%). The chlorine tolerance test shows that the hybrid membrane has much higher stability of flux than CA RO membranes.

C19 Magnetron Sputtering-based Synthesis of Pd film on UF membrane support for environmental catalysis, Michael J Detisch (University of Kentucky)*; DB Bhattacharyya (University of Kentucky); John Balk (University of Kentucky)

Composite membranes with metallic films as top layers were synthesized for applications in separations and catalysis. Commercial ultrafiltration membranes are used as a base for the production of nanostructured thin films to generate these composite membranes. Magnetron sputtering is used to deposit thin films of tantalum and MgPd alloy on top of polysulfone (PSf) membranes 100 nm thick. A dealloying step in water removes the Mg component of the film creating a nanoporous (np) film of Pd. This generates a sponge-like structure of interconnected nanowires with characteristic size of 5 - 10 nm. These nanostructures are imaged with high resolution SEM combined with FIB cross-sectioning. Surface chemistry is analyzed using x-ray photoelectron spectroscopy. The np palladium is used with H_2 gas for the degradation of chloro-organics from water. PCB's were dechlorinated while permeated in a water solution under pressurization with 5% H_2 gas (remainder Argon). Over 60% of PCB-1 was degraded in solution with a single pass through the composite membrane at 40 LMH. Np metals may be fabricated from a variety of precursor systems. Depending on the reaction desired, various metals may be incorporated into the membrane system with minimal difficulty. These membranes can also be tuned for selective separations in both water and solvent media. The membrane showed improved rejection properties and a corresponding drop in flux with the addition of metallic top layers. Flux dropped from 180 LMH/bar for the base UF PSf compared to 3 LMH/bar for the composite system. The molecular weight cutoff (MWCO) of the UF PSf was found to be above 70 kDa while for the composite membrane the MWCO dropped to below 10 kDa. This illustrates an extensive modification of membrane properties through the addition of a thin (100 nm) film. This research is supported by the NIH-NIEHS-SRC (Award no.: P42ES007380), and by NSF KY EPSCoR grant (Award no: 1355438).

C20 Influences of polymeric additives in different solvent systems on membrane performance and fouling resistance, Catharina Kahrs (Leibniz University Hannover)*; Jan Schwellenbach (Sartorius Stedim Biotech GmbH)

Phase separation of polymer solutions is a commonly applied technique for the production of porous membranes. One mechanism to induce the demixing of these solutions is the non-solvent induced phase separation (NIPS). By applying this technique it is possible to fabricate membranes with a broad range of different characteristics. A crucial variable during the NIPS process is the composition of the casting solution as it has an important impact on the features of the resulting membrane. Apart from the membrane-forming polymer, a number of different solvents, non-solvents, and additives are available which can be varied in order to affect the kinetic and thermodynamic parameters during membrane formation. Consequently, the improvement of the control of resulting membrane characteristics requires an enhanced understanding of the NIPS process and its influencing factors. In order to gain a better insight into the underlying mechanisms of polyethersulfone membrane preparation by NIPS, the effect of the additives polyvinylpyrrolidone (PVP) and polyethylene glycol (PEG) was studied in four different solvent systems. Apart from comparing the influence of the respective solvent features, another aim was to investigate the impact of concentration and molecular weight variations in dependence of the applied solvents. Therefore, membranes with varying concentrations and molecular weights of PVP and PEG were produced for each of the four solvents. Subsequently the membranes were characterized in respect of structure, performance and fouling resistance by scanning electron microscopy, permeability and protein retention measurements, as well as by fouling evaluation through static protein binding and dynamic filtration experiments. The results which were found indicate that both, solvents and additives, can significantly impact membrane performance, fouling resistance and structure which can be attributed to an alteration of the kinetics during the membrane formation process.

C21 Porous membranes with synergic solvent and thermal resistance from polyoxindole derivatives, Bruno Pulido Ponce de Leon (King Abdullah University of Science and Technology)*; Suzana Nunes (King Abdullah University of Science and Technology)

The expansion of membrane technology into new applications drives a constant demand for novel materials. Particularly, organic solvent ultra- and nanofiltration (OSN), has been a rapidly growing research field that aims for more efficient chemical separations. Polyoxindole derivatives constitute a new family of polymers and are promising candidates as material for polymeric membranes with solvent and thermal resistance. These high-performance polymers are synthesized by superacid polycondensation, an efficient one-step, one-pot, metal-free, room-temperature method. We have synthesized functionalized polyoxindoles and manufactured them into porous membranes by non-solvent induced phase separation (NIPS). The prepared membranes were crosslinked mainly using two different strategies: (1) by reacting with dibromides or (2) by incorporating a propargyl pending group, followed by a heat-induced reaction in hot glycerol. More recently, we've explored membranes prepared of poly(oxindole-2,2'-biphenol), which is made from the monomers isatin and 2,2'-biphenol. The material is insoluble in many common organic solvents, but it can be made into porous membranes from solutions in polar aprotic solvents. Additionally, this polymer has a high density of hydroxyl groups that make the polymer very attractive for further functionalization. Moreover, these polymer membranes undergo a thermal-rearrangement at 300 °C which leads to membranes with increased solvent and thermal resistance. The MWCO for these membranes ranges from 1 kg/mol to 20 kg/mol and can be tuned by casting conditions, with dimethylformamide permeation of 68 $L\ m^{-2}\ h^{-1}\ bar^{-1}$ at 120°C. The thermal rearranged membranes preserve their morphology and filtration performance. After treatment, all membranes have high solvent resistance, with minimum swelling and degradation temperature up to 400°C, allowing them to be used in demanding organic solvent filtration processes.

C22 Functionalized random zwitterionic copolymers as chlorine- and fouling-resistant nanofiltration membranes, Samuel J Lounder (Tufts University)*; Ayse Asatekin (Tufts University)

The purification of water is vital for many chemical separation processes. However, feedstocks containing foulants and/or chlorine represent harsh conditions in which many membranes are unable to operate. Accordingly, the development of membranes that can function effectively in such conditions is an important research endeavor. A zwitterion is defined as a molecule that contains an equal number of positive and negative charges. Random zwitterionic copolymers (RaZCs) are linear copolymers comprised of randomly distributed zwitterionic repeat units and non-zwitterionic repeat units. RaZCs self-organize to form a network of zwitterionic nanochannels that are surrounded by the non-zwitterionic repeat unit of the RaZC. The zwitterionic channels act as the effective ≈1 nm pores. Previous work shows that thin film composite membranes featuring these RaZCs as their selective layer exhibit high flux, exceptional fouling resistance, and size-based small molecule selectivity along with low salt rejection. However, for many applications, improved charged solute and divalent salt rejection would be beneficial. In this work, we show that RaZCs that combine zwitterionic, charged, and hydrophobic repeat units are versatile materials for the fabrication of robust membranes that exhibit high flux, divalent ion rejection, fouling resistance, and remain stable upon exposure to chlorine. Furthermore, the pore chemistry of these membranes can be tuned by a simple approach, thus rationally tuning membrane selectivity for desired applications. This approach represents a unique pathway to fabricating industrially relevant membranes with tunable separation properties.

C23 Deterioration of nanofiltration polyamide membrane by strong acid and its mechanism, Byung-Moon Jun (University of South Carolina); Hyung Kae Lee (Ulsan National Institute of Science and Technology); Young-Nam Kwon (Ulsan National Institute of Science and Technology)*

In this study, the effect of strong acids such as sulfuric acid and hydrogen halides (hydrochloric acid, hydrobromic acid, and hydroiodic acid) on polyamide NF membrane was systematically investigated using various analytical tools. The membrane exposed to sulfuric acid did not show noticeable difference in Scanning Electron Microscopy (SEM), Attenuated Total Reflectance-Fourier Transform Infrared spectroscopy (ATR-FTIR), X-ray photoelectron spectroscopy (XPS), contact angle, and zeta potential analysis compared to the virgin membrane. However, the performance changed after exposure to sulfuric acid, and the water flux and salt rejection of the membranes degraded at pH 0 and 1 were different. This was likely due to the distortion of hydrogen bonding or formation of tetrahedral structure caused by O- or N-protonation. On the other hand, the membrane exposed to hydrogen halides showed change of various physical and chemical properties along with permeability change after exposure to the acid. The degree of changes was in order of HCl, HBr, and HI. This was due to the halogenation of the polyamide membrane.

C24 Opening new doors: Epoxides as novel chemistry for interfacial polymerization, Rhea Verbeke (Katholieke Universiteit Leuven)*; Marijn Seynaeve (Katholieke Universiteit Leuven); Wouter Arts (Katholieke Universiteit Leuven); Elke Dom (Katholieke Universiteit Leuven); Ivo Vankelecom (Katholieke Universiteit Leuven)

TFC-membranes dominate the current nanofiltration (NF) and reverse osmosis market for water treatment and desalination. However, their applicability in more aggressive feeds (extreme pH, hypochlorite, solvents) is limited due to their limited chemical robustness. Provided the enormous growth potential of membrane technology in these industries, there is an ongoing quest to obtain solvent-, pH- and chlorine-stable (TFC) membranes, which are additionally performing well in terms of rejection and permeance. To comply with this demand, we successfully converted the well-known monophasic bulk epoxide polymerization (commonly used in e.g. the automotive and coating industry), into the synthesis of thin, yet cross-linked top-layers via interfacial polymerisation. Nanofiltration membranes with >90% Rose Bengal and 70% Methyl Orange rejection with water permeances above 2 L/m^2hbar were achieved in the proof-of-concept study. Optimization studies already resulted in permeances of 25 L/m^2hbar without compromising RB rejection. Additionally, the occurring reaction, a homopolymerisation of the epoxy monomer, results in a network of chemically stable polyether chains. As a consequence, unchanged membrane performance after immersion in oxidizing (500 ppm NaOCl, pH 4, 2.5 h) and acidic (pH 1, 48 h) environments was achieved. The correlations between IR, PALS, SEM and performance results further demonstrate the feasibility and reproducibility of this innovative membrane synthesis technique. The plethora of available epoxy monomers, reaction initiators and catalysts, as well as the use of other membrane supports, ensures high tunability in terms of final membrane performance, morphology and synthesis time. We therefore believe that this novel interfacial polymerisation chemistry may lay the foundation for a new generation of exceptionally stable (solvent resistant) NF TFC-membranes, opening immense possibilities with respect to chemical cleaning and novel membrane applications.

C25 Solution-diffusion with ion association: a non-mean-field model that successfully describes NF of multi-ion mixtures, Viatcheslav Freger (Technion - Israel Institute of Technology)*

Despite wide use of nanofiltration (NF), modeling NF separations and ion selectivity still presets a challenge, especially, for multi-ion solutions. Mean-field models of charged nanopores (e.g., Donnan or Poisson-Boltzmann) have been the standard for last two decades, yet they fail to reproduce correctly trends even for some single salts, which may be traced back to inherent flaws of mean-field theories in charged low-dielectric media. On the other hand, simple phenomenological models with adjustable ionic permeabilities work well, but require ad hoc composition-dependence and the physics behind this dependence remains obscure. Here we abandon the standard mean-field nanopore picture and introduce a simple solution-diffusion model of a homogeneous NF membrane that treats the strong departures from mean-field by adding ion association effects in a manner of classical Bjerrum theory. The model consistently explains our recent elaborate NF data for NaCl+$CaCl_2$ mixtures [1] and other published data. In particular, where existing models fail, the new model readily explains why (a) ion permeabilities rank Na+ > Cl- > Ca2+, (b) NaCl permeability increases with concentration while that of $CaCl_2$ does the opposite; (c) addition of a small fraction of Ca2+ sharply increases NaCl permeability, (d) effective membrane charge is much lower than independent estimates etc. The results emphasize the importance of non-mean-field effects in general and ion association in particular for correct modeling of ion separations in osmotic membranes, for which the new model offers a simple yet physically sound and transparent way to treat the problem. [1] N Fridman-Bishop, KA Tankus, V Freger, Permeation mechanism and interplay between ions in nanofiltration, J. Membr. Sci., 2018, 548, 449-458

C26 Relative permittivity properties of hydrated polymer membranes for desalination applications, Kevin Chang (University of Virginia)*; Hongxi Luo (University of Virginia); Geoffrey M Geise (University of Virginia)

Polymers are widely used in membrane-based desalination to address global water shortages. To be effective, these polymers must act as a selective barrier and prevent ion passage. One approach to engineering effective desalination membranes is to prepare membranes that suppress ion sorption, because low ion sorption can lead to favorable salt rejection and efficient desalination. Thus, it is important to understand how to engineer polymer chemistry that suppresses ion sorption and maximizes membrane desalination performance. Movement of an ion from solution into the membrane phase is associated with a change in free energy, which is sensitive to both solution and membrane properties. Describing this free energy change requires knowledge of the relative permittivity properties of the hydrated polymer. Few studies report this information for desalination membranes, so relative permittivity is often estimated using a linear relationship between polymer water content and relative permittivity. This relationship, which is based on a small data set (primarily Nafion® 117 data), may not be universally applicable to all hydrated polymers. We evaluated the functional relationship between permittivity properties and water content for a series of cross-linked poly(glycidyl methacrylate) (XL-pGMA) polymers. At equivalent water content, XL-pGMA has a lower relative permittivity compared to Nafion® 117. This result supports the view that polymer water content alone may be insufficient for estimating the relative permittivity properties of hydrated polymers. Since lower relative permittivity correlates with higher ion rejection, relative permittivity measurements provide important insight into ion sorption properties that can be used to inform molecular level polymer design. Thus, understanding the structure/property relationships between polymer chemistry and ion sorption, via membrane dielectric measurements, is critical to guide the design of advanced desalination membranes.

C27 Multi-solute transport behavior of aqueous mixtures through polyether-based membranes, Jung Min Kim (Auburn University)*; Bryan S Beckingham (Auburn University)

Molecular transport in polymer membranes is important for a wide variety of applications from traditional separations to innovative new energy devices including electrochemical cells such as direct methanol fuel cells and solar fuel devices for the photoelectrochemical (PEC) reduction of CO_2. In these applications, understanding and controlling the selective transport of ions and molecules is crucial. For instance, one of the major challenges in PEC cells for CO_2 reduction to liquid products is to understand and manage product crossover through the ion exchange membrane. While there have been numerous studies investigating single ion or molecular transport through hydrated dense membranes, there have been relatively few studies focused on concurrent multi-solute transport in these materials. Polyethers are promising materials for membrane applications due to their wide range of potential chemical functionalities and ease of membrane fabrication through UV-photopolymerization through diacrylate and acrylate end groups. Membranes prepared from poly(ethylene glycol) diacrylate (PEGDA) have been used as a model system for investigating fundamental aspects of ion and molecular transport where water permeability and selectivity over a solute were impacted by varying the PEGDA to water ratio of the pre-polymerization mixture due to variation in crosslink density. In previous work, our group utilized custom diffusion cells with in-situ ATR-FTIR spectroscopy for monitoring the simultaneous transport of multiple solutes through ion exchange membranes. Here, we leverage this methodology to examine this solution-diffusion transport of single and multicomponent alcohol mixtures through PEGDA membranes and PEGDA-based membranes incorporating various mono-acrylate units within the structure. Permeability values extracted from diffusion cell measurements are complemented by solute solubilities from absorption-desorption experiments in order to fully describe the transport behavior.

C28 Comparative analysis of novel covalent organic framework nanofiltration membranes synthesis, composition, thickness and substrate dependence, Gabrielle Levato (University of Illinois at Urbana-Champaign)*; David Burke (Northwestern University); Zhiwei Jiang (Imperial College London); Daniel Mosiman (University of Illinois at Urbana-Champaign); William Dichtel (Northwestern University); Andrew Livingston (Imperial College London); Benito Marinas (University of Illinois at Urbana-Champaign)

It is essential to start emphasizing sustainable and efficient water treatment technologies due to increased water scarcity and lack of safe drinking water. Pressure driven membrane systems can be used effectively to recycle seawater, wastewater effluent or industrial water to potable quality. Commercial membrane systems can remove a wide range of water contaminants with low use of chemical additives and no thermal input, but the relatively limited polymeric surface chemistry currently available restricts the range of water permeability and solute selectivity achieved. An alternative to conventional polymeric membranes is covalent organic frameworks (COFs). COFs have a crystalline structure created by strong covalent bonds made through synthetic reactions of organic building units. The structure provides a well-organized layer with pore size control based on chosen building units. Thin film composite nanofiltration membranes with a COF active layer have been developed via interfacial polymerization and exfoliation techniques, allowing active layer thickness, comparative synthesis feasibility and reproducibility, and COF-substrate interaction to be analyzed. This work begins to assess the composition of COF TFC membranes on ultrafiltration solvent-resistant polyacrylonitrile (PAN) supports of varying pore sizes, and specifically how COF active layer thickness and synthesis method influence permeation and rejection capabilities of organic and salt solutions. Advanced physicochemical characterization of the COF membranes using techniques including grazing-incidence wide-angle x-ray scattering (GI-WAXS), Rutherford backscattering spectrometry (RBS), atomic force microscopy (AFM), scanning electron spectroscopy (SEM), and transmission electron spectroscopy (TEM) provide valuable insight into COF membrane surface composition and support interaction. Combined, the two analyses of COF membranes elucidate clear pathways forward for further studies in this field.

C29 Copper-MOF Functionalized Nanofiltration Membranes for Enhanced Dye Removal: Effects of Dip-Coating and In-Situ Modification Techniques, Mostafa Dadashi Firouzjaei (University of Alabama); Zane Joseph Parkerson (University of Alabama); Milad Rabbani Esfahani (University of Alabama)*

Dyes are current threats to human and environmental health and pose a risk to the safety of groundwater and surface water. The main adverse properties of dyes are high toxicity, bioaccumulation, persistence in the environment, and are known to be endocrine disruptors and carcinogens. Presence of dyes in water decreases the gas solubility which consequently kills living species in water and also -has a mutagenic effect on humans. Effluent treatment and water-recycling can play a significant role in reducing discharge pollution and providing reusable process water. In this study, we developed a surface modified-polyamide nanofiltration (NF) membrane by coating polydopamine layer incorporated with the copper-based metal-organic framework (MOF). Two different methods, (1) dip-coating and (2) In-situ filtration processes applied for surface modification of polyamide nanofiltration membranes. The membranes and fabricated Cu-MOF nanoparticle structure characterized by scanning electron microscopy (SEM), transmission electron microscopy (TEM), Fourier-transform infrared spectroscopy (FTIR), X-ray powder diffraction (XRD), X-ray photoelectron spectroscopy (XPS), Atomic Force Microscope (AFM), Ultraviolet-visible spectroscopy (UV-Vis), and Zeta potential. A cross-flow filtration unit employed for dye removal, antifouling and permeation examination of the modified membranes. The flux-rejection trade-off was highlighted based on the fabrication techniques. However, the addition of hydrophilic Cu-MOF nanoparticles increased the water flux where the dip-coated membranes showed higher (up to 90% rejection compared to pristine polyamide NF membrane) dye rejection due to the thicker active layer.

C30 Advancing Conductivity-Permselectivity Tradeoff of Ion-Exchange Membranes with Sulfonated CNT Nanocomposites, Hanqing Fan (Columbia University)*; Yuxuan Huang (Columbia University); Ngai Yin Yip (Columbia University)

Ion exchange membranes (IEMs) are widely applied in energy and water technologies, such as reverse electrodialysis for sustainable power generation and electrodialysis for desalination. An IEM with lower ionic resistance and better selectivity for counterions can reduce the energy demand and raise the efficiency of IEM-based processes. However, an empirically-observed tradeoff between conductivity and permselectivity always constrains the performance of IEMs. The incorporation of rationally functionalized 1-dimensional nanomaterials as filler into the polymer matrix offers opportunities to depart from this tradeoff. When highly dispersed, the 1-D nanomaterials with large aspect ratio can form a percolating network to facilitate ion transport. In this study, we developed nanocomposite cation exchange membranes (CEMs) by incorporating sulfonic acid-functionalized carbon nanotubes (sCNTs) in sulfonated poly(2,6-dimethyl-1,4-phenyleneoxide), sPPO, polymer matrix. The fabricated nanocomposite IEMs exhibit improved conductivity while maintaining permselectivity. Intrinsic resistivity, the reciprocal of conductivity, is lowered with greater blending of sCNTs filler into the sPPO matrix, decreasing by approximately 25% with 20 w/w% incorporation of sCNTs. The variation of permselectivity is within 3% across the different degrees of sCNT incorporation. Therefore, compared with pristine membranes, the conductivity-permselectivity tradeoff line of the fabricated nanocomposite membranes are advantageously advanced, improving overall performance. Enhancement in conductivity is more pronounced for membranes with lower swelling degree. We posit that the interconnected network of sCNT reduces the tortuosity of the ion diffusion path, thereby increasing effective ionic conductivity. This study demonstrates the potential of rational utilization of nanomaterials to advance the conductivity-permselectivity tradeoff governing conventional IEMs.

C31 Ceramic-supported thin film composite membrane for organic solvent reverse osmosis, Mi Zhang (University of Connecticut)*; Lingling Xia (University of Connecticut); Jeffrey R McCutcheon (University of Connecticut); Marcus Weyd (Fraunhofer-Institut für Keramische Technologien und Systeme)

Membrane separations has often been sought as a replacement for thermal separations processes due to the potential 10X reduction in energy consumption. Reductions of this magnitude were realized when reverse osmosis replaced thermal distillation for desalination applications. It is widely believed that such benefits would translate to solvent separations as well. Organic solvent reverse osmosis (OSRO) is an emergent area of membrane separations that offers a membrane-mediated separation between miscible solvent mixtures. We use a recently developed polyamide thin film composite (TFC) platform to evaluate OSRO. These TFC membranes were made by forming the polyamide selective layer via interfacial polymerization onto a ceramic tubular ultrafiltration membrane. The TFC membrane was then tested in the mixtures of methanol (32.04 g/mol) with 1 w/w% decane (142.29 g/mol), and hexane (86.18 g/mol). These membranes exhibited decane and hexane rejections of 85% and 42%, respectively.

C32 Dynamic FO tests of commercial low pressure RO TFC membrane, Du Bai (University of Ottawa); Farhad Asempour (University of Ottawa); Boguslaw Kruczek (University of Ottawa)*

Reverse osmosis (RO) is already a mature membrane separation technology, which has dominated seawater and brackish water desalination industries. Nevertheless, the quest to improve the properties of thin film composite (TFC) and thin film nanocomposite (TFN) membranes for RO, forward osmosis (FO) and nanofiltration (NF) applications continues to be an important research topic. In turn, this requires appropriate testing tools to assess the effect of modifications of TFC/TFN membranes on their structure and performance. We have designed and built a novel testing system that allows characterization TFC/TFN membranes in a dynamic test, similar to the time-lag method commonly used to characterize gas separation membranes. The dynamic experiment in the new system is initiated by a step change in the draw solution concentration, which leads to a step change in the osmotic pressure across the membrane. Since the dynamic experiments are carried under zero hydraulic pressure gradient, they are referred as dynamic FO experiments. The capability of the new testing system and the dynamic FO experiments is demonstrated by characterizing commercial, low pressure TFC RO membranes (Toray Industries Inc.). In this presentation we will discus the effect of membrane orientation (active layer facing draw solution vs active layer facing feed solution) on the dynamic behaviour and the steady-state water and reverse salt fluxes. The effect of magnitude of the step change in draw solution concentration will also be discussed.

C33 New generation recycled polystyrene membranes for industrial water purification using membrane distillation, Machawe M Motsa (University of South Africa)*; Bhekie Mamba (University of South Africa)

Polystyrene is widely used as a packaging material this is poorly disposed after use. This material possesses several properties that could be ideal for membrane distillation membranes such as light weight, hydrophobicity, poor thermal conductivity and poor resistance to air permeance. Membrane filtration is not a common practise in most Southern African countries due to its heavy reliance on energy. Therefore, this work sought to develop membrane materials that will address both pollution by polystyrene and the costs incurred during buying commercial membranes. This was achieved by preparing new generation basic hydrophobic membranes from non-traditional polymers consisting of unique structural configurations for efficient mass transfer, thermal stability and good mechanical strength. Membrane characterization with scanning electron (SEM) and atomic force microscopy (AFM) revealed an asymmetric cross-sectional structure preceded by a thin top layer followed by a loosely arranged mesoporous layer largely characterised by microvoids. The membrane surface was found to be rough (SA(nm) = 297 ± 8.21) which enhanced it hydrophobicity. It also exhibited an average weak positive charge brought about by the modification with hydrophobic surfactants. Measured water contact angles and subsequent interfacial free energy confirmed the extreme hydrophobicity of the surface (CA> 117°). The membranes' separation properties were tested using a solar driven contact membrane distillation system for the purification of power generation cooling towers water with a total dissolved solute content of 1203 ppm. The recorded ion rejection was 97% at a temperature gradient of 50°C. The membrane exhibited good anti-scaling properties and permeate flux was stable over the 120 h experiment duration. The membranes presented a promising alternative to the treatment of industrial wastewater besides reverse osmosis and could be further optimised by various surface modifications.

C34 Advanced supported ionic liquid (SIL) membranes for Pi electron cloud mediated separation of aromatics, Mahmood Jebur (University of Arkansas); Arijit Sengupta (University of Arkansas); Ranil Wickramasinghe (University of Arkansas)*

Development of membranes for highly specific separations is challenging and will depend on specific surface interaction. Here we have investigated the feasibility of using Pi election cloud interactions. Imidazolium-based ionic liquids were synthesized using vinyl imidazole and three different alkyl halides (chlorobutane, bromohexane, and allyl bromide). A 0.45 micron pore size PTFE membrane was used as the support structure. The imidazolium-based ionic liquids were pumped into the membrane pores using a stirred cell at a pressure of 2 bar producing a supported ionic liquid membrane. The supported ionic liquid membrane could be used for specific Pi electron cloud-mediated separation of solutes. Here the separation of styrene, divinyl benzene and toluene in hexane (non-aqueous medium) and Ramazol brilliant blue R (RBBR), Congo red (CR) and Eichrome black T (EBT) from aqueous medium have been investigated. The specificity for separation of aromatics was found to be related to the affinity of the ionic liquid towards the Pi electron cloud density and extent of conjugation in the solute. The separation efficiency for all SIL membranes was observed to be divinyl benzene > styrene > toluene, while in case of dyes the complicated electronic configuration led to Congo red (CR) having the highest selectivity. It was also shown that the SIL membranes displayed antibacterial activity. The antibacterial activity was investigated using two-gram positive bacteria (Pseudomonas aeruginosa and Staphylococcus aureus). SIL membranes with longer carbon side chains on the imidazolium ring showed greater antibacterial activity (bromohexane > chlorobutane > allyl bromide). The results obtained here indicate that membranes for highly specific separations could be developed based on Pi electron cloud interactions.

C35 Graphene Oxide Membranes for Selective Molecular Separation of Lignin Model Compounds, Ashish Aher (University of Kentucky)*; DB Bhattacharyya (University of Kentucky)

In the quest for selective separation of value lignin monomers, the performance of Graphene Oxide-based (GO) membranes was investigated with model lignin monomeric compounds. GO membranes were fabricated on a commercially available polyvinylidene fluoride support (PVDF) by casting aqueous GO dispersion using wire wound rod. The degree of reduction of GO was controlled to tune the performance of GO membranes towards selective separation. GO membranes were thoroughly characterized to establish the chemical state of GO and to establish membrane performance parameters. Efficacy of these membranes towards rejection and separation of value lignin model compounds (dimers and trimers) was verified. Impressive performance with the rejection of over 70% for the model compound BMP (trimer-2,6-Bis[(2-hydroxy-5-methyl phenyl)methyl]-4-methylphenol) was achieved compared to only 20% rejection for GGE (dimer-Guaiacylglycerol-β-guaiacyl ether) with isopropanol-water (90%-10% by volume) mixture as solvent. This corresponds to an encouraging selective separation with selective permeation of GGE (dimer) 3.5 times compared to BMP (trimer). We hypothesize that the steric hindrance effect plays the most important role in the excellent molecular rejection and separation performance of our membranes. Moreover, the controlled reduction of GO membranes by varying thermal incubation time was achieved. A decline in the rejection rate with increasing extent of reduction was observed which was explained based on the higher partitioning of lignin-derived molecules on the more hydrophobic domain. The present contribution identifies the impressive performance of GO membranes for selective separation of value lignin monomers in polar organic solvent media.

C36 High Flux and Anti-fouling Electrospun Nylon 6 Ultrafiltration Membrane Coated with GO/N-carboxyethylchitosan (NCECS) Functionalized Polyvinyl Alcohol Hydrogel, Liuqing Yang (Concordia University)*; Tiantian Chen (Concordia University); Zhibin Ye (Concordia University); Benoit Barbeau (Polytechnique Montreal); Saifur Rahaman (Concordia University)

Highly hydrophilic polymers have been widely used to improve flux and anti-fouling properties of ultrafiltration (UF) membranes. In this study, a new nylon 6 UF membrane was fabricated by using electrospinning technique, coated with GO/N-carboxyethylchitosan (NCECS) functionalized polyvinyl alcohol (PVA) hydrogel. N-carboxyethylchitosan (NCECS) was used for the first time for the fabrication of UF membranes due to its high hydrophilicity. Electrospun Nylon 6 membranes coated with PVA containing different NCECS were fabricated and characterized via SEM, Raman spectroscopy, and contact angle measurements. Results showed that uniform pore size distribution and increased hydrophilicity were achieved. The flux and anti-fouling properties of fabricated membranes were investigated by comparing with the pristine UF membrane. The increased flux and anti-fouling performance exhibit the potential of using N-carboxyethylchitosan (NCECS) for the fabrication of UF membranes used in wastewater treatment.

C37 Development and Characterization of nZVI/TiO_2 Supported Membranes for Photocatalytic Treatment of 1,4-Dioxane in Water, Larissa L. S. Silva (Federal University of Rio de Janeiro)*; Wael Abdelraheem (University of Cincinnati); Minghao Kon (University of Cincinnati); Mallikarjuna Nadagouda (Wright State University); Ana Maria Rocco (Federal University of Rio de Janeiro); Cristiano Borges (Federal University of Rio de Janeiro); Fabiana Fonseca (Federal University of Rio de Janeiro); Dionysios Dionysiou (University of Cincinnati)

Using nano-zero-valent iron (nZVI) as a catalyst in Fenton processes for water treatment has recently gained much attention due to its environmentally friendly properties and excellent catalytic performance. However, the problem of aggregation of nZVI particles in aqueous solution limits its practical application in water treatment. It's believed that immobilization of nZVI on membrane followed by impregnating TiO_2 nanoparticles on top of them could improve the stability of nZVI in aqueous solution. Moreover, due to the photocatalytic properties of TiO_2, the overall degradation power of TiO_2/nZVI system will be enhanced. 1,4-Dioxane (DXN) is an emerging contaminant that is found in various water distribution systems at relatively high concentrations of 0.07-9.1 μg/L. Owing to its deleterious health implications, DXN has raised an urgent need for its removal from water systems. Thus, this work aims at applying an innovative photocatalytic treatment technology based on UV/TiO_2/nZVI/membrane for the treatment of the DXN in water. Hydrophilic membranes were synthesized and utilized as a support for nZVI and TiO_2 nanoparticles. Various methods of impregnation of TiO_2 onto the nZVI/membranes were also investigated. The modified membranes were characterized by Scanning Electron Microscope (SEM), Fourier-transform Infrared Spectroscopy (FTIR), X-ray Powder Diffraction (XRD), Energy Dispersive X-ray (EDX), mass gain and total iron lost. DXN was successfully degraded by the current hybrid photocatalytic system. Results revealed DXN degradation by TiO_2/nZVI/membrane was higher than that achieved by nZVI/membrane and TiO_2/membrane systems. Furthermore, the membrane's stability during the degradation process was enhanced due to the presence of TiO_2. The current study provides important mechanism information on the degradation of DXN by the suggested system and justifies the improved photocatalytic properties of TiO_2/nZVI membranes based on AOPs.

C38 Oxone Mediated TEMPO-Oxidized Nanocellulose Form-I and Form-II: Materials, Membranes and Modeling, John P Moore (University of Arkansas)*

Nanotechnology in science has shown small materials can be used for novel and versatile applications. In the past few years, nanocellulose (and new derivatives thereof) has become a growing interest in nanomaterial research. Nanocellulose has shown to have excellent hydrophilic and tensile properties allowing for the tuning of membrane filtration performance. While, improvements using nanocellulose materials has seen success in the membrane field, more work is needed to lead the advance of natural fiber-based nanomaterials. The focus of this publication is to identify the parameters of interest to characterize the performance enhancing effect and antifouling capabilities of novel nanocellulose material derivatives (Oxone mediated TEMPO-oxidized nanocellulose form I and form II) on ultrafiltration membranes. Membranes are characterized by means of separation performance, physical properties, and morphology. Modified membranes displayed increased hydrophilicity, increased tensile capacity, decreased membrane resistance, and decreased fouling at the surface due to adsorption/concentration polarization.

C39 Facile Fabrication of Amphiphobic Nanocomposite Membranes for Application in Membrane Distillation, Tiantian Chen (Concordia University); Liuqing Yang (Concordia University); Saifur Rahaman (Concordia University)*

Two superhydrophobic and amphiphobic membranes, which can repel both water and low surface tension liquids (oil), are fabricated via electrospinning technique and subsequent surface modifications. Highly hydrophobic nanofiber mats are first prepared by electrospinning a mixture of poly(vinylidene fluoride-co-hexafluoropropylene) and reduced graphene oxide. Surface superhydrophobicity and amphiphobicity are implemented by grafting fluoroalkylsilane of low surface energy or hydrophobic silica nanoparticles. SEM, XPS, liquid entry pressure, and contact angle measurements are applied to investigate the effects of the surface modification on the morphology and structure of the membranes. The analyses reveal changes in the surface roughness, elemental composition, and hydrophobicity of the electrospun nanofibers. Both membranes show excellent superhydrophobicity and amphiphobicity, clearly demonstrated by their wetting resistance with water and low surface tension organic solvents. These promising membranes are tested to desalinate water from 35 g/L sodium chloride solution in the presence of a surfactant (sodium dodecyl sulfate) through a direct contact membrane distillation (DCMD) process. Our membranes exhibit enhanced stability and durability of MD performance with high permeation flux (~42 kg/m^{2}h) and 99.99% salt rejection

C40 Graphene Oxide Nanoplatelets Embedded Polyamide Thin Films for Water Desalination, Mahsa Abbaszadeh (Mississippi State University)*; Santanu Kundu (Mississippi State University)

Polyamide thin film membranes are used in water desalination applications. Incorporation of graphene oxide nanoplatelets in the polyamide layer can alter the surface characteristic, enhance the anti-chlorination properties, and can improve the permeability and selectivity of the membranes. However, a standard and reproducible fabrication strategy to incorporate nanoplatelets/nanoparticles in the polyamide membranes is yet to be developed. In this study, a layer-by-layer synthesis technique has been used for embedding graphene oxide nanoplatelets (GONPs) in the polyamide layer. Polyamide layers with GONPs were constructed as alternating layers of GONPs and polyamide, and GONPs on top of the polyamide layer. Incorporation of GO resulted in an increase of surface hydrophilicity, as captured by the change of contact angle. The presence of GONPs in the synthesized membranes has been confirmed by XPS, SEM, TEM, and AFM. Water flux and salt rejection properties of synthesized membranes in chlorine and non-chlorine conditions have been investigated by using a dead-end cell. The GONPs containing membranes display significant improvement in the chlorine resistance properties in comparison to the pristine polyamide membrane. In addition, a small enhancement in the salt rejection properties has been observed for the GONP containing membranes.

C41 Development of Isoporous 0.2 Micron Nanoslit Silicon Nitride Membranes for Sterile Filtration of Biological Therapeutics, Evan Wright (McMaster University), Joshua Miller (Simpore), Andrew Gosselin (Simpore), Jared Carter (Simpore), James McGrath (University of Rochester), David Latulippe (McMaster University), James Roussie (Simpore)*

Current Good Manufacturing Practice (cGMP) standards for the production of biological therapeutics rely on sterile filtration processes using microporous polymeric membranes to remove microbial contaminants and product aggregates in order to ensure the safety of the final formulation. Existing polymeric sterile filters are not ideal for larger biological therapeutics such as antibody-particle conjugates, oncolytic viruses, or extracellular vesicles, as these often become entrapped within the high internal surface area of such polymer membranes. The resulting yield loss increases production costs of these therapeutics, and therefore, new process technologies are required to optimize the production of large-sized biological therapeutics. Here, we report on the development of nanoslit silicon nitride (NSN) membranes with isoporous sub-0.2 μm rectangular prism "pores" for addressing the need for improved biological therapeutic sterile filtration. We demonstrate fabrication processes that yield NSN membranes that are ultrathin (200 nm thick), exhibit high permeability (~33% porosity), and precision-patterned slit openings (0.19 μm X 10 μm). We characterized these newly developed membranes for their gas and hydraulic permeability, differential pressure tolerance, and model solute retention and transmission. Most importantly, we challenged these membranes with 10^7 B. diminuta per cm^2 of NSN membrane and obtained complete retention of bacteria, demonstrating similar bioburden reduction properties to those of commercially available 0.22 μm sterile filters. Given these results, we anticipate that NSN membranes will enable sterile filtration of emerging large-sized biological therapeutics, resulting in improved yield during cGMP manufacturing.

SESSION D - Innovation in Membranes for Energy Applications

D1 **A membrane to separate light olefins and paraffins; an update on pilot trials, module design and C4 separations,** Ken Loprete (Compact Membrane Systems)*; Bill Charlton (Compact Membrane Systems); Hannah Murnen (Compact Membrane Systems); Kenneth Pennisi (Compact Membrane Systems); Sudip Majumdar (Compact Membrane Systems)

The separation of light olefins from paraffins is one of the most energy intensive separations in the chemical processing industries. The state of the art separation method is distillation, a technology that has remained unchanged over decades. Membranes hold the promise of a more modular, energy efficient and capital efficient method for separation. However, they have historically been hampered by low performance and stability. Compact Membrane Systems has developed a novel membrane material using a silver embedded fluoropolymer material to transport olefins across the membrane. This material has demonstrated over 2 years of longevity in laboratory operation and has shown high selectivity and permeance under realistic operating conditions. During the past year, CMS has been operating a pilot utilizing this membrane material in a spiral wound element at the Delaware City Refining Company. We will share the latest results from that pilot operation as well as the update on the scale-up of our membrane modules. The Delaware City pilot is operating on a low concentration olefin stream (i.e., ~10-15% propylene balance propane in the feed) with the goal of achieving an HD5 quality propane product and an enriched propylene stream. We will discuss the improvements we have made to module design that have enabled high performance at realistic field conditions. In addition, we will demonstrate stability for over 60 days in the field. CMS will also discuss initial results on laboratory testing of C4 streams. This includes mixed gas separation of n-butane and 1-butene with permeances between 200-500 GPU and selectivities greater than 20. We will share results for a number of operating conditions and will highlight several exciting C4 separation applications including those with isomer streams. In addition, longevity results for >30 days of C4 operation will be shared demonstrating the stability of the membrane in realistic streams.

D2 **Cross-linked Polyethyleneimine (XLPEI) with superior H_2/CO_2 separation properties,** Sankhajit Pal (The State University of New York at Buffalo)*; Weiguang Jia (The State University of New York at Buffalo); Haiqing Lin (The State University of New York at Buffalo)

Membranes with superior H_2/CO_2 separation properties are needed for H_2 purification and CO_2 capture from the syngas derived from coal. Polyethylenimine (PEI) has been widely studied for CO_2 sorption, due to the strong interaction of amines with CO_2. This study demonstrates that the strong interactions between the CO_2 and PEI decrease the CO_2 diffusivity in the polymer, thus increasing H_2/CO_2 permeability selectivity. Specifically, PEIs were cross-linked with various amounts of hexamethylene diisocyanate (HMDI) to form free-standing films. The effect of cross-linking density on the properties (including gel content, density and glass transition temperature) is systematically investigated. As the HMDI content increases from 0 to 35 wt% in the film, the Tg increases from -52 to -4.8 °C. At a weight ratio of 65:35 for PEI: HMDI, the resulting film exhibits a H_2 permeability of 1.0 Barrers and H_2/CO_2 selectivity of 30 at 150 psig and 35 °C. Such selectivity is the highest reported for polymers for H_2/CO_2 separation. These results contradict the conventional wisdom that adding amine groups in polymers can increase CO_2 permeability.

D3 **Self-Cleaning Nanocomposite Membranes with Phosphorene-Based Pore Fillers for Water Treatment,** Joyner Eke (University of Kentucky)*; Isabel Escobar (University of Kentucky)

Photocatalytic membrane processes generate oxygen-reactive radicals under irradiation by light that could degrade pollutants in feed solutions. Phosphorene is a two-dimensional material made by exfoliating black phosphorus, a monotypic van der Waals crystal. Unlike other common two-dimensional materials, under light irradiation, phosphorene can generate reactive oxygen species (singlet oxygen, hydroxyl radicals and superoxide radicals). In this study, we exfoliated black phosphorus to synthesize phosphorene. Then immobilized the phosphorene nanomaterial into the membrane (sulfonated poly ether ether ketone and polysulfone blend). Studies were done under ultraviolet light irradiation. The dye, methylene blue was used as feed solution. Membranes were comprehensively characterized before and after modification. Results obtained showed enhanced flux recovery for the phosphorene modified membranes. Fluorescent studies also revealed lower coverage of methylene blue on these membranes as compared to the unmodified membranes. Mechanism of photocatalysis will be investigated and reaction equations proposed.

D4 Scale-up of A Catalytic Membrane Reactor for the Production of 5-Hydroxymethylfurfural from Lignocellulosic Biomass, Tanmoy Patra (University of Arkansas); Robert Beitle (University of Arkansas); Richard J. Ciora, Jr. (Media and Process Technology, Inc.); Ranil Wickramasinghe (University of Arkansas); Xianghong Qian (University of Arkansas)*

Significant efforts have been dedicated to developing economically viable processes for converting lignocellulosic biomass to biofuels, bio-based products and intermediates. 5-hydroxymethylfurfural (HMF) is a versatile platform chemical that can be produced from lignocellulosic biomass. The commercial viability for the production of HMF lies in its cost-effective production and separation from the reaction stream. This work explores the potential for development of a prototype catalytic membrane reactor for the production and separation of HMF using agricultural residues such as wheat and rice straws, as well as cornstover biomass. Our previous results have demonstrated the feasibility of producing HMF at the 0.01-0.1 L scale with over 70% yield from cellulose and pre-treated cornstover substrates in ionic liquid (IL), IL/gamma-valerolactone (GVL) and IL/H_2O mixed solvents. Our designed polymeric solid acid catalysts were immobilized on porous membrane substrates to increase the surface area and enable product separation. These catalysts consist of two adjacent polymer nanostructures, one sulfonic acid chain for catalyzing the biomass substrate, one poly (ionic liquid) chain for solubilizing cellulose and enhancing the catalytic activity. Here a semi-batch catalytic membrane reactor was designed and tested for the production and separation of HMF simultaneously in IL/H_2O mixed solvents at mild temperature of 110-140°C. Over 50% of HMF yield from cornstover biomass has been achieved.

D5 Thin-Film Composite Hollow Fiber Membrane for Post-Combustion Carbon Capture, Shouliang Yi (U.S. Department of Energy National Energy Technology Laboratory)*; James S Baker (U.S. Department of Energy National Energy Technology Laboratory); Ali Sekizkardes (Battelle/U.S. Department of Energy National Energy Technology Laboratory); Surendar Venna (U.S. Department of Energy National Energy Technology Laboratory); Victor A Kusuma (U.S. Department of Energy National Energy Technology Laboratory); Fangming Xiang (U.S. Department of Energy National Energy Technology Laboratory); Lingxiang Zhu (U.S. Department of Energy National Energy Technology Laboratory); Hyuk Kwon (U.S. Department of Energy National Energy Technology Laboratory); David Hopkinson (U.S. Department of Energy National Energy Technology Laboratory); Kevin Resnik (U.S. Department of Energy National Energy Technology Laboratory)

The National Energy Technology Laboratory (NETL) is developing the next generation of advanced CO_2 capture concepts. Post-combustion carbon capture from flue gas involves the separation of CO_2 (~10-16%) from nitrogen, the primary constituent of flue gas. Membrane and membrane-based processes have been considered as one of the most promising technologies for post-combustion carbon capture due to numerous advantages. Specifically, polymeric membranes have the additional benefits of being inexpensive and easy to manufacture. For industrial applications, asymmetric and/or composite membranes with a thin skin layer are typically used to achieve a high permeation flux. Among the various designs of membrane modules, hollow fiber membranes are preferred due to high packing density and self-supporting characteristics. In this presentation, Torlon, a commercial polyimide-amide polymer, was used to fabricate the highly porous substrate to develop high performance thin-film composite (TFC) hollow fiber membranes due to its exceptional mechanical, thermal and chemical resistant properties. Several advanced polymer materials including polymers of intrinsic microporosity (PIMs) and ultra-high molecular weight polyvinylamine were employed as a selective layer. The effects of dope compositions, spinning parameters, and dip-coating procedures were investigated in detail. The results showed that the CO_2 permeance of PIM-based TFC hollow fibers was 1500-2000 GPU with a CO_2/N_2 selectivity of 15-18 as measured using dry gas. The ultra-high molecular weight polyvinylamine-based TFC hollow fiber membranes demonstrated a CO_2 permeance of 50-60 GPU with CO_2/N_2 selectivity of ~50 under mixed gas with a composition of 14%CO_2/86%N_2 at 22 °C with humidity. We will review the effect of fabrication procedures for these composite hollow fiber membranes on the gas transport performance properties, mechanical properties, membrane morphology, durability, and applicability to future scale-up.

D6 Salinity gradient heat engine: optimising reverse electrodialysis for thermal-to-electric, Anna Hulme (Cranfield University)*

The Nano Membrane Toilet is a decentralised on-site sanitation system which requires an off-grid power supply to treat wastewater. Combustion of faeces within the system provides a reliable source of heat energy. However existing thermal-to-electric technologies are expensive and inefficient for conversion of heat below 100 °C. A salinity gradient heat engine consisting of a reverse electrodialysis (RED) stack and a thermal separation stage could offer improved efficiencies. RED utilises the Gibbs free energy of mixing solutions with a concentration gradient across ion exchange membranes to generate a voltage. Membrane distillation is a thermally-driven separation process which restores the concentration gradient of the RED feeds using low-grade heat to close the loop. Research on RED has focussed on large-scale power production using seawater and river water. For a small-scale heat engine, power output from a finite volume of feed solution must be maximised. The use of natural feeds with relatively low salinity gradients limits potential power densities. By contrast, use of artificial saline solutions enables optimal conditions for thermal-to-electric to be selected. A laboratory-scale five cell pair reverse electrodialysis stack was used to investigate the effect of process conditions on power density and energy efficiency in single pass and recycle. The effect of feed concentration, flow rate, temperature and current density on water and salt transport within the stack was quantified. Increasing residence time within the stack increased energy efficiency. Recycling feeds led to the highest energy efficiency. Despite increased water transport, high salinity gradients recovered the most work per volume of solution. This demonstrates the suitability of increased salinity gradients in reverse electrodialysis for thermal-to-electric applications. Selection of current density is critical to maximise energy efficiency and manage water transport.

D7 Use of desalination brine for low-energy concentration of orange juice via forward osmosis, Haley D White (Tennessee Technological University)*; Leif Templeton (Tennessee Technological University); Shelby Jones (Tennessee Technological University); Laura H Arias Chavez (Tennessee Technological University)

Membrane technology has the potential to enhance sustainability at the food-energy-water nexus by reclaiming both energy and materials from 'waste' streams. Forward osmosis (FO) is a selective separation process that could facilitate such reclamation. It requires minimal energy when draw solution regeneration is avoided. To do so, concentrated waste streams can be used as single-use draw solutions in certain applications. We investigated a possible implementation of this strategy: concentrating liquid commodities using seawater desalination brine as a draw solution. Seawater desalination produces ~50 million m^3/day of brine worldwide, representing 53 GWh/day in wasted chemical potential energy. Harnessing this energy for osmotic concentration can displace the use of prime energy for thermal concentration. Orange juice was selected as a model commodity due to the proximity of citrus production to seawater desalination in Florida. Juice concentration was conducted in a bench-scale FO system using commercial thin-film composite membranes and NaCl draw solutions simulating 50% or 60% recovered seawater desalination brine. Juice recovery reached 55% with water flux maintained at 5 to 8 L m^{-2} h^{-1}. Fouling was reversible, with up to 98% of initial water flux regained after physical cleaning. Juice concentrates met quality standards for recoveries of up to 32%. At higher recoveries, sugar levels declined, likely due to fermentation during long experiment times in the bench-scale system. Transport of both draw solutes and juice components through the membrane increased over time, perhaps due to stretching or tearing of the membranes, which were not supported by spacers. From our data, we predict that modest fluxes can be maintained up to the juice industry requirement of 80% recovery, with desalination brine as the only significant source of energy. Despite bench-scale limitations, this study shows the potential of 'waste' streams to power relevant separations.

D8 High-performance SSZ-13 membranes for CO_2 and N_2 removals from natural gas, Rongfei Zhou (Nanjing Technical University)*; Xinping LI (Nanjing Technical University); Bin Wang (Nanjing Technical University)

Removal of the undesirable impurities such as CO_2 and N_2 from natural gas is a significant processing because that the impurity gases decrease the energy content of the gas and CO_2 can corrode pipelines in the presence of humidity. Zeolite membranes possess superior permeances and chemical, thermal and high-pressure stabilities. A SSZ-13 zeolite (CHA) with a pore size of 3.80 Å × 3.80 Å is a promising membrane candidate for separating CO_2 (3.30 Å) and N_2 (3.64 Å) from slightly larger molecule CH_4 (3.80 Å) due to its molecular sieving ability when it is made in membrane. High quality SSZ-13 membranes with improved separation performance were prepared on the asymmetric α-alumina tubes by a single hydrothermal secondary step. Some modifications were made for membrane synthesis in this current study. The SiO_2/Al_2O_3 ratio in the membrane gel were improved from 40 in our previous study (J. Membr. Sci., 475, 303-310, 2015) to 200 in order to improve the hydrophobic performance against the humidity. Ball-milled SSZ-13 seeds were used in this study to improve the induction ability of seed layer. Moreover, membrane calcination in pure oxygen at lower temperature was carried out for the current membrane instead of in air at higher temperature for our previous SSZ-13 membrane. The current SSZ-13 membrane had a CO_2 permeance of 1.20×10^{-6} mol/(m^2 s Pa) (equal to 3600 GPU) with CO_2/CH_4 selectivity of 300 and a N_2 permeance of 1.10×10^{-7} mol/(m^2 s Pa) (equal to 330 GPU) with N_2/CH_4 selectivity of 13 at 298 K and 0.2 MPa for equimolar CO_2/CH_4 and N_2/CH_4 mixture, respectively. Such separation performances of the membrane were beyond those of most of the membranes. Membrane stabilities for a long-term test and in the wet mixture were also investigated. The current high-silica SSZ-13 membrane has excellent potentials for CO_2 and N_2 removal in natural gas purification as a facile separation process.

D9 Highly efficient CO_2 capture by mixed matrix membranes containing three-dimensional covalent organic framework fillers, Youdong Cheng (National University of Singapore)*; Dan Zhao (National University of Singapore)

Mixed matrix membranes (MMMs) have long been considered as promising membrane types for industrial energy-intensive gas separation processes. Current MMMs are still facing grand challenges of poor filler dispersion and poor polymer-filler interfacial compatibility. The present study demonstrates that these challenges can be addressed by fabricating MMMs containing three-dimensional (3D) covalent organic framework (COF) fillers with ultrasmall size-selective pores. Two different polymer matrixes, including glassy 6FDA-DAM and rubbery Pebax, are explored to validate the effectiveness of 3D COF fillers in improving the membrane separation performance. The pure organic nature of COFs confers them with high affinity with pure organic polymer matrixes, leading to good interfacial compatibility in the resultant MMMs. These porous COF-300 fillers can increase the membrane fractional free volume that enhances the membrane gas permeability. Besides, the ultrasmall pores of COF-300 fillers and the rigidified polymer chains at the filler surface can enhance the size discriminative processes, resulting in increased membrane gas pair selectivity. Moreover, the separation performance of COF-300 can be further improved by functionalizing it with polyethylenimine (PEI), enabling the design of advanced membranes suitable for industrial applications.

D10 Industrially relevant membrane testing and novel mixed-matrix membranes for a brighter membrane future, Raymond Thür (Katholieke Universiteit Leuven)*

In an attempt to pass the Robeson upperbound, mixed matrix membranes have been designed to combine the good separation potential of inorganic nanoparticles (e.g. MOFs) with cheap, well-processable polymers. Zirconium-based MOF are considered to be among the most promising materials for MMM synthesis, regarding their low cost, very good stability, easy synthesis in green conditions and functionalization. Two new Zr-MOFs were investigated in their potential for use in MMMs: (a) Bipyridine-based UiO-67 was used as filler Matrimid MMMs. The combination of bipyridine UiO-67 with Matrimid led to doubled CO_2/CH_4 mixed gas selectivity (from 38 to 75) while CO_2 permeability was enhanced by 63%. Characterization of fillers and membranes showed that the MOFs affinity for CO_2 was increased by incorporating bipyridine-linkers and that these Lewis basic sites function as CO_2 carriers. Gas permeation experiments suggested a facilitated transport mechanism causing these high selectivities. (b) A relatively new, extremely stable Zr-MOF (MOF-808), which can be tuned for enhanced CO_2 affinity was tested in 6FDA-based polymer membranes. It was noticed that controlling the colloidal stability of the MOF-polymer-solvent system via various strategies is key to obtain successful MMMs. Even though the perspectives for membrane technology are promising, there currently exists a discrepancy between academic and industrial testing of membranes. Testing of membranes under industrially more relevant conditions is therefore of high importance. Our group developed a second generation high-throughput gas permeation set-up, which allows simultaneous testing of 8 membranes for CO_2 separation at pressures up to 34.5 bar and for olefin/paraffin separation up to 25 bar, also in presence of water vapour. In this way, industrial streams like flue gas, natural or bio gas and olefin/paraffin mixtures can be mimicked to gain better insight in the real separation performance of the membranes.

D11 Torlon® Hollow Fiber Membranes for Organic Solvent Reverse Osmosis separation of Complex Aromatic Hydrocarbon Mixtures, Hye Youn Y Jang (Georgia Institute of Technology)*; Ryan Lively (Georgia Institute of Technology); JR Johnson (ExxonMobil); Dhaval Bhandari (ExxonMobil)

In this work, defect-free polyamide-imide (Torlon®) hollow fiber membranes were fabricated with the aim of investigating the potential for polymer-based organic solvent reverse osmosis (OSRO) separations of complex aromatic hydrocarbon mixtures. The quality of the synthesized membrane was probed using gas permeation experiments. The membranes were found to have He/N_2 permselectivities of ~200, whereas the dense film permselectivities are known to be ~ 310, suggesting that these membranes are nearly defect-free. These membranes were found to be stable in various hydrocarbon liquids. and did not experience significant swelling upon exposure to these liquids. Considering the high quality of the membrane and the stability of the polymer, these membranes were utilized in OSRO separation without any post-treatment procedures. The membranes were challenged with a mixture of aromatic hydrocarbons that are typically found in light fractions of crude oil. These multi-component separation experiments revealed a MWCO of 160-200 Da. To clearly demonstrate the OSRO separation modality, we also challenged the membrane modules with more concentrated mixtures such as a 90/10 (mol %) mixture of toluene and 1, 3, 5-triisopropyl benzene. The osmotic pressure for this system is estimated to be 25bar. At an upstream pressure of 70-80 bar, we find that the solvent concentration in the permeate is 99.4-99.6 mole %, and that the TIPB rejection coefficient is approximately 94.6-95.7%. The solvent permeance in these membrane modules is low (0.01 L/m^2-h-bar). Nevertheless, the ability to reject solutes as small at 160-200 Da using commercial polymers and commercially-inspired spinning techniques suggests that these membranes have the potential to enable challenging solvent-solute or even solvent-solvent separations found in the hydrocarbon processing industry. This work provides a benchmark for OSRO separation of complex hydrocarbon mixture using polymeric hollow fiber membranes.

D12 Performance Investigation of Polymer Derived Ceramic Composite Membrane in Microbial Fuel Cell (MFC), Vignesh Ahilan (University of Bremen)*; Camila Cabral de Barros (University of Bremen); Gourav Dhar Bhowmick (Indian Institute of Technology); Makarand M Ghangrekar (Indian Institute of Technology); Michaela Wilhelm (University of Bremen); Kurosch Rezwan (University of Bremen)

Polymer derived ceramic (PDC) membranes were developed as a Proton conducting membrane material for Microbial Fuel Cell (MFC). The chemical composition was altered by incorporating graphitic filler materials such as Graphene oxide and Carbon nanotube in the polysiloxane precursor. These distinct compositions were converted at 600°C and 1100°C to ceramic membranes. Physical characteristics of the synthesized membranes such as porosity, hydrophilicity, mechanical stability, ion exchange capacity, and oxygen mass transfer coefficient were determined to investigate the best membrane material for further testing in MFCs. The open porosity of samples ranged from 30 % to 40 %, and the pore size was attained in the range between 0.1-1 μm. The membrane ion exchange capacity was significantly improved after adding 0.5 wt.% of GO. It increased drastically corresponding to an increment of 78.6 % with respect to that of the ceramic membrane without GO, while the oxygen mass transfer coefficient of the membrane decreased by 52.6 %. The performance of synthesized membranes was assessed in dual chamber MFC system in terms of power density generated, normalized energy recovery, Coloumbic efficiency and chemical oxygen demand (COD) removal efficiency. The maximum volumetric power density and COD removal efficiency of MFC with filler materials incorporated ceramic separator were 7.23W/m3 and 89 %, respectively. The power generation is 1.5 times higher than for a MFC conducted with a Nafion 117 membrane tested under similar condition. This study reveals the feasibility of using PDC membrane as energy efficient membrane material for the application in MFC.

D13 PIM-1 based polymeric blend membranes for gas separation applications, Ashley Miles (Oak Ridge Associated Universities/U.S. Department of Energy National Energy Technology Laboratory)*; Ali Sekizkardes (Battelle/U.S. Department of Energy National Energy Technology Laboratory); Janice Steckel (U.S. Department of Energy National Energy Technology Laboratory); David Hopkinson (U.S. Department of Energy National Energy Technology Laboratory)

Polymers of intrinsic microporosity (PIMs) have been widely studied as membranes for gas separation research. PIMs are known for high permeability, high free surface volume, and processability with aprotic solvents. However, PIM-casted membranes show poor mechanical properties, as they suffer from highly brittle film formation and low selectivity due to pore size (~1 nm). Here, we functionalize PIM-based polymers and blend them with compatible rubbery polymers to obtain superior membranes. In our process, we create binding sites that improve gas permeability and miscibility properties for the blended polymers. The resulting membranes show improved film formation, high CO_2 permeability (2500 Barrer) and high CO_2/N_2 selectivity (28). A description of the fabrication and characterization of these blend membranes will be presented. In addition, gas transport properties of blended membranes were studied for potential use in post-combustion flue gas separation applications.

D14 The dynamic fate of Organic micro-pollutants in MBR and their accumulation in fish tissue, Gal Dagan (Ben-Gurion University of the Negev)*; Inbal Zaibel (Ben-Gurion University of the Negev); Dina Zilberg (Ben-Gurion University of the Negev); Moshe Herzberg (Ben-Gurion University of the Negev)

Despite the extensive application of municipal recycled water for irrigation (in Israel 85% of the municipal wastewater), currently such use is not permitted for aquaculture. Previous studies on the bioaccumulation of organic compounds in fish cultured in municipal recycled water have demonstrated that heavy metals and organic contaminant levels in tissue were below the FDA limits. Organic micro-pollutants (OMPs) impact on fish is a key factor in determining the potential of using treated wastewater for aquaculture. The fate of OMPs strongly depends on processes including adsorption and biodegradation and the related operational parameters. Hence, there is a great potential in using recycled municipal wastewater from membrane bioreactors (MBR) effluents, for aquaculture. MBR is a wastewater treatment process that combines conventional biological treatment with membrane filtration for biomass retention. MBR allows sewage treatment at high mixed liquor suspended solids (MLSS) concentration, fixed sludge retention time (SRT) and hydraulic retention time (HRT), decoupled from one another. Our objectives were to evaluate the dynamic fate of two model OMPs, carbamazepine (CBZ) and diclofenac (DFC), in the MBR effluents and their related accumulation in fish tissues grown in MBR effluents. Intriguingly, the transient retention of CBZ and DFC in the MBR was mainly due to adsorption to the sludge and the membrane: CBZ has shown greater adsorption capacity to the sludge, whereas, DFC showed higher adsorption to the membrane. Currently, we quantify effects of these adsorption processes on the dynamic appearance of CBZ and DFC in the MBR effluents. In addition, accumulation of heavy metals and 21 OMPs compounds of interest in tissues of fish that were cultured in MBR effluents for 2 months, were below the maximum concentration allowed by the FDA. This study provides a premise of possibilities for aquaculture using MBR effluents under proper restricted conditions.

D15 Improving H_2 Utilization of Solid Oxide Fuel Cell Using CO_2-Selective Membrane, Kai Chen (Ohio State University)*; Witopo Salim (Ohio State University); Yang Han (Ohio State University); Mike Gasda (Bloom Energy Corporation); Winston Ho (Ohio State University)

Polymeric membranes with a high CO_2/H_2 selectivity are desired to recover H_2 from the anode exhaust of a high-temperature solid oxide fuel cell (SOFC) system for H_2 utilization. At a high temperature of 120°C, it is very challenging to maintain the high CO_2/H_2 selectivity. The membrane reported here contained a polyelectrolyte with quarternaryammonium ions as the cations bound to the polymer backbone. Fluoride was chosen as the counterion, which catalyzed the reaction between CO_2 and water. The composition also included a boric acid to further catalyze the reaction and a hydroxide-based mobile carrier for the facilitated transport of CO_2. The membrane showed a CO_2 permeance of 100 GPU and a CO_2/H_2 selectivity of 116 at 120°C, making the membrane suitable for the abovementioned application. Hydration of the membrane was also found to be important as water is involved in the facilitated transport of CO_2. Next, a high-level techno-economic analysis was conducted for the integration of the membrane process into the SOFC system. Parameters studied were the membrane transport performances, H_2 recovery, and feed to sweep flow ratio. The CO_2 removal cost using the membrane with a CO_2 permeance of 100 GPU and a CO_2/H_2 selectivity of 116 was calculated to be $62.8/tonne. Furthermore, in order to improve the process, a vacuum was proposed to be pulled on the permeate side so that the removed CO_2 could be captured.

D16 Adapting a Blood CO_2 Removal Device for CO_2 Removal from Flue Gas, Katherine Hornbostel (University of Pittsburgh)*

The purpose of this project is to investigate the use of hollow fiber membrane contactors (HFMC) currently used for CO_2 removal in medical applications for CO_2 separation from flue gas. HFMC's are hybrid devices that combine the advantages of membranes (high surface area, gas permeability) with the advantages of solvents (CO_2/N_2 selectivity, high CO_2 absorption rate). HFMC's can be used with nearly any low-viscosity CO_2 solvent, ranging from water to amines. Because of the intrinsically high surface area of HFMC's, cheap/abundant solvents (e.g. water, carbonate solution) can be used in lieu of traditional solvents (e.g. amines). This proposal couples the unique expertise of two professors at the University of Pittsburgh: Prof. Katherine Hornbostel (MEMS) and Prof. William Federspiel (BIOE, McGowan Institute for Regenerative Medicine). Hornbostel brings experience with modeling CO_2 separation using encapsulated solvents, while Federspiel brings experience with experimental HFMC research for blood oxygenation and CO_2 removal. The ultimate application for this technology would be CO_2 separation from coal or natural gas power plants. Initial CO_2 separation experiments have been performed on a wearable artificial lung device that is commonly used to remove CO_2 from the blood of COPD patients. In these experiments at room temperature and pressure, simulated flue gas was flowed through the hollow fibers, and water or Instant Ocean flowed on the other side. The results demonstrated up to 50% CO_2 removal, with best performance occurring at high liquid flow-rates with water and a larger device. Future tests will be done using NaOH and Na2CO3 solutions as the sweep fluid, which should significantly boost CO_2 mass transfer rates. Future experiments will also study the impact of carbonic anhydrase coating, mechanical agitation, and higher temperature for raising CO_2 removal rates from simulated flue gas.

D17 Experimental study of dehumidification potential of PDMS hollow fiber membrane for HVAC applications, Michelle K Croal (Oakland University); Jonathan Maisonneuve (Oakland University)*

Energy consumption in buildings constitutes approximately 40 % of primary energy use in the USA, with heating, cooling and ventilation being sizable portions thereof. Furthermore, in hot and humid climates, latent cooling loads (i.e. dehumidification) can be almost equal to sensible cooling loads, if not greater. While desiccant systems allow for separating latent and sensible cooling loads, they still require thermal energy input to regenerate the saturated desiccant. Membrane dehumidification systems require minimal energy input, as they utilize humidity gradients to remove water vapor. In this study, a polydimethylsiloxane (PDMS) hollow fiber membrane module is tested under various isothermal flow conditions, with varying feed and sweep relative humidity. Latent energy removal varied widely, but was found at maximum to be 97% of psychrometric limit, presenting significant potential for energy savings in building applications. Net dehumidification efficiency ranged between 35 - 85%. Sensible heat transfer is also studied by considering fresh air intake at 35°C and exhaust sweep air at 20°C for additional energy recovery.

D18 The Perfluoropolymer Upper Bound, Albert Wu (MIT)*; James Drayton (MIT); Zach Smith (MIT)

Perfluoropolymers have fundamentally distinct thermodynamic partitioning characteristics compared to their hydrocarbon counterparts. The original upper bound theory for polymer gas separations assumes hydrocarbon solubility behavior for all polymers. In this study, the difference in solubility behavior for perfluoropolymers is explicitly considered to create a new level of design and predictability to transport performance in polymer membranes with diverse chemical characteristics. New theoretical perfluoropolymer upper bounds are established, showing a positive shift of the upper bound front as a result of improved solubility selectivity for certain gas pairs. Comparisons are made with current separation performance data. Using the framework of the solution-diffusion model, a deconvolution of transport and partitioning characteristics is evaluated. A new correlation is presented that relates penetrant properties and an adjustable parameter used in upper bound theory to predict the current upper bound position for a variety of gas pairs, including those not previously considered. Promising future applications for perfluoropolymers are discussed.

D19 Synthesis and characterization of catalytic PtNa-PdAu membranes for hydrogen production, Ana M Tarditi (Instituto de Investigaciones en Catálisis y Petroquímica)*; Laura Cornagllia (Instituto de Investigaciones en Catálisis y Petroquímica); Yohana Martínez Galeano Instituto de Investigaciones en Catálisis y Petroquímica)

Hydrogen is one of the most promising energy carriers because of being a zero-emission fuel and presents high efficiency. In this actual scenario, Pd-based membranes play an important role in hydrogen production and purification due to their excellent permeability and selectivity. Zeolites have been widely applied in catalysis, due to their small pore size capable of carrying out selective catalysis. We have reported the use of PtNaA zeolite as an efficient catalyst for the water gas shift reaction [1]. Considering that, the aim of this work was to develop catalytic PtNaA-PdAu membranes for hydrogen production. PdAu alloys were deposited by means of sequential electroless plating. The PtNaA zeolite was deposited on top of the PdAu alloys employing the encapsulated method reported in a previous publication [1]. Different synthesis variables were considered: suface modification with APTES and the number of the hydrothermal synthesis. The samples were characterized by SEM-EDS, XRD, and XPS and evaluated in streams containing H_2, CO, and H_2O. A compact and homogeneous zeolite coverage was successfully obtained by previous functionalization of the PdAu surface with APTES. XRD diffraction patterns of the PtNaA-zeolite deposited on the PdAu layer showed the characteristic reflections associated with the NaA zeolite phase. Upon exposure to CO, the membrane showed a reduction in the H_2 permeance, although the recovery was almost 100% after H_2 exposure at the same evaluated temperature. Likewise, no nitrogen permeation flux was detected, indicating that the PdAu membrane continues to be H_2 selective. Uniform and homogeneous PtNaA-PdAu membranes were obtained employing an optimized synthesis protocol. The membranes showed good stability and resistance to streams with contaminant gases and present themselves as promising materials for H_2 production. [1] Martínez Galeano, Y., Negri, F., Moreno, M.S., Múnera, J., Cornaglia, L., Tarditi, A.M., Appl. Catal., A 572 (2019) 176-184.

D20 Hybrid facilitated transport membranes containing 2D GO-based nanoplatelets and scaffolds for CO_2 separation, Saravanan Janakiram (Norwegian University of Science and Technology)*; Juan Luis Martín Espejo (Norwegian University of Science and Technology); Zhongde Dai (Norwegian University of Science and Technology); Luca Ansaloni (SINTEF Industry); Liyuan Deng (Norwegian University of Science and Technology)

Amine-based fixed-site carrier membranes transport CO_2 through reactional pathway in addition to solution diffusion under humid conditions. Incorporation of tailored fillers in facilitated transport membranes leads to enhanced separation performance and superior mechanical strength in the hybrid matrix. Anisotropically shaped thin nanosheets triggers nanoscopic changes in polymer chain packing, resulting in enhanced permeation performance at very low loading when compared to conventional fillers like Zeolites, MOFs. Additionally, incorporation of 2D nanofillers like graphene oxide have been found to improve both selectivity and permeability due to the molecular sieving effect while reducing CO_2 plasticisation. As a novel approach, the current study aims at resulting in a mechanically stable thin composite membrane with superior separation performance. GO-based nanofillers were successfully incorporated in facilitated transport sterically hindered polyallyl amine membranes to obtain novel membranes for CO_2 separation. Defect-free ultrathin selective layers with thicknesses less than 300 nm were successfully coated on porous polymeric supports. Good compatibility of GO-based fillers favoured uniform coating with no visible voids or interfacial defects. CO_2 permeance in hybrids increased as high as 610 GPU with a CO_2/N_2 selectivity of 37. Structural changes and chemical functionalization of GO nanosheets are also made to study the effect of filler modifications on the permeation performance. Furthermore, as a step closer to industrial realization, fabrication and testing of hollow fibers modules with high packing density are under investigation.

D21 Current Efforts in Harvesting Algae At-Scale For Biofuels to Meet the Dual Energy Challenge, Mark A Deimund (ExxonMobil Research and Engineering)*

Recently, ExxonMobil announced continuation of its partnership with Synthetic Genomics, Inc. (SGI) to begin a new phase of their joint algae biofuel research program, where the two companies will work to develop the technical ability to produce algae biofuels at large-scale. This endeavor will require innovation and scale-up efforts for many technologies across orders of magnitude in water purification/treatment, algae growth and harvesting, and lipid extraction to convert algae products to fuels. To date, larger-scale growth and harvesting pilot work has been primarily conducted at SGI's growth site in Calipatria, CA, known as the "California Advanced Algae Facility" (CAAF). Algae harvesting requires separation from dilute growth media (~1 g algae/L water) to allow water recycle and lipid extraction with minimal energy usage; thus, membranes are a natural choice for initial dewatering. A summary of ExxonMobil's test results for harvesting algae with membrane systems will be presented, as well as some key learnings from this larger biofuels scale-up effort. Membrane-based separation systems have been demonstrated to be very effective at algae separation during ExxonMobil pilot tests, providing nearly-pure permeate and a concentrated algae slurry. This slurry can then be sent for further processing and lipid extraction, with greatly reduced pumping capacity requirements and more efficient downstream handling. Additionally, the resulting permeate is virtually free of any potential contaminants, such as extracellular material or other undesirable species, enabling recycle of large amounts of growth media with minimal treatment. While parameters such as membrane fouling, cleaning protocols, and membrane lifetime are key considerations, membranes show promise as an initial dewatering step in the large-scale production of biofuels.

D22 Atomically-Ordered Intermetallic Hydrogen Separation Membranes, Casey O'Brien (University of Notre Dame)*

Palladium-alloy membranes are attractive for a wide range of hydrogen separation applications due to their rare combination of high permeability and high-selectivity for hydrogen separation from mixed gas streams. However, other reactive gases that are common to hydrogen-containing gas streams, such as CO, propylene, and H_2S, can severely inhibit hydrogen permeation by blocking hydrogen adsorption, or reacting with the membrane to form compounds with low permeability. Nearly all studies of Pd-based membranes have focused on substitutionally disordered alloys, where the non-Pd elements occupy lattice positions that are normally occupied by Pd. In this talk, I will discuss a new approach--intermetallic membranes--for creating high performance hydrogen separation membranes. In contrast to substitutionally disordered alloys, intermetallic compounds are solids composed of at least tow metals that form an atomically-ordered crystal structure where the different elements occupy definite positions with the lattice. The ordered crystal structure of intermetallics results in some unique properties that have been exploited in several fields, including hydrogen storage, and catalysis. Although intermetallic membranes have received minimal attention, they are very promising as hydrogen separation membranes. Our work with intermetallic PdCu suggests that intermetallic membranes perform better than disordered alloys. Intermetallic PdCu has higher permeability than all disordered PdCu alloys, including Pd. Intermetallic PdCu resists corrosion from H_2S under conditions that result in severe corrosion of disordered PdCu alloys. Intermetallic PdCu is also more resistant to poisoning from CO due to the isolation of active sites on the membrane surface. In this talk, I will also discuss a new operando spectroscopy tool that we developed to monitor the interactions of reactive gases with intermetallic membrane surfaces in-situ under realistic permeation conditions.

D23 Molten Hydroxide Dual-Phase Membranes for Intermediate-temperature Fuel Cells, Vedasri Vedharathinam (Lawrence Livermore National Laboratory)*; Anna Ivanovskaya (Lawrence Livermore National Laboratory); Maira Ceron Hernandez (Lawrence Livermore National Laboratory); Patrick Campbell (Lawrence Livermore National Laboratory)

Anion exchange membrane (AEM) fuel cells are sought after because they enable non-precious metals as electrocatalysts and enhanced reaction kinetics. Typical polymer membranes contain quaternary ammonium or phosphonium groups that conduct OH- ions at temperatures below 90°C, limiting their applications at elevated temperatures. Operation at elevated temperature allows for improved electrode kinetics, system efficiency, and simplified heat management. Hence, there is a need to develop hydroxide ion conducting membranes that are chemically and mechanically stable and can operate at intermediate temperature ranges (150 - 400°C) without compromising ionic conductivity. Here we report a novel hydroxide-ceramic dual-phase (HCDP) membrane technology that utilizes molten alkaline hydroxides (e.g. sodium, potassium, and lithium) impregnated into a porous zirconia matrix. In typical alkaline electrolytes, hydroxide ions react with CO_2 from air to form carbonate, leading to performance degradation of the fuel cell due to loss of OH-ionic conductivity. The CO_2 poisoning effects are rapid and severe, necessitating the use of pure oxygen instead of air as cathode feed. Our proposed technology overcomes the CO_2 poisoning issue when using air as the oxidant by shifting the chemical equilibrium of the undesired carbonate ions toward hydroxide and CO_2 in the presence of water at elevated temperatures, thus releasing CO_2 from the electrolyte in situ during fuel cell operation and preventing performance degradation. This contribution will highlight our recent developments in this area.

D24 High-Yield Recovery of Dissolved Methane Using Omniphobic Membranes, Xuesong Li (University of British Columbia); Abhishek Dutta (University of British Columbia); Qirong Dong (Fibracast); Sasha Rollings-Scattergood (Anaergia); Jongho Lee (University of British Columbia)*

Efficient recovery of dissolved methane from anaerobic effluents is crucial to renewable energy production via anaerobic wastewater treatment processes. In this study, we present a novel membrane process (solvent-based membrane contactor or SMC) for recovering dissolved methane through the normal range of domestic wastewater temperatures (15°C - 35°C) by utilizing an omniphobic (non-wetting) microporous membrane that traps air within its pores when submerged in liquids. Driven by solubility-gradient, dissolved methane is extracted from a methane-rich aqueous solution (feed), transported across the omniphobic membrane, and absorbed into a non-polar organic solvent (draw) that has an order of magnitude higher methane solubility than does water. We fabricated the omniphobic membrane by coating a microporous polymeric membrane with silica nanoparticles, followed by a surface fluorination with low surface energy materials. The omniphobic membrane exhibited excellent wetting resistance against both water and low-surface tension organic liquids (e.g. dodecane and ethanol). Using the omniphobic membrane and dodecane as the draw solution in the SMC process, we demonstrate nearly 90% or above recovery rates of dissolved methane from methane-saturated feed water at 15°C, 25°C and 35°C, simulating anaerobic effluents produced in the psychrophilic to mesophilic temperature range, while a negligible transport of water is observed due to the low water solubility in the draw. A comparative energy analysis suggests that this novel membrane process for methane recovery can enable a net energy production, with a higher value at a lower temperature (up to 0.46 MJ per m3 of methane-saturated feed at 15°C), which outperforms other dissolved methane recovery techniques.

D25 Aging minimization of CMSMs from the incorporation of copper pillars derived from metal organic polyhedra-18, Whitney Cosey (University of Texas at Dallas)*; Inga Musselman (University of Texas at Dallas); John Ferraris (University of Texas at Dallas); Kenneth Balkus (Univerisity of Texas at Dallas)

Membrane-based gas separations have made strides within the last two decades due to their lower operating cost and ease of manufacturing. While polymeric membranes are currently used commercially, they are limited by an inherent trade-off between selectivity and permeability. However, carbon molecular sieve membranes (CMSMs) are not bound by this trade-off and thus surpass polymeric membranes in terms of permeability and selectivity due to the presence of micropores (0.7-2.0 nm) and ultramicropores (<0.7 nm), which provide high flux and high selectivity. While several researchers focus on tailoring the pore aperture to improve performance, improving the stability of the CMSMs by reducing physical aging (collapsing of pores) has yet to be addressed. Highly soluble metal organic polyhedra-18 (MOP-18) was introduced into Matrimid® 5218 with a loading of 40 wt% to form a mixed matrix membrane

(MMM). Pyrolysis of the MMM at 550°C allowed for copper oxide to be reduced by carbon to copper metal nanoparticles. The nanoparticles act as pillars for the graphene sheets within the CMSMs, preventing the collapse of the micropores and thus minimizing the aging of the CMSM. Gas permeation at 35°C and 2 atm showed the membrane's permeability for CO_2 was 725 Barrer and was 39 Barrer for CH_4. After a period of 20 days, only a 2% decrease in CO_2 permeability and a 15% decrease in CH_4 permeability was observed, with retention of CO_2/CH_4 selectivity.

D26 (Cross-linked Poly(ionic liquid)-Ionic Liquid-Zeolite) Mixed-Matrix Membranes for CO_2/CH_4 Gas Separations Based on Curable Ionic Liquid Prepolymers, Collin A Dunn (University of Colorado Boulder)*; Zhangxing Shi (University of Colorado Boulder); Rongfei Zhou (Nanjing Technical University); Douglas Gin (University of Colorado Boulder); Richard Noble (University of Colorado Boulder)

A three-component, (cross-linked poly(ionic liquid) (PIL) - ionic liquid (IL) - zeolite) mixed-matrix membrane (MMM) platform based on curable IL prepolymers of controlled length has been developed for separating CO_2 from CH_4. Solutions of these curable prepolymers demonstrate increased resistance to support penetration compared to comparable solutions of analogous cross-linkable IL monomers. By adjusting the curable IL prepolymer chain length, it is possible to manipulate polymer susceptibility to support penetration, polymer solution gelation time, and gas separation performance in MMMs based on these materials. When a 50 wt% solution of the curable IL prepolymer with a degree of polymerization of 87 was cast on an ultrafiltration support membrane, only 3.7 wt% of the polymer penetrated into the support. As the degree of polymerization of the curable IL prepolymer increases, the CO_2/CH_4 gas separation performance of the resulting MMM performance also improves. An MMM synthesized using 64 wt% curable IL prepolymer, 16 wt% [EMIM][Tf2N], and 20 wt% SAPO-34 zeolite demonstrates a CO_2/CH_4 selectivity of 42±5, and CO_2 permeability of 47±1 barrers. This CO_2/CH_4 separation performance is comparable to the previous generation of MMMs based on curable small-molecule IL monomers with the same IL and zeolite. However, this new MMM system also exhibits faster curing gelation times and the ability to be solution-cast onto a porous support for formation of thin-film composite membranes without significant selective layer soak-in.

D27 Integral hollow fiber membrane with chemical cross-linking for pressure retarded osmosis operated in the orientation of active layer facing feed solution, Ye Li (Nanyang Technological University)*; Rong Wang (Nanyang Technological University)

Pressure retarded osmosis (PRO) is one of the promising techniques for harvesting salinity gradient energy. However, the performance of PRO is exacerbated by the severe membrane fouling occurring within the porous support. Changing the membrane orientation from active layer facing draw solution (AL-DS) to active layer facing feed solution (AL-FS) is a potential alternative for sustainable power harvesting due to its low membrane fouling propensity. In this work, a novel integral PRO hollow fiber membrane was fabricated using the phase inversion method followed by a facile cross-linking post-treatment. The structure and properties of the resultant membrane were systematically investigated using a series of characterization protocols. The PRO performance of the membrane was evaluated under various hydraulic pressures at the AL-FS orientation. The results reveal that the substrate with sponge-like structure and relatively dense inner surface can better withstand high pressure in PRO process. Two steps of cross-linking endowed the membrane with a smaller pore size on the outer surface that exhibits high rejection against various inorganic salts. The newly developed membrane can achieve a stable power density output around 4.3 $W \cdot m^{-2}$ at a pressure of 12 ~ 13 bar, using a real wastewater reverse osmosis (RO) retentate as the feed and a synthetic seawater brine (1.0 M NaCl) as the draw solution. This result demonstrates that the PRO operation in the AL-FS orientation offers a significant advantage over the AL-DS orientation by eliminating the pretreatment of the feed water or/and other measures associated with membrane fouling control in the AL-DS orientation. To implement the PRO operation with the AL-FS orientation, developing a robust membrane with an integral structure is critical. The newly developed integral hollow fiber membrane via chemical cross-linking presents potential for PRO operation in the AL-FS orientation.

D28 GHG separation membrane application in KOGAS LNG regasification facilities, Kyung Taek Woo (Korea Gas Corporation)*; Hyun Seok You (Korea Gas Corporation); Jong Tae Chung (Korea Gas Corporation); Jae Dong Kim (Korea Gas Corporation)

The Green House Gases (GHG) such as carbon dioxide (CO_2) and methane (CH_4) have been critical issue in terms of global climate change. The majority of GHG is emitted by combustion of fossil fuel to generate electric power. In order to overcome the issue, Carbon Capture, Utilization and Storage (CCUS) has been introduced one of the technologies to reduce GHG. The majority of GHG emission in KOGAS is generated in Submerged Combustion Vaporizer (SCV) one of the gasification facilities of LNG using combustion thermal of natural gas to boil up the water bath containing LNG pipelines. The flue gas from the SCV facilities is composed of CO_2 (7~8%), H_2O (13~15%), O2 (3~6%), CO (<100ppm), NO (<50ppm), and N_2 (balance). KOGAS has been considered the CO_2 separation technologies to SCV facilities in terms of the application site limitation and capture cost. Membrane gas separation process could be the solution to reduce CO_2 emission in KOGAS process. Therefore, the fundamental study of membrane gas separation process has been carried out to set basic operation conditions. KOGAS is emitted GHG not only CO_2 but also CH_4 from boil off gas (BOG) which is mainly generated in summer season. The BOG is treated by combustion process. As a result, CO_2 is generated by the combustion of BOG and it should be captured as well.

D29 High performance electrochemical structures for power generation and oxygen separation, Ralph A Bauer (Global Research and Development Inc.)*

Traditional inorganic ion conducting layer approaches to fuel cell power generation have hampered the solid oxide fuel cell's development and practicality in numerous applications. It has long been understood that thinning these layers is critical to overcoming the technical hurdles facing the technology. Recent work at Global Research and Development Inc. and the Ohio State University has developed a readily scalable fabrication methodology, that allows economic and reproducible formation of <200nm thick layers of metallic and oxide layers. Fabricated nano-layered solid oxide fuel cells have shown high power densities, long operational lifetimes while operating at <600C. Further a similar nano-layered electrochemical cell has been designed for the pumping of oxygen and has shown output ~100X above the state of the art electrochemical oxygen pumps.

D30 Techno-economic comparison of RO and MD for small-scale brackish water desalination, Haamid S Usman (Concordia University)*; Khaled Touati (Concordia University); Saifur Rahaman (Concordia University)

Membrane Distillations (MD) is a promising separation technology that can contribute to reduce the world-wide water-energy stress that our globe is facing. Its ability to use low-grade thermal energy is a potential advantage that attract the interest of research and makes of it one of the emerging technologies that may reduce the energy consumption of water desalination. Reverse osmosis (RO) is still considered as the best energy-efficient seawater desalination technique. However, for low scalability, RO is also criticized in term of cost-effectiveness. In the current work, we aim to compare the cost-effectiveness of MD and RO during brackish water desalination for a small-scale plant. For that, a techno-economic analysis was performed for both technologies to determine the cost of desalinated water using a desalination plant capacity of 100 m3/hr. It was assumed that MD is integrated with photovoltaic solar power technology, which provides an alternative source of energy to fossil sources. Capital cost (CapEX) and operation cost (OpEX) were calculated for each sub-system (i.e., intake, pre-treatment, RO separator, post treatment, and outfall) under real operational conditions. The cost breakdown results conclude that, even though the cost of water using RO system is still cheaper, MD can compete with RO for desalination plant with low capacity, if more technological improvements in membrane and heat recovery efficiency are brought to the system.

D31 Tailor-made amphiphilic copolymer membranes for efficient gas and water vapor transport, Faheem Hassan Akhtar (King Abdullah University of Science and Technology)*; Klaus-Viktor Peinemann (King Abdullah University of Science and Technology)

Membrane-based dehumidification is a critical, energy intensive and crucial process for several industries, e.g., air conditioning and flue gas dehydration. Various PEG-based membranes have been shown to have excellent water vapor/N_2 selectivity, but these membranes suffer from low permeabilities and mechanical strength. Herein, we demonstrate high-performance membranes for water vapor and CO_2 separation based on novel amphiphilic copolymers. We have synthesized ternary amphiphilic copolymers comprising PAN, PEGMA and PDMAEMA segments via an economical and facile free radical polymerization. These membranes are mechanically-robust, and water vapor permeability and selectivity are enhanced by the chain-length and content of PEGMA segments. In all the varying concentrations of PEGMA segments, a substantive improvement in water vapor permeability is observed with very high vapor/N_2 selectivities. Optimized PEGMA segments yield very high CO_2/N_2 and CO_2/CH_4 selectivities with an enhanced CO_2 permeability. The reason for this superior gas and water vapor separation performance will be discussed in detail during my presentation.

D32 High-Temperature and High-Pressure Permeability-Selectivity Upper Bounds for Gas Separation Membranes Prepared From High Performance Polymers, Edson V Perez (The University of Texas at Dallas)*; Kenneth Balkus, Jr. (The University of Texas at Dallas); John Ferraris (University of Texas at Dallas); Inga Musselman (The University of Texas at Dallas)

The performance of polymer membranes for gas separations is currently measured in terms of their ability to surpass the permeability-selectivity upper bound (Robeson plot). The current Robeson plot (L.M. Robeson, J. Membr. Sci. 2008, 320 (1-2), 390-400) uses data from membranes tested at 35°C and <5 atm and is, therefore, limited to membranes tested under these conditions of temperature and pressure. There is, however, the need to measure the performance of membranes for gas separations at higher temperatures and higher pressures, conditions at which there is not an available upper bound due to limited published data. In this project, several membranes from high performance polymers (e.g., VTECTM, Matrimid® 5218, polybenzimidazole, 6FDA-DAM) were cast, annealed at 300°C, and tested for H_2, CO_2, and N_2 gas permeation at temperatures of 35, 100, 200, and 300°C and at pressures of 5 and 30 atm. The high-temperature permeability-selectivity upper bounds for H_2/CO_2, defined from these data, are located above the 35°C bound (L.M. Robeson, J. Membr. Sci. 2008, 320 (1-2), 390-400). The bounds, with constant slope, show an increasing upward shift with increasing temperature. Increasing the feed pressure does not significantly affect the position or slope of the bounds. The results of the study suggest that higher temperatures increase membrane performance, and that there is marginal impact from a change in feed pressure. At the limits of the new bounds lie polybenzimidazole with the highest selectivity and smallest permeability and 6FDA-DAM with the smallest selectivity and highest permeability.

D33 Synthesis and Characterization of Anion Exchange Membranes based on Semi-Crystalline Poly(Ethylene-block-Styrene-block-Ethylene) Triblock Copolymer, Carrie Trant (Rensselaer Polytechnic Institute)*; Chulsung Bae (Rensselaer Polytechnic Institute); Sangwoo Lee (Rensselaer Polytechnic Institute)

The ability of polymer-based ion exchange membranes to efficiently transport ions is important in a wide range of electrochemical energy applications including fuel cells, batteries, and electrolyzers. To better understand the impacts of microstructured morphology on water uptake and the formation of hydrated ion conducting channels to aid ion conductivity, we investigated the effects of polydispersity and ion exchange capacity (IEC) on the microscopic and macroscopic structures and the performance of the semi-crystalline poly(ethylene-b-quaternized styrene-b-ethylene) (ESE) triblock copolymer-based alkaline anion exchange membranes (AAEMs). The quaternary ammonium group functionalization of the styrene middle block was quantitatively controlled by acid-catalyzed Friedel-Crafts bromoalkylation reaction followed by amination with trimethylamine to synthesize membrane materials with a range of IECs of 0.64 to 1.89 meq/g. Our results suggest that strongly-segregated microdomains without a long-range order are desirable for membrane processing and performance compared to the other membranes with anisotropic lamellar morphology.

D34 Effect of Polyvinylamine Content on the CO_2 Separation Performance of Facilitated Transport Membranes, James S Baker (U.S. Department of Energy National Energy Technology Laboratory)*; Victor A Kusuma (U.S. Department of Energy National Energy Technology Laboratory); David Hopkinson (U.S. Department of Energy National Energy Technology Laboratory)

Polyvinylamine (PVAm) is one of the most commonly studied facilitated transport polymers for CO_2 separation membranes due to its high-density of amine groups (carriers) and excellent film forming properties. The most common route to prepare PVAm is to first synthesize poly(N-vinylformamide) (PNVf), followed by acid or base hydrolysis to yield PVAm. The hydrolysis is typically not conducted to full conversion; therefore, the product polymer is a random copolymer of PVAm and PNVf repeat units. Since PVAm is a more effective CO_2 carrier than PNVf, it would seem logical to maximize conversion to PVAm to maximize the density of carrier sites within the polymer. However, several researchers have suggested that high degrees of conversion to PVAm can result in a decrease in CO_2 permeability, possibly due to formation of crystalline domains within the polymer film. To date there have been no systematic studies in the open literature confirming this trend. We are conducting a rigorous examination of the effect of PVAm content in the copolymer upon the CO_2 permeability and CO_2/N_2 selectivity for the polymer films. PNVf from a single batch of polymer was hydrolyzed to varying degrees of PVAm content, deprotonated to pH 11, and free-standing films of 20-30 microns dry thickness were cast for gas permeation and mechanical property testing. Gas permeability of the films was tested with humidified (95% RH), mixed feed gas (1.38 bar, 14: 86 CO_2: N_2) at 60°C. Moderate levels of PVAm content yielded the highest CO_2 permeability and selectivity, with 52 mole % PVAm films having a mean CO_2 permeability of 386 barrer and a CO_2/N_2 selectivity of 89. Polymers containing 0, 20, 42, and 76 mole % PVAm had CO_2 permeabilities of 161, 279, 349, and 249 barrer and selectivities of 36, 48, 79, and 49, respectively. We are using X-ray diffraction (XRD) of the polymer films to determine if there is a correlation between polymer composition and degree of crystallinity.

D35 Effect of organic ballast properties on the energy efficiency of a concentration gradient flow battery, Fei Liu (University of North Carolina at Chapel Hill)*; Ryan Kingsbury (Membrion, Inc.); Jeromy Rech (University of North Carolina at Chapel Hill); Wei You (University of North Carolina at Chapel Hill); Orlando Coronell (University of North Carolina at Chapel Hill)

A concentration gradient flow battery is a membrane-based system for storing clean energy (Kingsbury 2016, van Egmond, 2016). The concentration battery operates through a combination of an electrodialysis (ED) stage where energy generated from renewable resources is used to produce highly concentrated brine and dilute solution, and a reverse electrodialysis (RED) stage where the brine and dilute solutions are mixed to generate power. However, the non-ideal transport of water through the membrane by osmosis has a substantial negative impact on the power density and faradaic efficiency of the battery. A promising approach to mitigate osmosis is to use uncharged osmotic ballasts in the dilute solution to balance the osmotic pressure and significantly reduce faradaic energy losses. The objective of this study is to investigate the effect of ballast size, orientation (linear vs cyclic), and end group on the energy efficiency of a concentration battery. To accomplish this objective, we tested 7 sugars and 5 glycol-based compounds as osmotic ballasts in a closed-loop ED/RED cell. Our results show that the ballasts with higher molecular weight generally resulted in higher efficiency and lower water transport. We also observed that the glycol-based ballasts with asymmetric end groups had lower water transport compared to those with symmetric end-group. We also found that glycol-based ballasts resulted in lower open circuit voltage and water transport than sugar ballasts. Overall, the best performing ballast was triethylene glycol monomethyl ether, reducing osmosis by 128% at a faradaic energy efficiency of 53%. These findings contribute to the fundamental understanding of the effect of ballast physical and chemical properties on the performance of salinity gradient batteries and provide useful guidelines for optimizing osmotic ballasts. Our findings may also benefit other closed-loop energy processes such as RED systems for waste heat recovery and redox flow batteries.

SESSION E - Innovation in Membranes for Water Treatment

E1 Feasibility of long-term treatment of malting wastewater in a submerged anaerobic membrane bioreactor (SAnMBR) at psychrophilic temperature conditions, Esmat Maleki (Lakehead University)*; Baoqiang Liao (Lakehead University)

Malting wastewater that is discharged from the steeping step of malting plants has a low temperature ranging between 14°C and 18°C in Canada. Currently, most malting plants discharge their wastewater to the municipal water treatment plant and pay high discharge fees. Aerobic treatment of this wastewater doesn't recover its resources content. Malting wastewater is rich in nutrients and has high soluble COD content. Hence, it is a good candidate for the anaerobic treatment that converts the COD of wastewater to biogas. Anaerobic membrane bioreactor (AnMBR) technology is capable of recovering both energy and nutrients resources from this waste stream. However, operation of anaerobic bioreactors in low temperature slows down the biomass growth rate and hinders the hydrolysis reactions. Hence, most industrial plants consume energy to warm up their wastewater and operate their AnMBRs at mesophilic conditions which is costly. Operating the AnMBR at low temperatures could save energy and reduce the operational costs. It is necessary to evaluate performance of AnMBR technology at psychrophilic temperatures before its industrial scale implementation. Therefore, this study investigated the feasibility of long-term operation (376 days) of a submerged anaerobic membrane bioreactor for treating malting wastewater at psychrophilic conditions (21-16°C) and resource recovery in the form of biogas and nutrients. High COD removal efficiency of 92.58±1.44% was achieved, and the average biogas yield was 0.25±0.05 L at STP/g CODremoved. Moreover, the permeability of the fouled membrane was recovered up to 56.9% after physical cleaning, 80.0% after basic-chemical cleaning, and 89.8% after acidic-chemical cleaning. The results of this study proved the feasibility of treating malting wastewater at psychrophilic conditions in a submerged anaerobic membrane bioreactor.

E2 Biofouling in ultrafiltration process for drinking water treatment and its control by chlorinated-water and pure water backwashing, Shao Senlin (The University of Hong Kong)*

We investigated biofouling in ultrafiltration (UF) for drinking water treatment and its control by backwashing with chlorinated-water or pure water. By using sodium azide to suppress biological growth, the relative contribution of biofouling to total fouling was estimated, and its value (5.3-56.0%) varied with the feed water, and increased with the increases of filtration time and membrane flux. The biofouling layer could partially remove biodegradable organic matter and ammonia (32.9-74.2%). Backwashing using chlorinated-water partly inactivated the microorganisms (23.8%) but increased the content of extracellular polymeric substances (7.7%) in the biofouling layer. In contrast, backwashing using pure water led to a looser and more porous fouling layer according to optical coherence tomography observation. Consequently, the latter was more effective in reducing fouling resistance (33.41% reduction) compared to backwashing by chlorinated-water (8.6%). These findings reveal the critical roles of biofouling in pollutants removal in addition to membrane permeability, which has important implications for addressing seasonal ammonia pollution.

E3 Membrane Distillation of High Salinity Water by Induction Heated Thermally Conducting Membranes, Avner Ronen (Temple University)*; Arezou Anvari (Temple University)

Membrane Distillation (MD) has been gaining interest in the recent decade as a method for providing desalinated water from several water sources including high salinity water and produced water. The method has gained popularity due to the low required pressure, especially when the osmotic pressure makes membrane desalination such as RO uneconomic. MD is a thermally driven process in which the vapor pressure gradient over the porous hydrophobic membrane leads to the transport of water vapor. To ensure an efficient driving force, the feed solution is continuously heated, leading to a substantial energy requirement as most of the heat is not directly used due to a limited surface area and mass transport limitations. We suggest a novel approach which heats only the membrane-water interface through inductive heating (IH). Thus, significantly decreasing the required energy and increasing the permeate flux while mitigating inorganic scaling and the influence of high salinity such as found in produced water. Carbon nanotubes coated iron nanoparticles (CNT-Fe) membrane with high thermal conductivity are fabricated through spraying on a PTFE base support and the heating is done by exposing the CNT-Fe membranes with an iron loading of about 45 Wt.% to a high-frequency magnetic field. Preliminary results show the ability of heating water directly on the surface of the membrane to a boiling temperature in a matter of seconds by exposing the CNT-Fe membrane to an electromagnetic field at a fixed frequency (283 kHz). The heating ability was found to be related to the loading of CNT-Fe on the base support and the feed flow rate. In addition, permeate flux from the IH-MD system was found to be significantly higher than conventional MD systems and the flux increased with increasing feed salinity. Inorganic fouling (e.g. CaSO4), which typically leads to a flux decline in MD systems had almost no impact on the IH-MD system flux.

E4 Bromide Removal from Drinking Water Source Impacted by Energy Wastewater Discharge Using Electrically Conductive Membranes, Avner Ronen (Temple University)*; Mei Sun (University of North Carolina at Charlotte); Kartikeya Kekre (Temple University)

In recent years, wastewater discharges from the energy industry, including coal-fired power plants equipped with wet flue-gas desulfurization units and from shale gas extraction, have been identified as contributors to elevated bromide concentrations in several watersheds across the United States. Bromide reacts with drinking water disinfectants and dissolved organic matter to form DBPs in downstream drinking water system. Elevated bromide levels have been one important cause of compliance violation of DBP regulations across the US. In this research, we developed and evaluated a membrane-based electrooxidation process to remove bromide from impacted drinking water sources. The electrooxidation takes place on carbon nanotubes-based electrically conductive membranes (ECM) and uses an external potential applied on the membrane surface to oxidize bromide during filtration. We initially tested the oxidation using cyclic (CV) and linear voltammetry (LV) which allowed us to detect bromide oxidation at an anodic potential of 1-1.1V against a silver/silver chloride reference electrode. As these reactions occur when the contaminated solution is in contact with an electrode, which has a limited surface area and limited mass transport, bromide removal rates were relatively low. The anodic potential at that range did not impact membrane electrical conductivity which remained stable following 24 hours of applied potential. Next, a similar approach was evaluated using the membrane in a cross-flow filtration cell. As there is no diffusion limitation due to the convective transport in filtration, removal rates were very fast following a contact time of 1-3 seconds flow through the membranes. Applying a whole cell potential of 2.5V with the ECM serving as the anode allowed continues removal of about 60% of the bromide from the permeate stream in a single pass while additional removal was detected at the concentrate stream.

E5 A surface modification of forward osmosis (FO) membranes to enhance dye retention and fouling resistance for the treatment of textile wastewater, Sara Azari (Sterlitech)*; Mohammd Karimi (Amirkabir University of Technology)

Dye containing wastewaters are generated in large volumes in the textile industry and due to their toxicity to all forms of life, they are a major environmental challenge if not treated properly. Due to their high solubility in water and non- biodegradability, conventional treatment methods are unable to effectively remove or degrade the dye molecules. Membrane processes are alternative technologies that are commonly applied for tertiary treatment of dying wastewater. In this project forward osmosis membrane process was studied for treating/concentrating dye containing wastewaters. CTA and TFC FO membranes from HTI Water were tested in bench top FO test cells. Our initial results showed 10-15% and 12-20% flux decline due to fouling for CTA and TFC FO membranes, respectively. Moreover, diffusion of dye molecules into the membrane polymer structure and its penetration to the draw solution was noted. These observations may affect long term stability and efficiency of FO membranes to produce colorless permeates. Here we present a facile surface modification using L-Dopa to increase dye retention and to reduce membrane fouling. Membranes were coated on the active side using L-DOPA. Contact angle and zeta potential measurements showed higher hydrophilicity and higher negative surface charge of the modified membranes. While the membrane flux remained almost intact, surface modified membranes showed better antifouling properties, as indicated by flux decline with operating time. An improvement in dye retention by modified membranes was also observed. Such an improvement in the performance, can be explained by more hydrophilic nature of the modified membrane and electrostatic repulsion between charged molecules of the dye and the membrane surface. This surface modification can be performed in situ, i.e. in the original module assembly post-manufacturing process, making this approach very interesting from point of view of practicality in larger scale systems.

E6 Evaluation of surface, bulk and electrochemical properties of $Ti_3C_2T_x$ membranes used for selective ion removal from brine., Yousuf Z Bootwala (Georgia Institute of Technology)*; Wahiduz Zaman (Vanderbilt University); Kelsey Hatzell (Vanderbilt University); Marta Hatzell (Georgia Institute of Technology)

Transition metal carbides, carbonitrides, and nitrides or MXenes represent a large family of 2D materials derived from a class of ternary ceramic materials known as MAX described by the general formula Mn+1AXn where n=1,2,or 3 and M represents an early transition metal,A is a group A element,X is carbon and/or nitrogen.MAX phase materials are layered hexagonal materials with P63/mmc symmetry and are composed of metal carbide and/or nitride layers(Mn+1Xn) twinned with respect to each other and separated by an A-layer(Naguib et al).There are approximately 70 known MAX phases that have been experimentally produced and have numerous potential applications.More recently,they have attracted interest for both gas and aqueous phase separations due to their high stability in aqueous environment,flexibility,high surface area and conductivity(Ren et al).This includes organic contaminant removal from water, energy storage and gas separation(Ren et al).The layered structure of MXene allows the inter-layered spacing to be tuned to allow rejection of selective ions based on their size and charge.When compared to graphene oxide membranes,MXene membranes have higher water flux due to 2-3 layers of water molecules in the spacing between the layers(Ren et al).Here,we will explore the surface and bulk properties of pristine $Ti_3C_2T_x$ membranes,and $Ti_3C_2T_x$ membranes with ion based intercalants(K+, Li+).In situ surface science(ambient pressure XPS and FTIR)will be presented in various relevant conditions(e.g.temperature,relative humidity).In addition,the electrochemical properties of the membranes (conductivity, resistance, capacitance) will be presented. Finally,experimental insight into salt base separations will be presented. 1. Naguib, M. et al. 25th Anniversary Article: MXenes: A New Family of Two-Dimensional Materials. Adv. Mater. 26, 992-1005 (2014). 2. Ren, C. E. et al. Charge- and Size-Selective Ion Sieving Through $Ti_3C_2T_x$ MXene Membranes. J. Phys. Chem. Lett. 6, 4026-4031 (2015).

E7 Microwave-enhanced Membrane Filtration for Water Treatment, Wanyi Fu (New Jersey Institute of Technology)*; Wen Zhang (New Jersey Institute of Technology)

Sustaining high flux (permeability) and diversified pollutant rejection (selectivity) are two crucial benchmarks for membrane filtration processes. Here, we report a microwave-enhanced membrane filtration process that uses microwave (MW) irradiated and catalyst-coated ceramic membranes to achieve efficient removal of pollutants (i.e., 1,4-dioxane) and significant mitigation of fouling. MW irradiation was selectively absorbed by catalysts and hydrogen peroxide to produce "hotpots" on membrane surface that promoted generation of radicals and nanobubbles. These active species enhanced pollutant degradation and further prevented membrane fouling. In contrast to ultrasound and ultraviolet radiations, MW could efficiently penetrate membrane housing materials and selectively dissipate energy to membrane-impregnated catalyst nanoparticles. Our study of MW-assisted membrane filtration processes may open new avenues toward next-generation antifouling and high-efficiency separation techniques.

E8 Application of Nanofiltration (NF) Membranes for Perfluoroalkyl and Polyfluoroalkyl substances (PFAS) Removal, Daniel G Kulas (Michigan Technological University)*; Andre da Costa (Michigan Technological University); Ananya Ananya (Michigan Technological University); Sam Jacobs (Michigan Technological University)

Perfluoroalkyl and Polyfluoroalkyl substances (PFAS) are an emerging and extremely pervasive contaminant with a potentially high health risk. They are currently being found throughout the United States in ground and surface waters. At least one city's drinking water, Parchment MI, was found to contain PFAS above the EPA's limit of 70 ppt, forcing them to switch to an alternative water source. In all likelihood more cities will soon find themselves in the situation of having unsafe drinking water due to PFAS as testing increases. It is thus essential for sustainable solutions to be developed for the safe removal of PFAS from water. We will present an evaluation of membranes as a sustainable solution for the removal of PFAS from water in a drinking water treatment process. Past research has shown that nanofiltration (NF) membranes are capable of removing PFAS from pure water, however no study has looked at removing PFAS from a real water source. We are testing several commercial NF membranes in a range of MWCO from 200-1000 Daltons with sample water from Parchment MI containing a PFAS concentration of 30,000 ppt. All experiments are being performed in a stirred cell dead-end filtration system. We are also investigating the effect of contact angle and surface energy on PFAS rejection. The water's hardness is being varied to determine its effect on PFAS rejection was also tested in the presence of hardness and chlorine to determine potential placement points within the water treatment process. Results will be presented on both the PFAS rejection and flux of commercial NF membranes.

E9 Development of membrane synthesis and novel draw materials for forward osmosis process, Yusak Hartanto (Katholieke Universiteit Leuven)*

Thin-film composite forward osmosis (TFC-FO) membrane prepared by interfacial polymerization utilizes organic solvents which often have serious environmental and human health impacts. Furthermore, highly volatile nature of these solvents also increases membrane manufacturing cost due to significant solvent evaporation. Ionic liquid (IL) is proposed as organic phase to replace hexane in interfacial polymerization of polyamide layer. The resultant TFC-FO membrane prepared using IL has similar performance in terms of its water flux and reverse salt flux. This innovative method represents a step towards greener and more sustainable membrane preparation process. Another important aspect of forward osmosis (FO) technology is draw materials development. Thermo-responsive hydrogels are semi-solid FO draw materials which have potential to solve the problems of inorganic salts when used as draw solute materials. These hydrogels are often copolymerized with ionic comonomers to improve their water drawing ability. However, the water release ability of these hydrogels is hampered by the presence of ionic moieties within the hydrogel networks. Non-ionic copolymer microgels are recently proposed as novel FO draw materials which are able to overcome the ionic contribution of ionic thermo-responsive copolymer hydrogels/microgels

E10 3D Printed Self-Assembled Zwitterionic Copolymer Membranes, Xin Qian (University of Connecticut)*; Samuel J Lounder (Tufts University); Ayse Asatekin (Tufts University); Jeffrey R McCutcheon (University of Connecticut)

Zwitterionic copolymer self-assembled membranes have been shown to have impressive selectivity for nanofiltration applications. Conventional approaches used to fabricate these asymmetric membranes, however, typically result in thick selective layers being formed that can limit permeance. In this work, we evaluate the use of electrospray to print thin layers of amphiphilic zwitterionic selective layers onto UF membrane substrates. Dyes with different sizes are used to probe the selectivity of these membranes while their thickness is reduced by orders of magnitude compared to casting methods. As thickness is decreased, water permeance was found to proportionally increase while dye selectivity was maintained. While it was found that a threshold minimum thickness was required to maintain selectivity, we also found that the intrinsic permeability of the polymer films was changed as a function of thickness. Annealing the membranes was found to increase permeance as well without a significant loss of selectivity. Notably, one of our annealed ultra-thin TFC membranes was found to exhibit a water permeance value of 205LMH/bar and a Chlorophyllin rejection at 99.67%. Interestingly, this permeance is indistinguishable from the supporting UF membrane permeance, suggesting that even higher permeance without selectivity loss is possible.

E11 Antifouling and Antiwetting Janus Membrane for Treating Hypersaline Oily Wastewater by Membrane Distillation, Chenxi Li (University of British Columbia); Xuesong Li (University of British Columbia); Xuewei Du (Colorado State University); Tiezheng Tong (Colorado State University); Tzahi Cath (Colorado School of Mines); Jongho Lee (University of British Columbia)*

We present a novel approach to fabricate a robust Janus membrane with excellent wetting and fouling resistances for treating highly saline oily wastewater through membrane distillation (MD). An omniphobic membrane was first fabricated by grafting silica nanoparticles (SiNPs) on the nanofibers of a quartz-fiber mat, creating multi-level re-entrant structures, followed by surface fluorination to reduce the surface energy. The Janus membrane was then fabricated by grafting a zwitterionic polymer brush layer via surface-initiated atom-transfer radical-polymerization (ATRP) on the omniphobic substrate. Membrane characterization, including fluorescence microscopy, Fourier-transform infrared spectroscopy, and contact angle measurement, confirms that the surface hydrophilicity can be precisely controlled by tuning the ATRP reaction duration. The brush layer was confined on the top surface of Janus membrane, rendering the surface hydrophilic. Still, the rest part of the membrane remained omniphobic, resisting the wicking of low-surface-tension liquids including ethanol and hexane. A static oil fouling test using a crude oil showed that oil droplets were readily adsorbed on the omniphobic membrane surface in water. In contrast, oil droplets that were placed on the Janus membrane came off from the membrane surface when it was immersed in water, proving the effective oil fouling resistance. Finally, we performed direct-contact MD experiments using a surfactant-stabilized crude oil-in-saline water (1M NaCl) emulsion as a feed solution, which mimics industrial hypersaline oily wastewater. The results show that the Janus membrane exhibits superior wetting and fouling resistances compared to the omniphobic membrane, maintaining a stable water flux and nearly perfect salt rejection over a 12-hour MD test. Our work highlights the great potential of MD using antiwetting and antifouling Janus membranes for water reclamation from hypersaline industrial wastewaters.

E12 Treating Hydraulic Fracturing Produced Waters by Membrane Distillation, Ranil Wickramasinghe (University of Arkansas)*

Membrane distillation is an emerging technology for treating hydraulic fracturing flow back and produced waters (PW). These wastewaters are highly impaired often containing over 100,000 ppm dissolved salts as well as polar and non-polar organic compound and low surface tension compounds. Membrane stability is often limited for these highly fouling feeds. Typically, hydrophobic membranes are used. However, adsorption of organic compounds and low surface tension compounds must be minimized in order to minimize flux decline and breakthrough of the water. We have investigated the feasibility of using direct contact membrane distillation (DCMD) to treat real PW using commercially available polypropylene, polyvinylidene fluoride, and poly(ethylene chlorotrifluoroethylene) membranes. Two approaches have been used to improve membrane stability; pretreatment of the feed and modification of the membrane surface. Feed streams were pretreated using electrocoagulation. Electrocoagulation reduced the total suspended solids (TSS), total organic carbon (TOC) and turbidity by ~84, ~75 and ~89% respectively. This resulted in lower membrane fouling and consequently, lower flux decline during DCMD. Long-term experiments were conducted, concentrating the feed water, containing 135 g L^{-1} TDS, up to 265 g L^{-1}. Development of fouling-resistant membrane surfaces is essential to minimize pretreatment costs and suppress the breakthrough of water. However, the development of economically viable membrane surface modification protocols will be essential. Here we have modified base membranes by grafting a variety of polymers (hydroxyethylmethacrylate, acrylic acid, polyionic liquids, zwitterionic groups) using UV initiated polymerization. In addition, surface oxidation of the membrane has been conducted using KMnO4 and NaOH. Our results indicate that by modification of the membrane surface the overall fouling resistance can be enhanced leading to lower electrocoagulation requirements.

E13 Synthesis of Polystyrene-b-poly(solketal methacrylate) (PS-b-PSMA) Polystyrene-b-poly(glyceryl methacrylate) (PS-b-PGMA) Block Copolymers for Isoporous Membranes, Sarah Saleem (Helmholtz-Zentrum Geesthacht); Sofia Rangou (Helmholtz-Zentrum Geesthacht); Clarissa Abetz (Helmholtz-Zentrum Geesthacht); Brigitte Lademann (Helmholtz-Zentrum Geesthacht); Volkan Filiz (Helmholtz-Zentrum Geesthacht)*; Volker Abetz (Helmholtz-Zentrum Geesthacht)

Block copolymers are well known for the formation of many microphase separated morphologies depending on the architecture of the block copolymer(s) involved, molecular weight, composition and thermodynamic properties. Anionic living polymerization provides a way to the creation of well-defined block copolymers by the sequential addition of monomers. It is mostly used to synthesize block copolymers with controlled compositional and structural parameters such as molecular weight, narrow molecular weight distribution, copolymer composition, branching, and other architectural parameters. In this work; we compare double hydrophobic polystyrene-b-poly(solketal methacrylate) (PS-b-PSMA) and amphiphilic polystyrene-b-poly(glyceryl methacrylate) (PS-b-PGMA) diblock copolymer membranes which are prepared by combining the block copolymer self-assembly in solution with a non-solvent induced phase separation (SNIPS)1. Diblock copolymers (i.e., PS-b-PSMA) were synthesized by sequential living anionic polymerization, whereas polystyrene-b-poly(glyceryl methacrylate) (PS-b-PGMA) were obtained by acid hydrolysis of the acetonide groups of the polysolketal methacrylate (PSMA) blocks into dihydroxyl groups (PGMA). Membrane structures and bulk morphologies were characterized by scanning electron microscopy (SEM) and transmission electron microscopy (TEM); respectively. The resulting PS-b-PGMA diblock copolymers produce an ordered hexagonal cylindrical pore structure during the SNIPS process, while membranes fabricated from the double hydrophobic (PS-b-PSMA) do not under similar experimental conditions. Membrane performance was evaluated by water flux and contact angle measurements.

E14 Uranium isolation and concentration using reactive membranes for nuclear forensics applications, Abenazer W Darge (Clemson University); Timothy DeVol (Clemson University); Scott Husson (Clemson University)*

This presentation describes the development of reactive membranes for the isolation and concentration of uranium from circumneutral pH solutions by ultrafiltration (UF). Polyethersulfone (PES) UF membranes were modified by grafting uranium-selective polymer films from surface of the membrane via UV-initiated polymerization. The resulting uranium-selective membranes (USMs) were used as a substrate for direct alpha spectrometric analysis of bound uranium isotopes. To guide membrane synthesis, we studied growth kinetics of the selective polymer film by measuring thickness as a function of UV-irradiation time using ellipsometry with a model PES substrate that simulates the surface of USMs. Alpha spectroscopy pulse height spectra were analyzed to determine the optimal selective layer thickness that would result in the highest uranium uptake without sacrificing energy resolution. Batch uptake experiments were conducted to determine uranium binding capacity of USMs at different concentrations. Flow through uptake experiments were conducted to identify the residence time for optimum detection efficiency at selected pH values. Results from this work are being used to develop a detection method for the rapid analysis of radionuclides in water using a field portable device, which would be a powerful nuclear forensics tool.

E15 Development of Nano-enabled Membrane Technology for Water Reuse in Agriculture, Xingmao Ma (Texas A&M University); Yinghao Wen (Texas A&M University)*; Jieming Yuan (Texas A&M University)

Due to water contamination and increased water demand for clean fresh water, conventional water supply is insufficient to meet all societal needs. Alternative sources such as treated effluents from wastewater treatment plants, brackish groundwater must be explored to augment current water supply. The need for alternative sources is particularly strong for the agricultural sector due to its vast water needs. Desalination can be a powerful technology to treat polluted water for water reuse in agriculture. Challenges in water reuse for agriculture include adequate removal of pollutants to ensure food safety, but not to over-treat the source water to remove essential nutrients. The technology for such purpose also needs to be cost effective and is suitable for large scale use. Therefore, there is a critical need for the membrane community to develop both highly permeable and selective membranes for water desalination. We investigated the effectiveness of nanoparticles enhanced ultrafiltration for water reuse in agriculture and compared the effectiveness of the newly synthesized membrane with several conventional ultrafiltration and nanofiltration membranes, using a synthetic wastewater effluent as our source water. The results demonstrate strong potential for the applications of membrane desalination in water reuse for agriculture.

E16 Detection of Electrically Conductive Membrane Fouling by Impedance Spectroscopy, Nan Zhang (McMaster University)*; Charles-François de Lannoy (McMaster University)

Monitoring membrane fouling is critical for the onset of fouling detection and effective fouling removal in ultrafiltration process. This work investigates the use of electrochemical impedance spectroscopy (EIS) on the surface of electrically conductive membranes to precisely measure the onset of surface fouling. An electrochemical cell was developed in which an electrically conductive membranes acted as a working electrode and a graphite electrode acted as the counter electrode. Conductive membranes were fabricated by coating single-walled/double-walled carbon nanotubes (SW/DWCNT) on polyethersulfone (PES) supporting membranes. Membrane fouling was simulated by pressure depositing different amounts of latex beads onto the surface of the membrane in a dead-end filtration cell. Changes in membrane permeance were measured to correlated fouling amount. Clean membranes had a permeance of 392 LMH/bar. Reduction to membrane permeance of 9.1%, 9.2%, 17.1% and 20.6% were observed for membranes covered with 0.028mg/m^2, 0.28 mg/m^2, 1.40 mg/m^2 and 2.80 mg/m^2 (approximately equivalent to 1/10, 1, 5, and 10 latex layers on the membranes), respectively. These small changes were statistically resolvable in Nyquist plots. It was observed that the higher frequency (103 ~ 106 Hz) diameter of the semicircle in Nyquist plots increased with greater fouling, while the impedance in the diffusion region (1 ~ 103 Hz) decreased with increased latex fouling. These observations were hypothesized to correspond to exposure area of conductive layer decline and negative-charged fouling particles accumulation at the interface. The interpretation for fouling layer formation was proposed by fitting EIS results with a proper equivalent circuit. Morphology of fouled membranes, surface conductivity, surface hydrophilicity, and pore size were measured by SEM, four-point probe conductivity analysis, contact angle measurements, and MWCO experiments, respectively.

E17 Stability of Polyamide Nanofiltration Membranes with Peracetic Acid/Hydrogen Peroxide Disinfection, Mohsen Ghafari (State University of New York at Buffalo)*; Boya Xiong (MIT); Ning Dai (State University of New York at Buffalo)

Chlorine is a conventional and effective disinfectant for biofouling control, but it reduces the lifetime of polyamide membranes. Peracetic acid/hydrogen peroxide (PAA/$H_2$02) mixture has been or proposed as a low-cost alternative disinfectant in wastewater treatment plants, but yet the impact on polyamide membranes has not been investigated at a mechanistic level. In this work, we compare the chemical mechanism of polyamide incubated with PAA/$H_2$02 mixture and chlorine. We demonstrate that PAA/$H_2$02 as a disinfectant with comparable anti-microbial activity to chlorine is capable of maintaining chemical properties and performance of polyamide membranes. The water flux and salt rejection of commercial flat sheet DOW NF90 membranes were monitored using a dead-end filtration cell after the exposure to PAA/H_2O_2 and NaOCl (100-24000 ppm·h) at pH=6.5. Results showed that PAA/H_2O_2 mixture was less detrimental to the membrane than chlorine. Water flux decreased by 18% after ~18000 ppm·h exposure to PAA/$H_2$02 mixture compared to 67% decrease after ~4000 ppm·h exposure to chlorine. In FTIR analysis, disappearance of the amide II and aromatic amide vibration peaks at 1541 and 1609 cm^{-1}, respectively, indicated the cleavage of amide cross-links in the membrane treated with chlorine. In contrast, no such change was observed for PAA/$H_2$02 treated membrane. Membrane treated with ~18000 ppm·h PAA/H_2O_2 increased surface hydrophilicity based on water contact angle measurements (62 to 46), while those treated with chlorine showed an initial increase and then a decrease in water contact angle, likely due to initial chlorination of -NH- and aromatic ring and subsequent degradation of amide groups to amine and carboxyl groups with continued chlorine exposure. Further characterizations including XPS, SEM, and AFM will be carried out to elucidate the mechanism. Findings will inform appropriate conditions to enable an effective biofouling control with maximized polyamide membrane lifetime.

E18 Fouling-Resistant Two-Dimensional (2D) Covalent Organic Framework Membranes for Industrial Water Reuse, Phuoc Duong (University of Wyoming); Valerie Kuehl (University of Wyoming); Mohammad Afroz (University of Wyoming); John Hoberg (University of Wyoming); Bruce Parkinson (University of Wyoming); Katie D Li-Oakey (University of Wyoming)*

Membranes with high permeability, selectivity, and fouling resistance provide significant advantages in water reuse and purification such as low consumption of energy, equipment and cleaning costs, and prolong membrane life span. In this study, a new highly porous two-dimensional (2D) covalent organic framework (COF) with carboxylic nanopores of approximately 2.8 nm was synthesized and used to form thin selective layers of composite membranes for shale gas produced water treatment. It is well-known that water transports predominantly through graphene oxide (GO) membranes via an interlayer spacing. In contrast, the high density of hydrophilic nanopores of the COF reported in this study allows for facile water transport primarily through the pores, resulting in a far superior water permeance of ~2260 L m^{-2} h^{-1} bar^{-1}, as compared to ~7.6 L m^{-2} h^{-1} bar^{-1} for the GO membranes under the same conditions. The COF membrane exhibits excellent performance for treatment of shale gas produced water with an oil rejection of 100% and a total dissolved solid (TDS) removal rate of 25%. In addition, the COF membrane shows enhanced fouling resistance compared to commercially available polymeric membranes.

E19 Reduced-holy graphene oxide(r-HGO) membranes with enhanced water permeance for water purification, Xiaoyi Chen (The State University of New York at Buffalo)*; Liang Huang (The State University of New York at Buffalo); Zhihao Feng (The State University of New York at Buffalo); Janavi Gohil (The State University of New York at Buffalo); Haiqing Lin (The State University of New York at Buffalo)

Reduced-graphene oxide (r-GO) membranes have emerged as a potential candidate for next-generation nanofiltration (NF) applications. However, previously reported r-GO membranes have lower water permeance than polyamide-based NF membranes because of the small 2D channels required to achieve good size-sieving ability. We report here an effective approach in enhancing the water permeance by creating nano-pores on the r-GO nanosheets. Specifically, GO sheets are chemically etched using H_2O_2 to form holey GO sheets, which are then deposited on porous supports before being reduced by HI. The prepared membranes were thoroughly characterized using SEM, Raman Spectroscopy, FTIR, and XRD. The effect of the etching and reduction conditions on the membrane performance is systematically investigated. Increasing the etching time increases the water permeance with slight decrease in the salt rejection rate, while increasing the reduction degree decreases the water permeance and increases the salt rejection. For instance, increasing the etching time from 0 h to 4 h improves the water permeance from 0.4 L m^{-2}·h^{-1}·bar^{-1} to 6.6 L m^{-2}·h^{-1}·bar^{-1} by over 15 times, while the Na_2SO_4 rejection remains at 98.5%. The r-HGO membranes exhibit NF performance comparable with the commercial polyamide-based membranes. For instance, when r-HGO with 4 h etching and a reduction using 34 wt.% HI was fabricated into a 20 nm-thick membrane, it exhibits water permeance of 12 L m^{-2}·h^{-1}·bar^{-1} and Na_2SO_4 rejection of 91.4% at 4 bar and 22°C. We have also prepared multi-layer r-HGO membranes with various degrees of etching and evaluated for NF applications. We will also present the long-term stability test results of these membranes using a crossflow system.

E20 A Comparison of PolarClean, Gamma-Valerolactone and their Mixture as Bio-derived Solvents for Polysulfone Membrane Fabrication, Xiaobo Dong (University of Kentucky)*; Halle Shannon (University of Kentucky); Isabel Escobar (University of Kentucky)

Petroleum-derived solvents used for membrane fabrication through the nonsolvent induced phase separation (NIPS) process are known for their toxicity. One big challenge is that by 2020, some common solvents, N-Methyl-2-Pyrrolidone (NMP) and Dimethylacetamide (DMAc) will be banned by Registration, Evaluation, Authorisation and Restriction of Chemicals (REACH) in Europe[1]. Therefore, bio-derived and low-toxicity solvents are a focus of increasing investigations. Examples of bio-derived, low-toxicity solvents include methyl-5-(dimethylamino)-2-methyl-5-oxopentanoate (PolarClean) and gamma-valerolactone (GVL). PolarClean has been previously used to fabricate polysulfone (PSf) membranes but having membrane pores collapse during membrane cleaning. To address this issue, GVL was investigated as a sole solvent and a co-solvent with PolarClean to fabricate PSf membranes[2,3]. Membranes prepared using GVL as a sole solvent were observed to be gelatinous, hence not ideal for filtration[2]. GVL and PolarClean were used as co-solvents, viable membranes were cast with surface charge and hydrophilicity not being significantly different from membranes made using PolarClean alone. Specifically, the solvent mixture produced PSf membranes with zeta potential values of about -20mV and contact angles at around 65°. Furthermore, the average pore size of membranes decreased from 48nm to lower than 1nm as adding GVL in dope solutions. The membranes prepared using solvent mixtures showed pure water flux at 300-500 LMH and over 99% BSA rejection. Therefore, the use of PolarClean/GVL as co-solvents showed promise for the fabrication of PSf membranes. 1. https://chemicalwatch.com/22940/reach-clp-hub-authorisation-vs-restriction 2. Dong, X.; Shannon, H.D.; Escobar, I.C. Green polymer chemistry: New products, processes, and applications, American Chemical Society: 2018; Vol. 1310, pp 385-403. 3. Wang, H.H.; Jung, J.T.; Kim, J.F.; Kim, S.; Drioli, E.; Lee, Y.M. J. Membr. Sci. 2019, 574, 44-54.

E21 Filterability of crude oil emulsions stabilized by COREXIT 9500 dispersant, Seyma Kucuk (Michigan State University); Charifa Hejase (Michigan State University)*; Iryna Kolesnyk (National University of Kyiv-Mohyla Academy); Jia Wei Chew (Nanyang Technological University); Vlad Tarabara (Michigan State University)

During the 2010 Deepwater Horizon oil spill, nearly 800 million liters of oil were released into the Gulf of Mexico polluting the deep ocean and more than 1600 km of the Gulf's shoreline. A large amount (1.84 million US gallons) of dispersant, Corexit 9500A, was applied to the spill; as a result, the oil was emulsified into microdroplets that remained suspended in the water column. Even though small droplets facilitate biodegradation, they can pose significant environmental risks. Membrane filtration is one of the most cost-effective remediation technologies that can remove emulsified oil. Despite extensive studies, mechanisms of membrane separation of emulsified oil remain poorly understood. In the present work, we investigate the impact of salinity on the properties and filterability of crude oil-in-water emulsions stabilized by Corexit 9500A. To our knowledge, this is the first study on membrane filtration of Corexit-stabilized crude oil emulsions. Direct Observation Through the Membrane (DOTM) technique was employed to visualize the droplets in real time on the surface of a microfiltration membrane. Corexit altered both the emulsion stability and the wetting properties of the membranes. Corexit decreased the oleophobicity of the membrane in both DI water and seawater. Emulsion characterization and constant pressure filtration tests indicate that salinity affects the stability of emulsions and can enhance microfiltration performance. Almost compete rejection of oil was achieved, but was accompanied by a precipitous flux decline. The low values of the steady state permeate flux indicate that microfiltration is suitable as a polishing step that follows an extensive pretreatment by large throughput deoiling unit processes such as hydrocyclonic separation or flotation.

~~**E22 Polyamide/CNT Thin Film Nanocomposite (TFN) Seawater RO Membranes with Improved Boron Rejection,**~~ ~~Aysa Guvensoy (Istanbul Technical University); Süer Kurklu-Kocaoglu (Istanbul Technical University); Cansu Yildirim (Istanbul Technical University); Sadiye Velioglu (Istanbul Technical University); S. Birgül Tantekin-Ersolmaz (Istanbul Technical University)*~~

~~Thin film nanocomposite (TFC) RO membranes have remarkably improved in terms of salt removal; however, removal of boron from seawater is an emerging problem. According to World Health Organization (WHO), boron concentration should not exceed 2.4 ppm for potable water and 0.5 ppm for irrigation water used for crops such as citrus fruits and nuts. Since boron is present in seawater as boric acid and its effective diameter is close to hydrogen-bonded cluster of water molecules, conventional TFC membranes fail to achieve effective boron removal. Incorporating nanomaterials into TFC membranes has potential for improving selectivity as well as water flux and anti-fouling properties. However, this requires optimizing the conditions for loading, size, type, and functional groups of the nanomaterials as well as the interfacial polymerization and post-treatment protocols. In this study, we investigate the influence of embedding carbon nanotubes functionalized with carboxyl groups into polyamide RO membranes on their boron rejection performance. Deposition and alignment of SWCNTs on a commercial polysulfone UF membrane was carried out by vacuum filtration. The polyamide layer was prepared by interfacial polymerization using m-phenylene diamine (MPD) and tri-mesoyl chloride (TMC) as monomers. Membranes were characterized by SEM, FTIR, XPS, polarized Raman spectroscopy, and crossflow filtration tests. Results showed that concentration of CNT significantly affects dispersion and alignment of CNTs inside the polymeric matrix as well as homogeneity, thickness and crosslinking ratio of polyamide layer. TFN membranes exhibited increased salt and boron rejection relative to neat polyamide membranes. To the best~~

~~of our knowledge, this is the first assessment of the performance of TFN desalination membranes for more effectively removing boron from seawater. This work is supported by the Scientific and Technological Research Council of Turkey (TUBITAK) through Project No. 114Y165.~~

E23 New generation of patterned membranes for water treatment, Asad Asad (University of Alberta); Dan Sameoto (University of Alberta); Mohtada Sadrzadeh (University of Alberta)*

Surface patterning is a promising method to enhance membranes performance without the need to apply chemical approaches. The addition of patterns simply increases the effective surface area and thus increases the water permeate flux without sacrificing selectivity. It also generates local streamlines at the vicinity of the pattern which might help reducing the attachment of fouling materials. In this study, a novel method is developed to prepare patterned membranes based on hydrogel facilitated phase separation (HFPS). HFPS method relies on the high-water content in hydrogel, which initiate phase separation once in contact with polymer solution resulting in a dense skin layer at the patterned side of the membrane. This method offers a facile approach for direct fabrication of various patterns with different dimensions on the membrane surface. The filtration performance of the HFPS patterned and unpatterned membranes were examined using a cross-flow filtration system. Results showed that the pure water flux through HFPS patterned was doubled compared with the unpatterned one. Moreover, fouling test with Bovine Serum Albumin (BSA) solution showed 76% increment in the flux results for the patterned membrane with almost the same rejection after 90 minutes of filtration demonstrating the significant advantages of patterned membranes for water treatment.

E24 Enhancing boron rejection on electrically conducting reverse osmosis membranes through local electrochemical pH modification, Bongyeon Jung (UCLA)*; David Jassby (UCLA)

As a result of growing population, demand on seawater desalination as a source of drinking water and irrigation water is growing. Among various desalination technologies, reverse osmosis (RO) takes up a large portion of world's desalination capacity due to its high salt removal efficiency. However, RO membranes show relatively low rejection rate on molecules having small size and neutral charge. Boron is one of the examples. Boron is not particularly harmful to mammals but can be lethal to highly boron-sensitive crops such as citrus. Thus, boron concentration should be monitored and strictly limited when desalinated water is used for agricultural purposes. A common practice in RO plants to reduce boron concentration of produced water is to elevate pH using alkaline chemicals. This result in transformation of boron from non-charged to charged and fully hydrated form, which can be easily removed by RO membranes. Although this method can successfully produce water containing small amount of boron, its application is limited by scaling. Thus, typically a double-pass design is employed, which raises water production cost. In this study, electrically conducting RO membranes were exploited to eliminate alkaline dosing and the double-pass design while achieving high boron removal efficiency. When membrane acting as a cathode was connected with an external potential source, hydroxide ions were generated along the membrane surface by water electrolysis. Therefore, pH in the permeate stream was elevated as potential applied to the membrane increased. In addition, boron elimination efficiency was significantly improved compared to membrane with no voltage applied.

E25 Cellulose Triacetate Membrane Parameter Estimation Under Osmotic and Osmotic-Assisted Process Conditions, Jacob L Weidman (U.S. Department of Energy National Energy Technology Laboratory/Oak Ridge Institute for Science and Education)*; Sara Osipi (Federal University of Rio de Janeiro); Alexander V Dudchenko (Carnegie Mellon University); Meagan S Mauter (Carnegie Mellon University); Nicholas Siefert (U.S. Department of Energy National Energy Technology Laboratory)

Power plants and oil & gas wells that produce high salinity brines as effluent pose a technological challenge for implementing waste treatments to achieve zero liquid discharge (ZLD). While mechanical vapor compression is a commercially-available option, emerging membrane technologies have the potential to lower the capital cost and energy costs associated with ZLD. Commercially-available reverse osmosis (RO) membranes are limited by the maximum pressure they can withstand, and this restricts their use to concentrations < 90 g/L. To overcome these limitations, an osmotically-assisted reverse osmosis (OARO) process can be used to produce potable water by subsequently stepping down salt concentration through a series of cellulose triacetate membranes, a commonly-used membrane material for seawater desalination. In this study, cellulose triacetate membranes were tested with sodium chloride solutions in a lab-scale OARO configuration to determine critical transport parameters of the membrane, including water permeability (A), salt permeability (B), and the membrane structural parameter (S). Various conditions were tested, including a range of pressures and concentrations, as well as different configurations of RO, FO, PRO, and OARO. Analysis of the results provided baseline values for the membrane parameters and demonstrated an expected high rejection of sodium chloride ions. However, the data also suggests that the A and S parameters are dependent on operating conditions, as compression of the support layers and possible dehydration of the membrane impact performance. These lab scale results provide a promising baseline for potential OARO applications, necessitating consideration of scale-up effects that would impact performance on the power plant scale. Further studies on OARO membranes will focus on utilizing hollow fiber modules and designing novel membranes with optimal parameters for more effective performance.

E26 Porous hydrophobic-hydrophilic composite flat and hollow fiber membranes for direct contact membrane distillation, Aishwarya Puranik (New Jersey Institute of Technology); Lydia Rodrigues (New Jersey Institute of Technology); John Chau (New Jersey Institute of Technology); Lin Li (New Jersey Institute of Technology); Kamalesh K Sirkar (New Jersey Institute of Technology)*; Ashok Sharma (Applied Membrane Technology Inc); Stephen Conover (Applied Membrane Technology Inc); Adam Juelfs (Applied Membrane Technology Inc); Connor Colling (Applied Membrane Technology Inc); Saket Sharma (Applied Membrane Technology Inc)

Among a number of innovations taking place in direct contact membrane distillation (DCMD)-based desalination, composite membrane structures are of increasing interest. Specifically, in terms of membrane structure, an integral composite of a thin hydrophobic porous coating plasma polymerized over a porous hydrophilic layer or a stacked composite of a thin porous hydrophobic layer over a more conventional hydrophilic porous membrane are of interest. Integral composite membranes in flat form or hollow fiber form have been investigated for desalination by DCMD. A class of such membranes of both types yielding quite high water vapor fluxes have been characterized using a variety of characterization techniques: Contact angle, liquid entry pressure (LEP), bubble-point pressure, scanning electron microscopy (SEM), energy-dispersive X-ray spectroscopy (EDS), atomic force microscopy (AFM). Novel conditions created lead to very high water vapor fluxes compared to those from conventional hydrophobic membranes supported on a mesh support.

E27 Polyamide-imide: a new platform for fabrication of oil/water separation membranes, Nusrat Helali (University of Alberta); Masoud Rastgar (University of Alberta); Mohtada Sadrzadeh (University of Alberta)*

Hydrophilic membranes with underwater oleophobic property can potentially enhance both water permeability and antifouling properties for oily wastewater treatment. In this study, polyamide-imide (PAI) was used to make porous hydrophilic membranes via non-solvent induced phase inversion technique. To increase the hydrophilicity of membrane further, hydrophilic additives PEG (0.4 and 6.0 kDa) and PVP (10 and 360 kDa) were added to the polymer solution. Oil rejection experiments were conducted for three consecutive cycles at a very low transmembrane pressure i.e., 7 psi to evaluate the reusability of the synthesized membranes. Results showed that the use of additives resulted in a wide range of membrane morphology, porosity, pore size and surface chemical property, which influenced the water permeation and flux recovery. However, all prepared ultrafiltration membranes showed above 98% rejection. Membranes modified with PEG 0.4 and 6 kDa and PVP 10 kDa had underwater oil contact angle (OCA) above 155°, and showed lower flux decline and higher flux recovery (FRR) in all three cycles. In the case of additive PVP 360 kDa, a lower flux recovery observed due to decrease in oleophobicity (OCA=124°).

E28 Monolayer Graphene transfer onto polyvinyl alcohol (PVA) microfiltration membrane for water desalination., Mansour Saberi (Clemson University)*

Fresh and clean water is a primary concern for all the people around the world. So far, seawater desalination has been mainly performed via multistage flash distillation and reverse osmosis. Reverse osmosis (RO) is of interest in brackish water desalination because of near-resistance-free extremely-fast transport of water molecules through a membrane while obstructing all ions. One of the most famous and recently studied candidates for use in RO membranes is graphene which has excellent thermal, chemical, and mechanical stability and it is the most prominent thinnest possible membrane with its one atom thickness acting as the membrane. A crucial stage in the fabrication of an efficient graphene RO membrane is a suitable substrate that can offer the essential support to its single atom thick layer while also allowing for rapid water transmission. This poster presentation will focus on poly (vinyl alcohol) that is cross-linked to make a hydrogel as a support for graphene sheet as a RO membrane. The graphene is transferred onto PVA support by a novel method to decrease the chance of defect formation. Experimental work combined with membrane characterization methods (FTIR, SEM, and TEM) and filtration studies using a home-made RO system to examine the graphene as a RO membrane will be presented. These results will be compared with the other synthesized membrane and different type of supports.

E29 Membrane Distillation-Crystallization (MDC) for Maximum Water Recovery in Inland Desalination, Evangelos Balis (University of Nevada, Reno); Sage Hiibel (University of Nevada, Reno)*

Alternative freshwater demand is increasing around the world, and many coastal regions rely on seawater desalination for potable water. For inland regions brackish ground waters are an alternative water source, and desalination can potentially provide potable water. Unfortunately, most desalination techniques developed for coastal regions cannot be directly translated to their full extent inland due to water chemistry differences between seawater and brackish waters. Concentrated brine disposal is also a challenge in inland areas as the ocean cannot serve as a repository for the brines. Thus, zero liquid discharge (ZLD) approaches that minimize the concentrated waste streams while maximizing water recovery are desired. This work couples membrane distillation (MD) and crystallization to effectively treat RO brine resulting from desalinating inland brackish waters with the ultimate goal of ZLD. MD is a non-pressure-driven process that uses a hydrophobic, microporous membrane. Compared to conventional distillation, MD requires small temperature differences and is well suited for coupling to other processes. Because the driving force of MD is the vapor pressure of the brackish water, which is relatively independent of salt concentration, it is an ideal technology because it can treat brines to near saturation. When coupled with crystallization, the MDC system can remove salt as solid crystals away from the membrane surface to maximize membrane longevity. The MDC system was tested using a hyper saline feed solution (323 g/L NaCl). After 9 hours of operation, the MDC system was able to remove 2.5 times more water from the feed solution than MD alone. This was achieved by capturing 226 g of solid salt in the crystallizer. This work validates MDC as an effective method for reducing feed volumes beyond the threshold of the saturation limit, and also offers the potential of a value-added byproduct in the solid salt that can be sold to offset operational costs.

E30 Insights regarding thermomechanical bonding between porous membranes and thermoplastic polymers, Masoud Aghajani (MAST Center, University of Colorado Boulder)*; Adrienne Blevins (University of Colorado Boulder); Jason P. Killgore (National Institute of Standards and Technology); Ryan Sylvia (MilliporeSigma); Christina Carbrello (MilliporeSigma); Alan R. Greenberg (MAST Center, University of Colorado Boulder); Rong Long (MAST Center, University of Colorado Boulder); Yifu Ding (MAST Center, University of Colorado Boulder)

During filter device fabrication, a porous membrane needs to be intimately bonded with a thermoplastic polymer under controlled thermomechanical conditions. Under mechanical loading, the viscous polymer flows and infiltrates the membrane pores to create both mechanical interlock and interfacial adhesion. Failure to do so can result in product failure and thus increase the overall cost of manufacturing. Because of its practical importance, there is a need for greater understanding of the underlying mechanisms that drive bonding process. Fundamentally, the complexity of this bonding process rises from the interplay between polymer pore-flow and membrane densification, both of which are functions of temperature, pressure and time. In this study, we aim to understand the nature of this interplay so that successful bonding can be carried out in a predictive way. We systematically investigate the thermomechanical bonding between model membranes (with varying pore size and chemistry) and selected polymers. We utilize a suite of AFM-based nanomechanical measurements that enable us to precisely characterize not only the degree of infiltration but also local interfacial structures. In this presentation, we will discuss the relationships between thermomechanical bonding conditions, properties of both the polymers and the porous membranes, and the effectiveness of the bonding.

E31 Ion exchange polymer coatings enhance solute rejection of polyamide thin-film composite membranes, Mikayla D Armstrong (University of North Carolina at Chapel Hill)*; Ryan Kingsbury (Membrion, Inc.); Kasia Grzebyk (University of North Carolina at Chapel Hill); Orlando Coronell (University of North Carolina at Chapel Hill)

Significant research efforts to enhance the water permeability of desalination membranes has substantially improved their energy efficiency. However, a recent review made a strong case for the importance of improving water-solute selectivity (i.e., solute rejection) over water permeability (Werber et al, 2016, Environ. Sci. Technol. Lett., 3, 112-120). Our recent work shows that ion exchange (IX) polymers are better at excluding inorganic solutes from the membrane matrix than polyamide active layers of thin-film composite (TFC) membranes, while solutes are less mobile in polyamide than in IX polymers. Accordingly, to leverage the strengths of different polymers for enhanced water-solute selectivity, we coated polyamide TFCs with an IX polymer, yielding membranes with the high solute exclusion properties of IX polymers and low solute mobility properties of polyamide. The polyamide TFCs that were coated included three commercial membranes and membranes prepared in-house by interfacial polymerization of trimesoyl chloride with m-phenylenediamine (MPD) or piperazine (PIP). We evaluated solute rejection (i.e., NaCl, trivalent arsenic As(III) and boron (B)) and water permeability of uncoated and coated membranes in a dead-end filtration system with simulated groundwater. All coated membranes rejected solutes better than, or similar to, their uncoated controls, but displayed lower water permeability due to uncontrolled coating thicknesses. Notably, NF90 solute rejection increased from 89% to 91.7% for NaCl, 59.2% to 83.5% for As(III), and 66.0% to 87.8% for B. Solute rejection by the TFC-MPD membrane increased from 69.9% to 86.3% for As(III), and 63.6% to 74.7% for B, and remained unchanged at 95% for NaCl. Coating polyamide membranes with an IX polymer resulted in hybrid membranes with improved ability to reject solutes from simulated groundwater. Future work will focus on increasing water permeability by minimizing coating thickness.

E32 Mechanical behaviour of the porous support of TFC membranes and its influence on the overall resistance, Masoud Aghajani (University of Colorado Boulder)*; Alan R. Greenberg (University of Colorado Boulder); Yifu Ding (University of Colorado Boulder)

It is commonly believed that the overall resistance of thin film composite (TFC) membranes is dictated by the crosslinked, ultrathin polyamide barrier layer, while the porous supports merely serve as the mechanical support. However, this understanding is based on the "separate" rating of the resistance of the porous support under high trans-membrane pressure (TMP), which is distinctively different from the situation during high-pressure reverse osmosis (RO). In this study, we used two different methods to study the mechanical behavior of the porous support and how it affects the resistance of TFC membranes. For the first method, we used nanoimprint lithography to achieve precise compressive deformation in commercial TFC membranes. Based on DI water permeation measurements, the intrinsic membrane resistance showed negligible changes at strain levels lower than 30-40%, but increased exponentially at higher strain level of 40-60%. These results suggested substantial contribution of the porous support in the overall resistance at high strain values. Therefore, in our second method the goal was to measure the true resistance of the porous support during high-pressure RO. To this purpose we developed a new measurement method that uses a series arrangement of a TFC membrane on top of a typical porous support. Our results show that, under high overall TMP (> 5 MPa), the plastic deformation of the additional porous support is as much as 3X larger than that of the same support subjected to identical TMP but without the TFC membrane on top. Moreover, DI water permeation experiments revealed as much as an order of magnitude increase in resistance for the deformed porous supports in comparison to the pristine (undeformed) ones. Most importantly, our analysis indicates that the resistance of a representative commercial porous support used to fabricate TFC membranes can account for up to 33% of the overall TFC membrane resistance during high pressure permeation (5.5-8.2 MPa).

E33 Development of Membrane Distillation through Tuning Membrane Surface Properties and Module Design, Siamak Nejati (University of Nebraska-Lincoln)*; Mahdi Mohammadi (University of Nebraska-Lincoln); Mona Bavarian (University of Nebraska-Lincoln)

Membranes with special wettability have a variety of applications. Our focus here is on the membrane distillation (MD) process. Designing membranes for the MD process is tightly bound to feed composition and properties. Because the MD process is accompanied by phase change, interfacial and surface properties of materials play a key role in determining the performance of the membranes. The other critical factor in MD operation is thermal management; this concept is the bottleneck of the MD process and better thermal management, even at the membrane level, can dramatically enhance membrane performance. Here we present the predicted performance of the MD membrane, using first principle multiphysics modeling; we will demystify the underlying phenomena governing the MD process for different membrane structures. We will design composite membranes based on the modeling results and fabricate them through a combined phase inversion, spray-on, and state-of-the-art vapor phase polymerization. The experimental result will be compared to our model prediction. A few examples of these membranes, e.g., Janus membranes and omniphobic membranes, will be demonstrated. Particularly, the omniphobic properties of the created surfaces will be highlighted. We will demonstrate that the omniphobic structure of the fabricated membranes hinders the wetting by large droplets, (ca. 1.5 mm in diameter), and how low surface energy liquids come into contact with the surface with an impact velocity of around 2 m/s. Additionally, we will demonstrate that through proper design, we can achieve permeate flux exceeding what has been reported in the literature for MD membranes. The fabricated membranes were challenged by complex feed compositions, and accelerated scaling and oil wetting tests were performed. Lastly, we demonstrate the integration of a wide-band IR absorber into the module design to realize the development of a solar desalination technology.

E34 Integration of Forward Osmosis in the Treatment of Sewage by Chlorella vulgaris: Comparison between External and Immersed Systems, Xue Jin (Oregon State University)*; Mathieu Larronde-Larretche (University of Glasgow)

Microalgae have attracted increasing attention due to their promising application in sustainable biofuel generation, wastewater remediation, carbon dioxide sequestration and pharmaceuticals production [1]. Despite the promise, one technical challenge remaining to be overcome is the high energy cost of algae harvesting and dewatering that accounts for 20-30% of the total operating cost [2]. In our previous study, the feasibility of microalgae dewatering using forward osmosis (FO) has been demonstrated. The key fouling mechanisms, involving the excretion of carbohydrate by the microalgae under stress conditions, have also been described [3-4]. In this study, we further developed the integration of FO dewatering with wastewater treatment by microalgae. An immersed and an external FO cell were compared based on nutrients recovery, biomass recovery, filtration performances, and long term efficiency. Our findings reveals that the wastewater treatment efficiency (nutrients removal), as well as the production of biomass were greater with the immersed FO system due to a greater microalgae growth. However, these may not be sustainable in a long term operation of the immersed system. The external FO system was found better in terms of salinity build-up and FO dewatering performances. Overall, an external FO dewatering is recommended due to its better flexibility and sustainability. Reference: [1] S.A. Razzak, et al., Renew. Sust. Energ. Rev. 27 (2013) 622-653. [2] C. Gudin, C. Thepenier, Adv. Biotechnol. Process. 6 (1986) 73-110. [3] M. Larronde-Larretche, X. Jin, Algal Research, 23 (2017). 12-19. [4] M. Larronde-Larretche, X. Jin, Algal Research, 15 (2016). 1-8.

E35 Nanoengineered condensation surfaces for air gap membrane distillation, Yashwant S. Yogi (Purdue University); Sina Nejati (Purdue University); Akshay K. Rao (Purdue University); Rishav Roy (Purdue University); Abhimanyu Das, Longnan Li (University of Illinois, Urbana-Champaign); Soumyadip Sett (University of Illinois, Urbana-Champaign); John Lienhard (MIT); Nenad Miljkovic (University of Illinois, Urbana-Champaign); Justin A. Weibel (Purdue University); Jaichander Swaminathan (Massachusetts Institute of Technology); David Warsinger (Purdue University)*

The thermal desalination technology air gap membrane distillation (AGMD) is an energy efficient configuration, being especially dominant at high salinity. however, the large mass transfer resistance of the air gap dramatically reduces permeate flux, impairing performance. Higher condensation performance can be achieved through smaller air gap size, but with normal condensate flow pattern (film type) the air gap gets flooded before reaching the optimal gap size (<1mm). It was found through system level numerical modeling and experimentation that condensing flow regimes like dropwise and jumping droplet achieved using hydrophobic and superhydrophobic condensing surfaces respectively were able to both improve the mass transfer coefficient in the airgap, and improve droplet shedding, allowing for thinner gap sizes. This allowed increases of GOR (Energy efficiency) by 2.1x while avoiding flooding even at gap size of 0.2mm. More novel condensing surfaces like a combination of the superhydrophobic and superhydrophilic pattern created pathways for faster permeate removal producing more complex flow regimes.

E36 Fouling of Anion Exchange Membranes in Ferric Chloride Solutions: mechanism and the role of ion exchange capacity, water content, and pore structure, Michael J. McGrath (University of Colorado Boulder)*; Nicholas Patterson (UC San Diego); Bryce Manubay (University of Colorado Boulder); Hans Funke (University of Colorado); Samantha Hardy (University of Colorado Boulder); Andrew Basalla (University of Colorado Boulder); Xiujun Yue (UC San Diego); Ping Liu (UC San Diego); Douglas Gin (University of Colorado Boulder); Richard Noble (University of Colorado Boulder)

Anion exchange membranes (AEMs) have potential applications in redox flow batteries (RFBs) because of their potential to prevent electroactive metal ionic species crossover, and therefore prevent self-discharge. The all iron flow battery (AIFB) has been evaluated as a prime example thanks to the economical advantages stemming from the abundance of iron. However, fouling of AEMs has previously been observed in chloride-based AIFBs due to the presence of the ferric tetrachloride anionic complex and subsequent dehydration of the membrane. A phase-segregated AEM made from a polymerizable surfactant with positively charged discrete 1-nm pores was hypothesized to sieve iron species from the pores to prevent fouling. However, the membrane still rapidly fouled. Impedance measurements indicate the conductivity of AEMs drops rapidly upon exposure to ferric chloride solutions. Raman spectroscopy confirms iron tetrachloride is able to enter the membrane. SEM/EDX examines the distribution of iron across the membrane during the relevant time-scale of fouling. Iron tetrachloride is likely less hydrated compared to Cl-, supporting the observation that iron tetrachloride leads to dehydration of the pores. The fouling of the phase-segregated, polymerizable surfactant AEM is compared to that of three commercial homogeneous AEMs with different ion exchange capacities and water contents. We show that fouling occurs at different iron chloride concentrations for different AEM's, depending on their ion-exchange capacity and water content. Discrete pore modification of the phase-segregated AEM affects its fouling kinetics, demonstrating the importance of membrane structure and water domain environment on the fouling behavior of AEMs. Research of AEMs for AIFBs should recognize the need to prevent iron tetrachloride from entering the membrane to prevent fouling; suggestions for modification of AEMs to prevent fouling are made.

SESSION F - Innovation in Modeling of Membrane Fundamentals and Processes

F1 Molecular Dynamics Processing Tool for Transmembrane Simulations, Jia Lin Cheoh (Purdue University)*; Abhimanyu Das (Purdue University); David Warsinger (Purdue Unversity)

Molecular dynamics has become a widely used simulation method to investigate the properties of membranes in advancing technologies for a variety of fields including but not limited to renewable energy, water processing, and smart health applications. Advancement in high- performance computing and molecular dynamics simulations make it possible to perform simulations that closely models real phenomenon on transmembrane channels. However, performing molecular dynamics simulations on transmembrane channels is both time-consuming and difficult to comprehend for non-trained scientists in this field. We develop a simplify molecular dynamics simulations tool that sets the only requirements for end users to be that they have the high-performance computing power and the relevant input files without having to perform all the intermediary molecular dynamics processing steps. Our research will present implications for future interdisciplinary research at the intersection of membrane separation methods and high-performance simulations techniques, and further advance the pace of membrane-related research with the explicit solvent simulation method.

F2 Low energy-consumption pre-treatment system for integrated forward osmosis - reverse osmosis desalination process, Yunchul Woo (Korea Institute of Civil Engineering and Building Technology)*; June-Seok Choi (Korea Institute of Civil Engineering and Building Technology); Kwang-Duk Park (Korea Institute of Civil Engineering and Building Technology)

Forward osmosis (FO) process has been attracting attention for its potential applications such as industrial wastewater treatment, wastewater reclamation, and seawater desalination. Particularly, in terms of fouling reversibility and operating energy consumption, the FO process is assumed to be more preferable to the RO process. Moreover, hybrid FO-RO membrane systems should have an economic advantage compared to currently available technology for desalination such as a standalone RO model, and comparable costs with wastewater treatment and recovery system. The objective of this study was to evaluate the best economical, viewpoint of operating energy consumption and fouling, pretreatment system for FO-RO hybrid desalination process. Various processes were used for FO pretreatment of real secondary wastewater effluent. SWRO was operated with a diluted draw solution, such as treated secondary wastewater effluent by FO system. With applying above fouling control strategies, FO and RO fouling tests were conducted for 12 to 48 h and then flux decline rates were compared based on the initial flux of each case. Energy consumption during each FO-RO with the pretreatments was calculated by operating pressure and measured feed water flux. Our results suggest that membrane fouling can be effectively controlled by the pretreatment process. Moreover, the osmotic dilution process by the FO system is a non-pressurized membrane system and, thus, the energy consumption of the overall desalination system was the lower than standalone RO system. In addition, the structure of the fouling layer on the FO membrane is relatively weak and reversible enough to be disrupted by simply hydrodynamic washing. Therefore, the FO-RO process with low salinity feed water through a suitable pretreatment system is possible as a less energy consuming desalination system with efficient membrane fouling control.

F3 Arrested mobility effects on the spinodal decomposition of ternary polymer solutions for immersion precipitation, Jan Garcia (UC Santa Barbara)*; Douglas Tree (Brigham Young University); Kris Delaney (UC Santa Barbara); Glenn Fredrickson (UC Santa Barbara)

Many polymer membranes are made by immersion precipitation: a polymer solution film is immersed in a nonsolvent bath, inducing phase separation of the film into a polymer-rich phase that becomes the membrane matrix and a polymer-poor phase that becomes the membrane pores. Microstructure formation of these membranes is still not fully understood, given the interdepedendent nature of several physical processes: the mass transfer of solvent and nonsolvent through the film-bath interface, the ensuing phase separation of the film, the coarsening of domains, and finally the glass transition of the polymer-rich phase that arrests the membrane microstructure. In this work, we use phase-field models of the polymer-nonsolvent-solvent system to solve the coupled convection-diffusion and momentum equations that describe the immersion precipitation process. We model the glass transition using contrasts in the viscosity and mobility of the polymer-rich and polymer-poor phases. We report how the glassy dynamics changes the microstructures formed by bulk spinodal decomposition. We also study how the mass transfer kinetics between the bath and film are changed by the formation of a glassy interface.

F4 Hydro- and perfluoro-carbon sorption, diffusion and permeation modeling in poly(dimethylsiloxane) using a PC-SAFT Equation of State, Liang Liu (The University of Melbourne)*

The transport behavior of gases and vapors in poly(dimethylsiloxane) (PDMS) has been investigated extensively due to its easy processability and good separation performance. As a rubbery polymer, the membrane selectivity is mainly determined by penetrant solubility. The Flory-Huggins model is widely used to model this sorption behavior. Although this model works well for small gases such as N_2, CO_2 and CH_4, it lacks predictive ability for some other penetrants, for example, hydro- and perfluoro-carbons, as a concentration dependent interaction parameter must be used to achieve reasonable results. It is of great interest to develop a fundamental model to investigate the thermodynamic properties of these penetrant-polymer systems, so that they can be incorporated into an appropriate diffusion model to determine the transport properties of the membrane. In this work, the Perturbed Chain Statistical Associating Fluid Theory (PC-SAFT) equation of state was applied to investigate the sorption of small gases, hydrocarbons and perfluorocarbons in PDMS. All model results agreed well with experimental data and only one fixed interaction parameter over the entire experimental pressure range was used for each penetrant-polymer pair. The sorption behavior of gas mixtures in the polymer matrix was also investigated, without any adjustable parameter, indicating the versatility of the model. Moreover, permeation data could be fitted to a solution-diffusion model incorporating this approach. The diffusivities and concentration profiles of the penetrants across the membrane were also explored.

F5 The Potential of Low-Temperature Forward Osmosis for Water Recovery, Jeffrey Martin (University of Toronto); Georgios Kolliopoulos (University of Toronto); Vladimiros Papangelakis (University of Toronto)*

Industrial wastewater treatment poses significant challenges, for it is necessary to recover clean water for reuse while reducing the volume of the effluent. Compared to traditional water recovery technologies, forward osmosis (FO) is an attractive solution due to its low fouling potential and favourable energy consumption. Forward osmosis uses a concentrated draw solution to spontaneously pull water across a semipermeable membrane from an aqueous process stream, rejecting and concentrating solutes in the process water. There is significant potential for FO to recover water from low-temperature effluents such as mining activities in the winter in the north, yet the effects of temperature on the performance of FO have only been investigated in the literature for temperatures greater than 20°C. There is a lack of understanding of FO performance at temperatures between the solution freezing point and room temperature. To address this gap, we have experimentally studied the performance of a Sterlich forward osmosis membrane cell at low temperatures, by measuring the water flux and reverse draw solution flux for different draw solution concentrations. It was found that the water flux across the membrane can be reduced by 50% due to the lowering of the osmotic driving force, the decreasing draw solute diffusivity and increasing draw solution viscosity. However, this increased mass transfer resistance also inhibits the reverse draw solute flux, which provides a greater water flux selectivity at lower temperatures. Based on this work, the potential of FO for low-temperature applications can be optimized through a detailed understanding of its performance, elucidating the water flux selectivity and determining the optimal sub-room temperature operating window for the FO process.

F6 Fouling mechanisms in constant flux crossflow ultrafiltration, Alon Kirschner (The University of Texas at Austin)*; Yu-Heng Cheng (The University of Texas at Austin); Donald Paul (The University of Texas at Austin); Robert Field (University of Oxford); Benny Freeman (The University of Texas at Austin)

Four fouling models due to Hermia (complete pore blocking, intermediate pore blocking, cake filtration and standard pore blocking), have long been used to describe membrane filtration and fouling in constant transmembrane pressure operation of membranes. A few studies apply these models to constant flux dead-end filtration systems. However, these models have not been reported for constant flux crossflow filtration, despite the frequent use of this mode of membrane operation in practical applications. We report derivation of these models for constant flux crossflow filtration. Of the four models, complete pore blocking and standard pore blocking were deemed inapplicable due to contradicting assumptions and relevance, respectively. Constant flux crossflow fouling experiments of dilute latex bead suspensions and soybean oil emulsions were conducted on commercial poly (ether sulfone) flat sheet ultrafiltration membranes to explore the models' abilities to describe such data. A model combining intermediate pore blocking and cake filtration appeared to give the best agreement with the experimental data. Below the threshold flux, both the intermediate pore blocking model and the combined model fit the data well. As permeate flux approached and passed the threshold flux, the combined model was required for accurate fits. Based on this observation, a physical interpretation of the threshold flux is proposed: the threshold flux is the flux below which cake buildup is negligible and above which cake filtration becomes the dominant fouling mechanism.

F7 Hitz zeolite membrane separation system (HDS) for liquid separation, Akira Hamasaki (Hitachi Zosen Corporation)*

Hitachi Zosen Corporation (Hitz) has actively developed and commercialized our zeolite membrane separation systems for several kinds of industry use in liquid- and gas-separation. Hitz membrane elements are 16 mm in diameter and 1130 mm in length. The element is composed of a Al_2O_3 porous support tube with a closed end and a zeolite thin film deposited on outer surface of the tube by a hydrothermal synthesis process. The element constitution is only inorganic material, which enables the membrane to be used under high temperature and high-pressure conditions. We have developed six varieties of zeolite membranes (HZM-1 to HZM-6) with different composition, pore size and chemical affinity and the optimized zeolite membrane is selected depending on the separation use and the customer requirement. Regarding the liquid separation use, HZM-1 and HZM-2 membranes are used for several solvents dehydration and HZM-3 to HZM-5 membranes are used for the dehydration of the solvents with higher water content and/or acidity. The performance of the membrane elements has been tested in pervaporation (PV) and vapor permeation (VP) process. For example, the VP dehydration flux and separation factor of HZM-1 membrane element for the feed of ethanol (90 wt%)/water (10 wt%) at the temperature of 130 degree centigrade was 50 kg m^{-2} h^{-1} and several thousand or more, respectively. These performance is 6 times higher than that of the conventional zeolite membrane. Hitz dehydration system (HDS) was adopted for the first commercial bio-ethanol plant with the capacity of 50,000 L/d in Japan. The system has been operated successfully for about six and half years without any trouble and membrane replacement and it is found that the bio-ethanol product purity is 99.9 to 99.95 vol% at the capacity of 50,000 to 53,000 L/d.

F8 Effect of short-term contact with C1-C4 alcohols on the water permeance and salt rejection of MPD-TMC thin-film composite reverse osmosis membranes, Jaime A Idarraga-Mora (Clemson University)*; Michael Lemelin (Clemson University); Steven Weinman (Clemson University); Scott Husson (Clemson University)

This poster discusses the effect of C1-C4 alcohols on the transport properties of thin-film composite reverse osmosis membranes. Five commercial membranes were studied using direct-flow filtration to quantify the changes in water permeance and sodium chloride rejection before and after contact with five different C1-C4 alcohols. Membrane performance was correlated with active layer characteristics. Young's modulus measurements of the surface showed decreased stiffness of the active layer after contacting the membranes with alcohol, suggesting a plasticization effect. Water permeance generally increased without decreasing rejection after short-term alcohol contact. The extent of these changes depends on the membrane and alcohol used. Analysis of the data using a dual-sorption model found correlations between the change in water permeance after alcohol contact and both the initial water permeance of the membrane and the change in free energy of mixing of water and the alcohols. We suggest that the mixing of water with the alcohols facilitates penetration of the alcohols into the active layer, likely by disrupting inter-chain hydrogen bonds, thus increasing the free volume for water permeation. Our studies provide a modeling framework to estimate the changes in transport properties after short-term contact with C1-C4 alcohols.

F9 Whey recovery using forward osmosis with hollow fiber configuration - Evaluating the factors limiting the flux performance, YiNing Wang (Nanyang Technological University)*; Rong Wang (Nanyang Technological University)

The focus of whey processing has been changed from waste treatment to the production of valuable products owing to the advancement of membrane technology. Producing high quality whey products at low energy consumption entails novel techniques for whey concentration. This study proposes forward osmosis (FO) as a potential approach to concentrate whey protein solutions, which is superior to the concentration via reverse osmosis (RO), mainly owing to the low hydraulic pressure during FO process. In particular, this work validated the feasibility of FO-based whey concentration using high-performance hollow fiber (HF) membranes that were fabricated in-house. The investigation focused on the effects of various operating conditions including different modes of physical cleaning on the concentration performance. The experimental results reveal that optimal concentration efficiency would be achieved by appropriately choosing the cross-flow velocity, draw solution concentration and the maximum attainable concentration of whey protein solution. It is worth noting that a rapid formation of a polarized or gel foulant layer at the membrane surface takes place (fouling accumulation with time is of minor importance in this case). This proposed mechanism is evidenced from the flux decline behaviour at different whey protein concentration. It is further supported by the results of physical cleaning with water, showing that the post-process cleaning after one concentration cycle was able to recover the membrane performance, whereas the intermediate cleaning during one concentration cycle had little impact on the concentration efficiency enhancement. This study has practical significance for the application of FO in the concentration of dairy streams and other streams with high solid content.

F10 Bimetallic Nanoparticles Incorporated Poly(methacrylic acid) Functionalized Membranes for Water Remediation: Synthesis, Advanced Characterization and Reactive Properties, Hongyi (Derek) Wan (University of Kentucky)*; Md. Saiful Islam (University of Kentucky); Nicolas Briot (University of Kentucky); Anthony Saad (University of Kentucky); Lindell Ormsbee (University of Kentucky); DB Bhattacharyya (University of Kentucky)

The dechlorination of toxic chlorinated organics in groundwater has been reported using bimetallic (Fe/Pd) nanoparticle methods. However, the aggregation as well as the difficulties to recycle for the particles hinder the application in field. The integration of catalytic particles and functionalized commercial membranes was achieved (in both commercial flat-sheet membranes and spiral membrane modules) to solve these issues. The Fe/Pd nanoparticles were synthesized inside the pores of membrane through in situ reduction after ion-exchange with the carboxylate groups (methacylic acid (MAA) was polymerized inside the pores). The functionalized MAA controlled the size of particles and recapture the Fe after dechlorination treatment. The reactive membranes were tested in the treatment of lab synergic polychlorinated biphenyls and field groundwater (from a Superfund site, which includes trichloroethylene, carbon tetrachloride, etc.). At 2.2 s residence time, 90% of chlorinated organic species (10-1000 ppb) in the groundwater were dechlorinated in the convective flow mode. The pilot studies with functionalized commercial spiral membrane modules are in progress. Advanced characterization methods, such as FIB, TEM, XPS were applied to understand the correlation between nanoparticle properties and depth inside membrane pores. The depth profile of functionalized membrane indicated that particle size was uniform inside membrane pores (24±6 nm) but slightly smaller than those nanoparticles located on the surface (39±9 nm). Furthermore, the use of TEM and XRD enables the calculation of Pd monolayer coverage on the Fe surface. The effects of Pd surface coverage and particle compositions on the H_2 production (Fe corrosion in water) and corresponding dechlorination rate were evaluated to find the optimal Pd content for the dechlorination. This research is supported by the NIEHS-SRP grant P42ES007380. Partial support is also provided by NSF KY EPSCoR grant (Grant no: 1355438).

F11 Membrane compaction in forward osmosis process, Daniel Yee Fan Ng (Nanyang Technological University); Yunfeng Chen (Nanyang Technological University); Rong Wang (Nanyang Technological University)*; Zhili Dong (Nanyang Technological University)

Membrane compaction is commonly observed in polymeric membranes when subjected to elevated hydraulic pressure. Such phenomena is thought to be the main cause of irreversible reduction in membrane permeability for most hydraulic pressure-driven membrane separation process. In forward osmosis (FO) process, it is commonly assumed that there were little hydraulic pressure exerted on the membrane, and therefore the highly porous membrane should experience negligible compaction. In the current investigation, three types of TFC membranes were fabricated from different hollow fiber substrates that possess different water permeability. The TFC membranes were continuously tested in FO experiments for 24 hours by using DI water as feed and NaCl solutions of different concentration as draw, and their performance were checked by with fresh feed solutions. At the end of FO experiments, the water flux of the TFC membrane with the most permeable support layer decreased by 9% and 21% , whereas the water flux of the TFC membrane with the less permeable support layer decreased by 32% and 52%, when 0.5 M NaCl and 2.0 M NaCl were used as draw solution, respectively. Furthermore, unexpected "plummeting" water flux pattern was also observed on the TFC membrane with the least permeable support layer when switching to higher draw concentration in the FO experiments. Visible changes in the cross-section morphology and surface topography of the TFC membranes were also observed under SEM and AFM after the FO experiments, suggesting that these signs of membrane compaction could be driven by "negative hydrostatic pressure" build-up within the support layer of the TFC membranes. The current study could shed light on the important considerations that should be taken into account during membrane fabrication and when establishing the testing protocols for characterization of forward osmosis membrane.

F12 Novel Isothermal Acid Sweep Membrane Distillation for Selective and Energy Efficient Removal and Recovery of Ammonia, Stephanie N McCartney (Columbia University)*; Natalie Williams (Columbia University); Chanhee Boo (Columbia University); Ngai Yin Yip (Columbia University)

In membrane distillation (MD), the permeation of volatile components is traditionally driven by a partial vapor pressure gradient set up by a transmembrane temperature differential. Conventional MD can be employed for the separation of volatile components, e.g., soluble gases such as ammonia, from aqueous feed streams and collected on the sweep side, but the process suffers from undesired simultaneous water permeation that dilutes the sweep stream and incurs large energy losses due to the enthalpy of vaporizing water. In this study we present a novel operation of MD, termed isothermal acid sweep membrane distillation (IASMD), for the separation and recovery of ammonia from aqueous solutions, while suppressing water transport. The innovative operating features of IASMD that distinguish it from conventional MD are: i) same temperature on both the feed and sweep sides and ii) acidic solution on the sweep side. High ammonia flux with minimal water permeation for isothermal membrane distillation was realized with 500 mM total ammonia nitrogen (TAN) to simulate urine as the feed, and either 100 mM acetic acid or deionized (DI) water as the sweep. Traditional MD operating at feed and sweep temperatures of 40 and 20°C, respectively, exhibited ammonia flux of 9.3 mol/m^{2}h with water flux at 8.3 kg/m^{2}h, while IASMD was enhanced ammonia flux to 13.9 mol/m^{2}h, while diminishing water flux to 0.15 kg/m^{2}h. Ammonia permeation in isothermal MD is enhanced 1.5-fold when acidic sweep solution is used compared to DI water. IASMD operation from 20 to 60 °C demonstrates that ammonia flux increases with increasing temperature, from 5.6 to 22.8 mol/m^{2}h, while suppressing water flux to below 0.5 kg/m2h across all temperatures. Furthermore, IASMD is substantially more energy efficient than traditional MD, achieving energy savings of 92%. IASMD has promising potential for the selective and energy efficiency removal and recovery of volatile contaminants from aqueous waste-streams.

F13 In-situ micro-rheology of a foulant layer, Jose Agustin Epstein (Technion - Israel Institute of Technology); Guy Z Ramon (Technion - Israel Institute of Technology)*

Fouling remains a major problem in membrane-based water purification and desalination processes, leading to increased operational costs and necessitating extensive pre-treatment. In particular, characterization of organic matter deposits and bio-fouling, their resulting properties and the ways in which they conspire to determine process losses requires further physical insight, from which better control strategies may be developed. This research demonstrates an in-situ approach for foulant monitoring, aimed at measuring the mechanical properties of the foulant layer at there membrane surface, during the separation process. We employ fast confocal microscopy and particle tracking algorithms for the analysis, with alginate as a model foulant used to create a hydrogel deposit on a nano-filtration membrane (NF270). Fluorescent particles of 0.5 μm diameter are embedded in the deposit, and particle trajectories are utilized to calculate the storage (G') and loss (G'') modulus using the generalized Stokes-Einstein relation (GSER). Confocal microscopy enables the accurate monitoring of known positions within the layer, resolving depth variation of its structure and mechanical properties. Result demonstrate the ability to characterise the deposit mechanics and heterogeneity. Specifically, the alginate layer is shown to be a frequency/dependent viscoelastic material with mechanical depth variations. Notably, different permeate fluxes generate changes in the material response as our result show that the foulant-layer mechanical properties are dependent on the permeate flux, with responses varying from viscoelastic to elastic-solid-like. This research demonstrates a first analysis of this type of characterization, opening new avenues for studying foulant properties and response to operating conditions. Such physical insight can lead to better fouling control practices.

F14 Mixed-gas permeation in thermally rearranged poly(benzoxazole-co-imide) polymer membranes, Adele Brunetti (Institute on Membrane Technology of Italian National Research Council); Elena Tocci (Institute on Membrane Technology of Italian National Research Council); Jong Geun Seong (Hanyang University); Young Moo Lee (Hanyang University)*; Enrico Drioli (Institute on Membrane Technology of Italian National Research Council); Giuseppe Barbieri (Institute on Membrane Technology of Italian National Research Council)

The mutual influence of each component of a gas mixture on the other gas permeation was systematically investigated for mixtures including CO_2, N_2, CH_4, H_2, and CO. The mixed gas permeation was measured at 35°C at up to 5 bar of feed pressure in thermally rearranged poly(benzoxazole-co-imide) (TR-PBOI) membranes. How and how much CO_2 (the most soluble gas) and H_2 (the fastest diffusing gas among the gas species tested) affect each other permeation as well as the permeation of the other gases were specifically addressed by a discussion based on sorption and diffusion contributions. For quantifying the sorption contribution to the permeation, sorption isotherms of binary mixtures were calculated using the Grand Canonical Monte Carlo approach. CO_2 concentration in the polymeric matrix was not affected by other gases except for CH_4 because of its competitive sorption. CO_2 permeance measured feeding mixed gases was independent of the other gas in the mixture and did not deviate from its single gas value. However, CO_2 significantly affected the other gases permeation, reducing H_2, N_2, CO, and CH_4 permeance by 18, 14, 12, and 1.3%, respectively. H_2 permeance did not change in mixtures but H_2 influenced N_2, CO, and CH_4. Consequently, in all CO_2-containing mixtures the mixed gas selectivity was greater than the one based on single gases, whereas the opposite trend occurred in all H_2-containing mixtures. Both H_2 and N_2 gases in mixture showed linear sorption isotherms, confirming a very low interaction between them and with the polymer. The permeance of CO_2 and H_2 remained unchanged, whereas the permeance of the other gases was reduced by the presence of CO_2 or increased by the presence of H_2.

F15 Plug flow reactor IEMB - performance and modelling, Shalom Fox (Ben-Gurion University of the Negev); Kristina Stadnik (Ben-Gurion Univ.); Amit Thakur (Ben-Gurion University of the Negev); Zeev Ronen (Ben-Gurion Univ.); Yoram Oren (Ben-Gurion University of the Negev); Jack Gilron (Ben-Gurion University of the Negev)*

Ion exchange membrane bioreactors have been used to remove oxyanions from process streams and contaminated groundwaters. The process involves Donnan dialysis to drive the contaminating ions from the water side to the bioside where bioreduction takes place. In the past most work has been done using reactors with high water recycling rates on the water side which caused them to operate at low fluxes since the membrane was exposed to the lowest exit concentration. This group has designed two forms of plug flow reactors, one plate and frame and one spiral wound to drastically increase the throughput possible. The reactors have been tested on perchlorate, nitrate, and mixed perchlorate and nitrate feeds at different flow rates. The design equation for the reactor incorporated the flux equation developed by Velizarov and coworkers, but modified it to reflect the ion adsorption isotherms obtained with different membranes for the same oxyanion. The results of the design equation are compared to the PFR results and future directions are suggested.

F16 Isolation of Extracellular Vesicles from a Biofluid Using Ultrathin Microslit and Nanopore Silicon Membranes, Jared Carter (SiMPore inc.)*; Cassandra Walinski (SiMPore Inc.); Akash Patel (SiMPore Inc.); Joshua Miller (SiMPore Inc.); Jon-Paul DesOrmeaux (SiMPore Inc.); James A Roussie (SiMPore Inc.)

Extracellular vesicles are emerging as a potentially powerful class of next-generation diagnostic analytes. In particular, Urinary extracellular vesicles (UEVs) are abundantly present in urine, carry protein and nucleic acid cargoes from their cells of origin, and derive from all cell types of the nephron and urinary tract. Because urine can be obtained non-invasively, UEV-based urinalysis liquid biopsies have received a great deal of attention. Isolating UEVs out of urine and away from matrix factors [e.g., urinary cells and Tamm-Horsfall protein (THP) filaments] remains a challenge especially for microfluidic-based approaches. Without integrated sample preparation capabilities, these microfluidic UEV detection systems may find little utility as next-generation research and diagnostic tools. Here, we report on the development of a new three-step UEV isolation procedure that has a total processing time of less than 1hr for 500µL human urine samples. The process relies on chip-format nanomembranes that can be integrated into microfluidics, including microslit silicon nitride (MSN) membranes for urine matrix factor removal and nanoporous silicon nitride (NPN) membranes for UEV capture. We demonstrate that MSN membranes with 0.5x50 µm slit rectangular prism "pores" can yield a filtrate that is cell-free, protein-depleted (4X reduction), and ≥95% of initial volume, in under 5-minutes at 900xG centrifugation. We further demonstrate that NPN membranes with 50 nm mean cylindrical pores can capture UEVs with nearly quantitative efficiency from the MSN membrane's filtrate in 20 minutes at 600xG centrifugation. We verify our results by multiple metrology and molecular biology techniques. Together, these results demonstrate ultrathin silicon nanomembranes offer unique capabilities for UEV sample preparation and may provide the sought-after solution for UEV isolation in microfluidic analysis platforms.

F17 High-Pressure Reverse Osmosis for Energy-Efficient Desalination of High Salinity Brines, Douglas M Davenport (Yale University)*; Akshay Deshmukh (Yale University); Jay Werber (Yale University); Yuhao Du (Yale University); Menachem Elimelech (Yale University)

Water scarcity, expected to become more widespread in the coming years, demands renewed attention to freshwater protection and management. Critical to this effort are the minimization of freshwater withdrawals and elimination of wastewater discharge, both of which can be achieved via zero liquid discharge (ZLD), an aggressive wastewater management approach. Reverse osmosis (RO) is widely accepted as the most energy-efficient desalination process for solutions with a concentration less than ~70,000 mg/L total dissolved solids. However, the high brine salinities encountered in ZLD necessitate the use of thermal desalination processes. Because of the high energetic cost of thermal desalination, ZLD is particularly challenging for high-salinity wastewaters. In this presentation, we highlight the potential of high-pressure reverse osmosis (HPRO) (i.e., reverse osmosis operated at a hydraulic pressure greater than ~100 bar) to efficiently desalinate hypersaline brines. We will first discuss the inherent energy efficiency of membrane processes compared to that of conventional thermal processes for brine desalination. We then present process-scale energetic modeling to showcase the opportunity of HPRO to reduce energy requirements for the desalination of key high-salinity industrial wastewaters. The current state of membrane materials and processes for hypersaline brine desalination is also presented, emphasizing several process design considerations unique to HPRO. The most pressing research needs for the development of HPRO are also discussed, notably the development of membranes and modules suitable for high pressures. Lastly, we will share recent insights into the nature of compaction and salt transport at pressures and salinities relevant to HPRO.

F18 Computationally Efficient Methods of Simulating Membrane Spacers, Jacob Johnston (Colorado School of Mines)*; Jincheng Lou (Colorado School of Mines); Nils Tilton (Colorado School of Mines)

Membrane spacers play an important role in fluid flow, concentration polarization, temperature polarization, and mineral scaling in membrane separation processes, including nanofiltration, reverse osmosis, and membrane distillation. To date, computational fluid dynamics studies of spacers have been limited by the fact that most commercial computational fluid dynamics software uses body-fitted grids to simulate spacer surfaces. Such grids are time consuming to build, and limit the number of spacer geometries that can be considered. Thus motivated, we leverage recent advances in immersed boundaries to develop a method of simulating spacers without body-fitted girds. Rather, the method uses simple Cartesian grids that nevertheless recover the same order of accuracy as traditional grids. We show that the method accurately simulates not only the no-slip condition for the fluid flow, but also the advection of heat and salts near the spacer surface interface, which are often tricky to simulate. Using our method, we simulate heat and mass transport in a direct-contact membrane distillation system with spacer elements in both the feed and distillate channel.

F19 Dissipative Particle Dynamics Simulations of PES membrane via NIPS, Eric Ledieu (University of Arkansas)*; Yuanhui Tang (University of Arkansas); Rosario Cervellere (University of Arkansas); Paul Millett (University of Arkansas); David Ford (University of Arkansas); Xianghong Qian (University of Arkansas)

Polymeric membranes produced by Non-solvent Induced Phase Separation (NIPS) have been widely used across a wide array of microfiltration and ultrafiltration processes. However, despite their wide use, current fabrication processes are largely determined by trial and error. Better insights on the factors affecting membrane morphology would be of great use. To that end, Dissipative Particle Dynamics (DPD), a mesoscale simulation technique, was used to simulate membrane formation process involving polymer (polyether sulfone, PES), solvent (methyl-2-pyrrolidinone, NMP) and nonsolvent (H_2O) with several million beads and over several microseconds. The interactions between these beads are determined according to Flory-Huggins thermodynamic theory and continuum transport equation. Two configurations of the periodic unit cell were set up: (1) symmetric unit cell with water on both sides of the polymer/solvent solution; (2) asymmetric unit cell with polymer solution surrounded by water on one side and a hard wall on the other side. Both synthetic models were found to form pores in the nanometer scale at the skin layer consistent with experimental data, although symmetric unit cell produced much larger inner pores resembling the finger-like structure found deeper within the membrane, which warrants further study.

F20 Leveraging Surplus Energy from Nuclear Power Generation for Seawater Desalination: A Process Modelling Investigation, J. Hunter Himes (Tennessee Tech University); Mary Adkisson (Tennessee Tech University); Laura H Arias Chavez (Tennessee Tech University)*

Nuclear power plants represent a significant, untapped supply of low-grade waste heat and intermittently available surplus electricity that could be used for seawater desalination. The majority of the thermal energy generated at these facilities is discharged to the environment as low-temperature heat; only one third is converted into electricity. In addition, nuclear power production in the United States cannot be ramped up or down to match the fluctuating demand for electricity, forcing generation facilities to sell electricity at a loss at times of low grid demand. If an appropriate partner process can be identified, this surplus energy can be used to desalinate seawater, enhancing coastal water supplies without significantly increasing energy demand while also improving the economics of nuclear power generation. To effectively leverage both surplus electricity and waste heat, a wide range of desalination technologies were investigated for stand-alone use or in combinations. These include reverse osmosis, membrane distillation, multi-pass nanofiltration, and others. Processes were ordered according to their unique advantages and disadvantages with respect to maximum achievable water recovery, energy utilization, tolerance of fluctuations in available energy, and susceptibility to scaling. Material and energy balances were used to model potential process combinations in MATLAB in order to predict the quantity and quality of water that could be produced from a given amount of energy. The feasibility of achieving zero liquid discharge in this unique energy use context was also investigated. Results were considered together with water-use data in southeastern Florida to estimate the cost and value of desalinating and supplying water alongside a hypothetical 300 MW (thermal) nuclear power plant at Turkey Point, Florida. These metrics were used to identify the most promising directions for future research into surplus-energy-powered seawater desalination.

F21 Maxwell-Stefan characterization of surface diffusion flow through defect containing silicalite membranes., David Carter (University of Ottawa); Shaaima Al Akwaa (University of Ottawa); Boguslaw Kruczek (University of Ottawa); Handan Tezel (University Ottawa)*

Although zeolite membranes can withstand temperatures in excess of 500°C and harsh chemical environments, their widespread industrial adoption for separating gas mixtures has been impeded by two main factors. These are firstly the ability to fabricate zeolite membranes containing as few defects as possible, and secondly, the means to evaluate the performance of novel membranes, and therefore design of membrane processes. Our group and others in the literature have investigated the former challenge. The present study addresses the latter challenge of characterizing novel membranes and utilizes the Maxwell-Stefan (MS) model to determine characteristic diffusivities and exchange coefficients, which describe only the surface diffusion contribution to overall membrane permeance for defect containing membranes. Since surface diffusion flow is subject to synergistic adsorption and diffusion phenomena, these diffusivities and exchange coefficients can be used to relate the pure gas permeances of novel membranes with gas mixture separation performance predictions. To accomplish this goal, it is necessary to discount the defect contributions to overall membrane permeance, which has been accomplished by utilizing parameters determined using a simple model that is based on the results of in-situ pure gas permeation experiments. For two defect containing silicalite membranes, MS diffusivities of CO_2 ($ĐCO_2$) and N_2 ($ĐN_2$) of 7.26 - 8.33 x 10^{-8} $m^2 \cdot s^{-1}$ and 1.17 - 1.95 x 10^{-8} $m^2 \cdot s^{-1}$, respectively, were determined. The corresponding exchange coefficients $ĐN_2/CO_2$ and $ĐCO_2/N_2$ were also determined, and $ĐN_2/CO_2$ was found to be at least 1.38 times greater than $ĐCO_2/N_2$ for both membranes. These characteristics are similar for both membranes, and can be used to predict the CO_2/N_2 gas mixture separation performance of silicalite membranes at untested conditions.

F22 Self-Regulating FO-RO Hybrid Systems, Noah Ferguson (University of Connecticut)*; Nicole Beauregard (University of Connecticut); Jeffrey R McCutcheon (University of Connecticut)

Though reverse osmosis (RO) is an industrially mature and widespread desalination technology, its susceptibility to fouling makes it challenging to operate using more difficult feedwaters. Pretreatment is one approach to reducing fouling in RO membranes, and such systems often comprise a majority of a desalination plant's total footprint. Hybridization of forward osmosis (FO) with RO has been considered as an alternative configuration to more conventional pretreatment schemes, but the lack of understanding of how these systems might operate and the potential for complex control schemes to balance fluxes between the FO and RO systems create uncertainty. In this work, we demonstrate self-regulating behavior of FO-RO hybrid systems that can lead to drastically reduced complexity of these systems. We show this behavior using a module-scale test bed that can mimic the behavior of larger scale systems. In short, the system shows volumetric flux convergence between the FO and RO modules when perturbed by a change in RO module pressure and therefore does not need to have complex controllers in place to ensure even flows between the two processes.

F23 Estimating salt diffusion coefficients in polyamide active layers of reverse osmosis membranes using microscale continuum modeling, Riley Vickers (University of North Carolina at Chapel Hill)*; Jingbo Wang (UCLA); Timothy Weigand (University of North Carolina at Chapel Hill); Casey Miller (University of North Carolina at Chapel Hill); Orlando Coronell (University of North Carolina at Chapel Hill)

Analysis of water and solute transport through reverse osmosis (RO) membranes in the literature shows that water and solute permeability of RO membranes depend on geometric (thickness, pore structure, surface roughness) and intrinsic transport (partition and diffusion) properties of their polyamide (PA) active layers. Developments in characterization techniques through the past decade have yielded experimental methods to quantify almost all geometric properties and water and solute partition coefficients of PA active layers. However, there is no method to experimentally quantify diffusion coefficients of salts and other solutes of common interest in PA active layers. Atomistic modeling techniques are the primary method for estimating diffusion coefficients, but they do not account for the morphology and topology of PA active layers (Ding et al., 2014, Desalination, 343 48-53). Our previous work (Lin et al., 2018, J. Membr, Sci., 564, 935-944) estimated water diffusion coefficients in RO membranes during permeation of pure water by using finite element microscale continuum modeling. In this study, we build upon that work to estimate both water and solute diffusion coefficients in RO membranes during permeation of salt solutions. To achieve this, we experimentally determine membrane water and salt permeability, water and salt partition coefficients, active layer thickness, pore sizes, and pore volume fraction. We then use this geometric characterization, water flux, and salt rejection data as model inputs to estimate water and salt diffusion coefficients. Using these salt diffusion coefficients, we evaluate which geometric or intrinsic transport properties have the strongest correlation to salt permeability. This modeling approach constitutes a tool for membrane scientists and developers to estimate the diffusion coefficient of solutes in PA active layers, and aids in targeting structural and intrinsic transport properties for optimization of membrane performance.

F24 Energy-efficient Biogas Upgrading Membrane Process, Chung-Seop Lee (Airrane Co. Ltd); Jinhyuk Lim (Airrane Co. Ltd); Sang Hoon Han (Airrane Co. Ltd.)*; Sejong Kim (Airrane Co. Ltd); Seong Yong Ha (Airrane Co. Ltd); Won Seok Chang (Korea District Heating Corporation); Moon Sei Oh (Korea District Heating Corporation)

Biogas is a kind of gas mixture produced by the anaerobic digestion or fermentation of organic matters such as municipal solid waste, sewage sludge, feedstock waste or any other biodegradable waste. Biogas from the anaerobic digestion has been mostly combusted for heating and power generation; however, the purification of biogas is getting popular in that the cost of methane by biogas purification is more economic than the cost of power generation by the combustion. Biogas purification has been mostly performed using pressure swing adsorption (PSA) process with solid adsorbent, but more recently the application of gas separation membrane is enlarged due to its simple process, compact footprint, low maintenance, high energy saving and low operation expenses, etc. The dehydrated biogas is consisted of mainly methane (60~70 vol%) and carbon dioxide (30~40 vol%) with few hydrogen, hydrogen sulfide, oxygen, nitrogen and siloxane, etc. Its low oxygen and nitrogen concentration enables the separation of methane from biogas more efficient than the separation from landfill gas; the state-of-the-art polymer membrane is the most economic process for biogas purification. Membrane-based biogas purification is performed in the order of dehydration, pretreatment, gas compression, heat exchange unit and membrane skid, etc. Pretreatment removes siloxane and acid gas which may reduce the long-term operation performance of membrane separation due to the sorption on the surface or corrosion of polymer membrane. The pretreated gas is compressed up to 10-20 barg. The multi-stage of membrane modules is tested with small membrane area at the lab-scale study, and the methane concentration of retentate stream was over 96-98 vol% enough to be put in the liquefied natural gas grid while the carbon dioxide enriched gas stream was permeated at the downstream.

F25 Multi-stage Membrane Process for CO_2 Separation from LNG flue gas, Chung-Seop Lee (Airrane Co. Ltd)*; Sejong Kim (Airrane Co. Ltd); Chung-Seop Lee (Airrane Co. Ltd); Won Seok Chang (Korea District Heating Corporation); Moon Sei Oh (Korea District Heating Corporation)

The enormous amount of greenhouse gas emission by the combustion of fossil fuels from 19th centuries causes the man-made global warming with the increase of atmospheric CO_2 concentration over 400 ppm. There are the six major greenhouse gases: carbon dioxide (CO_2); methane (CH_4); nitrous oxide (N_2O); hydrofluorocarbons (HFCs); perfluorocarbons (PFCs); and sulphur hexafluoride (SF6). CO_2 has the lower greenhouse index, but the effect to the global warming is more than 50% due to the worldwide fossil fuel for power generation, automotive, and petrochemical processes, etc. One of the substantial alternatives against the global warming is carbon capture, utilization and storage (CCUS) technologies. CCUS can reduce the emission of greenhouse gases into the atmosphere and also create the additional beneficial products. Membrane-based gas separation process is one of the most feasible gas separation processes for CCUS competing with amine-based solvent-scrubbing and sorbent-based capture. We developed a bench-scale multi-stage CO_2 separation process inside a 3×6 m^2 of movable container box. The process is equipped in the order of flue gas blower, knockout drum, heat exchanger, gas compressor, gas dryer, multi-stage membrane cascade and three vacuum pumps connected to each membrane permeate flow. The membrane process was installed at Pangyo Division in Korea District Heating Corporation (KDHC) which operated the combined heat and power plant (CHP) and supplied both heat and electricity. The flue gas from CHP is composed of the lower CO_2 concentration than the one from the pulverized carbon (PC) power plant, cement and steel manufacturing plant. The parametric experiments were focused to increase the CO_2 concentration from the lower CO_2 feed stream to the level for the appropriate CO_2 utilization process.

F26 Cost Optimization of Pressure Retarded Osmosis, Alexandra N Newby (Carnegie Mellon University)*; Timothy V Bartholomew (Carnegie Mellon University); Meagan S Mauter (Carnegie Mellon University)

Pressure Retarded Osmosis (PRO) is a technology which mixes two water sources of different concentrations and generates electricity, and is a potential solution to the growing need for low carbon energy sources. Previous analyses on the PRO membrane show that electricity can be generated. However, these studies are limited by not considering the operation, inefficiencies, and cost of a full system needed for PRO, nor is the capacity factor for PRO considered to account for the total cost of this technology. This study identifies the cost optimal design and operation using a nonlinear programming model to minimize the levelized cost of electricity (LCOE) of an entire PRO system for a specified feed salinity and process and financial parameters. Our model includes a detailed finite difference module-scale model that explicitly considers the pressure drop along the module and non-constant solution properties, mass and heat transfer coefficients, and equipment inefficiencies. Additionally, the LCOE and capacity factor for PRO is compared to other low carbon energy sources to determine the competitiveness of this technology. Preliminary results show that PRO should not be pursued low carbon energy solution until after membrane technology and equipment efficiencies can be improved significantly.

F27 Tearing propagation of an airship envelope fabric based on FEM and X-FEM methods, Chen Yonglin (Shanghai Jiaotong University); Wang Fengxin (Shanghai Jiaotong University)*

This paper studied tearing properties of an airship envelop fabric by numerical analysis. The finite element method (FEM) and the extended finite element method (X-FEM) were carried out to simulate the crack tip propagation on such fabric. ANSYS and MATLAB were applied as FEM and X-FEM software, respectively. To simulate the discontinuity on the specimen, both of the Galerkin's and the Heavisde functions were introduced during calculating strengthening function in X-FEM. The strain contours and the stress fields were obtained by two methods. Meanwhile, the stress field propagation in membrane crack tip was contrasted between such two methods. Then, the stress intensity factor of the airship envelope was solved. It is found the stress distribution of such fabric was proved to be nonlinear. And the X-FEM is preferable to simulate crack propagation on membrane material.

F28 Modelling of the effect of current density and contact time in electro-coagulation on membrane fouling, In-Soung Chang (Hoseo University)*

Electro-coagulation process has been gained an attention recently because it could overcome the membrane fouling problems in MBR (membrane bio-reactor). Effect of the key operational parameters in electro-coagulation, current density (ρ_i) and contact time (t) on membrane fouling reduction was investigated in this study. A kinetic model for ρ_i and t required to reduce the membrane fouling was suggested under different MLSS concentration. Total 48 batch type experiments of electro-coagulations under different sets of current densities (2.5, 6, 12 and 24 A/m^2), contact times (0, 2, 6 and 12 hr) and MLSS (mixed liquor suspended solids) concentration (4500, 6500 and 8500mg/L) were carried out respectively. After each electro-coagulation under different conditions, a series of membrane filtration was performed to get information on how much of membrane fouling was reduced. The membrane fouling decreased as the ρ_i and t increased but as MLSS decreased. Total fouling resistances, R_t (=R_c+R_f) were calculated and compared to those of the controls (R_o), which were obtained from the experiments without electro-coagulation. A kinetic approach for the fouling reduction (R_t/R_o) rate was carried out and three equations under different MLSS concentration were suggested: ρ_i^n *t = constant. Those equations state that the product of ρ_i and t needed to reduce the fouling in certain amounts (in this study, 10% of fouling reduction) is always constant.

F29 Characterization and Molecular Dynamics Simulation of Poly(arylethersulfone) Ultrafiltration Hemodialysis Membrane, Arash Mollahosseini (University of Saskatchewan); Amira Abdelrasoul (University of Saskatchewan)*

A significant number of patients around the world with end-stage renal diseases (ESRD) are receiving hemodialysis treatment as an alternative therapy for kidney transplant. Polymeric membranes, as the main element of the hemodialysis systems, still lack an optimum biocompatibility profile. Consequently several research approaches are currently focusing on biocompatible polymeric membranes with low- or non-activation behaviour. This study aimed to assess the effects of blood on hemodialysis membrane. As part of the experiment, before and after assessments were conducted on blood contacted membranes. To increase the relevance of the study, poly (arylethersulfone-vinyl pyrrolidone) (PAES-PVP) hemodialysis membranes were chosen. Fourier transmission infrared (FTIR) spectroscopy and Raman spectroscopy were used to assess the initial chemical structure of the membrane along with the subsequent variations with the blood infections. Primary chemical interactions were discussed based on the variations in spectroscopy peaks. Scanning electron microscopy (SEM) and Transition electron microscopy (TEM) were implemented in order to assess the morphological patterns in the membrane's surface structure and cross section. The research helped visualize the structural membrane variation, as well as blood aggregations and blood clots on the membrane's surface. Molecular dynamics (MD) simulation was performed using material studio software to evaluate the interaction of PAES-PVP with major human blood proteins, specifically in terms of interaction energy. The macromolecules were chosen from plasma protein component since they are considered to be the main elements responsible for blood constituent fouling on hemodialysis membranes. This is the first time that an MD technique was applied for the analysis of blood-hemodialysis membrane interactions.

F30 Humidified gas permeabilities in polybenzimidazole gas separation membranes, Joshua D Moon (The University of Texas at Austin)*; Benny Freeman (The University of Texas at Austin); Donald Paul (The University of Texas at Austin)

Polybenzimidazoles (PBIs) have recently been investigated for use as gas separation membranes due to their thermal stability and good H_2/CO_2 separation performance at elevated temperatures. However, PBIs are hydrophilic, and effects of water vapor on PBI separation performance are generally unexplored. Recent studies have shown that water binds strongly to PBIs and that the high water uptake of PBIs (up to 25 wt%) causes significant swelling and changes in volumetric properties such as polymer fractional free volume. The effect of humidity on gas transport in both commercial PBI (Celazole®) and recently synthesized sulfone-containing PBIs has been examined using single gas humidified permeation measurements. Samples were pre-equilibrated with water at relative humidities ranging from 2 to 60% at 35°C prior to permeating a single gas species (e.g., H_2 or CO_2). Gas permeabilities decreased by up to an order of magnitude at low humidities, which is believed to be due primarily to competitive sorption. Gas permeabilities increased slightly at higher humidities where more significant swelling and plasticization occurred. Humidified gas permeabilities can be correlated with changes in fractional free volume that result from water uptake and swelling.

F31 Hollow Fiber Countercurrent Dialysis for Continuous Buffer Exchange of High Value Biotherapeutics, Christopher Yehl (Penn State University)*

There is growing interest in the development of continuous biomanufacturing processes for the production of high value biotherapeutics. These continuous processes have the potential to provide lower cost operation, greater manufacturing flexibility, and enhanced product quality. Buffer exchange, desalting, and formulation of high-value biotherapeutics are currently performed using batch diafiltration (DF); however, this type of tangential flow filtration process would be difficult to implement as part of a fully continuous biomanufacturing process. The objective of this study was to explore the potential of using countercurrent dialysis for continuous protein formulation and buffer exchange. Experiments were performed using concentrated solutions of immunoglobuin G (IgG) with commercially available polysulfone hollow fiber dialyzers having 1.5 and 1.8 m^2 membrane surface area. More than 99.9% buffer exchange was obtained over a range of conditions, as determined from the removal of a model impurity (vitamin B12). The dialyzers were able to process more than 0.5 kg of IgG per day in an easily scalable low-cost process. In addition, buffer requirements were less than 0.02 L of buffer per gram IgG, which is several times less than that used in current batch DF processes. These results clearly demonstrate the potential of using low-cost hollow fiber dialyzers for buffer exchange and product formulation in continuous bioprocessing. Further studies will be performed to explore potential opportunities for using hollow dialyzers for simultaneous buffer exchange and product concentration by employing a transmembrane pressure to drive the ultrafiltration. Higher flux / permeability dialyzers are being generated in our laboratory by chemical modification of commercially-available dialyzers. This provides exciting new opportunities for the use of membrane systems in continuous downstream processing of high value biotherapeutics.

F32 Multiscale modeling of microporous polymer blends for CO_2 separation, Alec Kaija (U.S. Department of Energy National Energy Technology Laboratory)*; Ali Sekizkardes (Battelle/U.S. Department of Energy National Energy Technology Laboratory); Samir Budhathoki (U.S. Department of Energy National Energy Technology Laboratory); Janice Steckel (U.S. Department of Energy National Energy Technology Laboratory)

Polymers of intrinsic microporosity (e.g. PIM-1) are promising materials for gas separations - particularly in post-combustion CO_2 capture - due to their ultra-high permeability. Unfortunately, PIM-1 suffers from brittleness and low gas selectivity. Recently, the high permeability of PIM-1 has been leveraged by incorporating it into polymer blends for use in gas separation membranes. Several polyphosphazene polymers have been identified for blending with PIM-1 because they have better mechanical properties and higher CO_2/N_2 selectivity than PIM-1. Here we discuss approaches to modelling different PIM/polyphosphazene blends with the goal of predicting miscibility, phase separation, and gas permeation behavior. While atomistic simulations are effective for predicting certain chemical and physical properties (e.g. gas diffusivity) they do not lend themselves to the kinds of temporal and spatial scales necessary for evaluating miscibility. We discuss some of the challenges of modelling these particular systems as well as the development of a coarse-grained model for the consideration of chemical and physical properties not feasibly predicted by atomistic models alone.